Disabling Obesity

Paolo Capodaglio · Joel Faintuch
Antonio Liuzzi
Editors

Disabling Obesity

From Determinants to Health Care Models

Springer

Editors
Paolo Capodaglio
Istituto Auxologico Italiano IRCCS
Piancavallo, VB
Italy

Antonio Liuzzi
Ospedale San Giuseppe
Piancavallo (Oggebbio VCO)
Italy

Joel Faintuch
Department of Gastroenterology
Sao Paulo University Medical School
Sao Paulo
Brazil

ISBN 978-3-642-35971-2 ISBN 978-3-642-35972-9 (eBook)
DOI 10.1007/978-3-642-35972-9
Springer Heidelberg New York Dordrecht London

Library of Congress Control Number: 2013933016

Printed on acid-free paper

Springer is part of Springer Science+Business Media (www.springer.com)

Preface

Obesity is currently regarded as one of the major health challenges of the developed world. Excess body weight is an important risk factor for morbidity and mortality from cardiovascular diseases, diabetes, cancer, musculoskeletal disorders, and even psychiatric concerns. Obesity is estimated to cause nearly three million deaths per year worldwide. Several epidemiological studies in the past two decades have shown that adiposity, as measured by body-mass index (BMI), has increased in many populations and it has been argued that this could reverse life-expectancy gains in high-income nations. Unlike many other conditions, obesity is hard to define quantitatively. BMI, the most commonly used measure of obesity, fails to actually measure the amount and distribution of fat mass, the crucial factor for the risk of developing some of the obesity-related comorbidities. High-quality studies point out that patients classified as simply overweight according to BMI are in fact obese when we take into consideration the actual measure of fat mass by air plethysmography. Also, obesity is not necessarily associated with comorbidities: there are indeed metabolically healthy obese individuals. Thus, we need to consider separately individuals presenting simple obesity from those at risk of developing or who have already developed complex clinical situations potentially leading to disability. Comorbidities can tip the balance of independence in patients who already have functional limitations mainly due to the excess of mass itself (reduced tolerance to effort, reduced muscle strength and power, reduced balance, sleep apnoea) or who develop conditions where an abnormal metabolism of adipose tissue prevails (diabetes, cardiovascular conditions, nonalcoholic fatty liver disease). Those patients, like others suffering from chronic conditions characterized by relapses and sudden precipitation of the clinical-functional picture (i.e., neurodegenerative and rheumatic conditions), may enter phases of disability, which require a "here and now" multidisciplinary approach to avoid or minimize the progressive worsening of obesity, its comorbidities, and ultimately, disability.

Morbid obesity with comorbidities leading to disability represents the real social and economic burden for the National Health Systems worldwide. Also, the presence of multiple and associated comorbidities often represents an obstacle to being admitted to clinical settings for the treatment of metabolic diseases. On the other hand, clinical units with optimal standards for the treatment of pathological conditions in normal-weight patients are often structurally and technologically inadequate for the care of patients with extreme obesity. The evaluation and treatment of patients with disabling obesity requires clinical facilities where these complex patients can be treated with appropriate therapeutic and rehabilitative protocols carried out by specially trained operators and within an environment, which is ergonomically adequate and safe for both patients and staff alike.

Given the magnitude of the problem, it is not surprising that thousands of books have been published on the epidemiological, genetic, molecular, pathophysiological, and obviously therapeutic aspects of obesity. The aim of this book is to focus on the pathophysiological and rehabilitative aspects of disabling obesity, highlighting multidisciplinary rehabilitation interventions as key to counteracting the disabling aspects of complicated obesity. After an overview of the current knowledge of epidemiology and of genetic and environmental determinants of obesity, the first chapters intend to define the general features of obesity-related disability and how it can be measured. The final chapters will deal with the most advanced treatment protocols for the related specific clinical conditions and the existing evidence of effectiveness.

We do hope that this book will help to spread seminal ideas about the relevance of creating clinical units adequately equipped to face the various clinical pictures of disabling obesity and provide a whole range of intensities of rehabilitation care, from intensive in-patient interventions to the planning of long-term out-patient treatments. At present, we are very aware that obesity cannot be cured, but especially because of that, we should target our efforts on preventing the progression of the disease, its comorbidities, and the related disability.

Gerold Stucki
Paolo Capodaglio
Antonio Liuzzi
Joel Faintuch

Contents

Chapter 1
An Up-to-Date Vision on the Aetiology and on the Epidemiology of Obesity and Morbid Obesity

Antonio Liuzzi and Anna Maria Di Blasio

Abstract Obesity is a pathological condition characterized by an excess of body fat. Both environmental and biological factors have been considered to explain the worldwide diffusion of the obese phenotype. Most likely, the abnormal accumulation of body fat can be seen as a complex interaction between the two factors. In particular, several studies have focused on the hypothalamic mechanisms controlling food intake and energy expenditure; extensive genetic investigations have demonstrated a low but appreciable percentage of obese patients bearing a single gene mutation underlying the clinical picture. Moreover, Genome-wide association studies have shown a relevant number of loci bearing Single Nucleotide Polymorphisms that increase obesity susceptibility by interacting with environmental conditions. Obesity is now considered as a social health problem since excess of fat per se or through complex metabolic activities of adipocytes may allow the development of several chronic comorbidities such as type 2 diabetes, cardiovascular disease, liver steatosis potentially evolving to cirrhosis and sleep apnea. As a consequence, obesity must be considered a serious disease leading to poor quality of life and, potentially, severe disability.

1.1 Introduction

Obesity is a clinical condition, or a disease, whose prevalence sharply increased in the last decades in western countries and, later on, worldwide. Dramatically, obesity has been defined as a pandemia and as one of the major health problems

A. Liuzzi (✉)
Department of Endocrinological Medicine, Istituto Auxologico Italiano IRCCS, Piancavallo, Verbania, Italy
e-mail: liuzzi@auxologico.it

A. M. Di Blasio
Laboratory of Molecular Biology, Istituto Auxologico Italiano IRCCS, Milano, Italy
e-mail: a.diblasio@auxologico.it

P. Capodaglio et al. (eds.), *Disabling Obesity*, DOI: 10.1007/978-3-642-35972-9_1,

of the western countries. Although prevention of obesity is prospectively very important, everyday we have to face the problem of the patients who are now obese, since the incidence of obesity-related diseases is increasing in both adults and children (Yanovski and Yanovski JA 2011). The metabolic consequences of fat excess such as non-alcoholic fatty liver disease, type 2 diabetes, hypertension are already present in obese children and are associated with even more severe health consequences in adulthood. From a clinical point of view, the challenge is not to prevent obesity but rather the progression the disabilities already present. However, whether obesity may be disabling per se or only when associated to specific pathologies is still lively debated. In the US, obesity has been listed among the disabling conditions up to 1999. Thereafter, Social Security decided to consider the effects of obesity under other listed impairments such as respiratory, cardiovascular, and joint diseases. Nowadays, in the US an obese subject weighing 200 kg is not considered disabled unless other pathological conditions are associated (Lutkenhaus http://www.geraldlutkenhaus.com).

The aim of this chapter is to provide an up-to-date vision of the evidence regarding the relevance of environmental, genetic, and epigenetic factors in determining the prevalence of subjects with an excess body fat. Particular attention will be paid to the factors leading to disabling obesity. Obesity is a too generic term that includes patients with Body Mass Index (BMI) only slightly over 30 as well as individuals with a BMI of 60 or over who obviously have different problems and need specific approaches to their social and clinical problems. Actually, not all obese people develop comorbidities, which may be related to different metabolic activities of fat tissue compartments rather than to the mere body fat amount.

1.2 Aetiology

The regulation of energy balance is a very complex mechanism only partially clarified. Energy stores as fat mass are kept relatively constant in normal subjects by systems controlling both food intake and energy expenditure.

The main regulatory centres are located in the hypothalamus where specific neurons integrate afferent long-term and short-term signals of energy status conducted by nervous or hematic pathways and respond with signals to key regulatory sites. These include the anterior and posterior pituitary gland, the cerebral cortex, premotor and motor neurons in the brainstem and spinal cord, and autonomic (parasympathetic and sympathetic) preganglionic neurons. Briefly, leptin produced by adipocytes and other peptides such as CCK, GLP1 send to the hypothalamus inhibitory signals on food intake, whereas other gastrointestinal peptides such as ghrelin trigger sensation of hunger. Leptin is the most important hormone in long-term regulation of energy stores. When fat mass increases, secretion of leptin is also increased and, in the normal subjects, this leads to decreased food intake. Neuropeptides produced by specific hypothalamic nuclei react to peripheral stimuli: in particular the melanocortin family, derived by the cleavage of proopiomelanocortin, acts on specific receptors that modulate caloric intake. Other neuropeptides such as Agouti (Joly-Amado et al. 2012) may control the relative balance between lipids and carbohydrate metabolism acting on the autonomic output of liver, muscle and pancreas. Whereas hypothalamus is the site of homeostatic control, the rewarding mechanisms involved in food intake are located in the amygdala. The frontal cortex has been identified as the voluntary controller of the whole system. Overall, the integrated action of these centres results in endocrine, behavioural, and autonomic responses influencing energy balance. Although derangements of these mechanisms in obesity are still partially uncleared, genetic and functional studies have provided information on some differences between normal and obese. In particular, functional magnetic resonance and positron emission tomography evidenced peculiarities in the activation of frontal cortex, amygdala and hypothalamus obese subjects (Le et al. 2006; Del Parigi et al. 2005). Due to the complexity of the mechanisms involved, it is not surprising that the nosographic identity of obesity is still under discussion. This condition is defined as a disease by the World Health Organisation (WHO) but its specific etiologic agents have not been identified yet. An epidemiologic and a biology models have been suggested to explain the worldwide prevalence of obesity. According to the first one obese, as people affected by an infectious disease, are healthy but live into an obesogenic environment (Swinburn et al. 1999): abundance of food, poor physical activity, chemicals, some adenoviruses would determine a pathological increase of fat tissue. In this model, the sociocultural influences are stressed. Obese people are often considered guilty and laws acting on food industry and on the single patient are necessary in order to control this phenomenon. Diseases derived from obesity should not be covered by medical insurance, whereas patients have to be rewarded when they loose weight and punished by laws when they do not. In the second model, obesity is a medical problem to be approached by a biological point of view evaluating all the possible mechanisms that lead the single individual to accumulate an excessive fat amount. In the last years, scientific research has evidenced a relevant number of genetic, neuroendocrine and metabolic alterations in obese patients so that the biological basis of obesity cannot be denied. It seems obvious the definition of obesity as the percentage of fat tissue exceeding normality. However, the upper limit of normality of this parameter is at present poorly defined due to the lack of large population studies employing reliable methods to measure the actual quantity of fat tissue. BMI, the ratio between weight and height, does not measure fat stores and its limit of 30 to separate overweight from obese individual is only a statistical value over which the risk of cardiovascular disease and mortality increases. The conceptual value of this index has been challenged by studies showing that patients classified as overweight by BMI are actually obese if fat mass is measured by reliable methods as air plethysmography (Gomez-Ambrosi et al. 2012). Moreover, adipose tissue is functionally heterogenous so that consequences of obesity may be quite different if visceral or gluteo femoral adipose tissue is increased. The first one is dangerous since it synthesizes cytokines involved in the development of insulin resistance. In contrast, the second one is even protective for the cardiovascular diseases risk since it sequestrates the excess of triglycerides preventing the phenomenon of lipotoxicity i.e., the deposition of ectopic fat into visceral organs such heart, liver, muscles (Manolopoulos et al. 2010). These differences have an impact on the kind of disabilities associated to obesity and probably reflect different evolutionary pattern. Gluteo-femoral fat is a compartment with a slow turnover, relevant in conditions as pregnancy, lactation or famines. The visceral compartment reflects the need for a fast energy availability in stressful events as war or hunting. The meaning of obesity in the evolution of our species is still largely speculative: is it a genetic advantage, an extreme of normality, or a pathology? Nowadays we are worried because the obesity phenotype has an increased prevalence and seems harmful for health; however, the obesogenic environment derives from the success of the human species in the long way aiming to reduce physical effort and increase food availability. Humans have been able to expand and prevail over the other species just due to favorable genetic mutations such as erect position and brain development. Moreover, mankind has developed evolutionary traits toward palatable and fatty foods to be consumed in community (Krebs 2009). Obesity may be a condition entirely genetically determined by selection of individuals capable of providing food by hunting and agriculture, with the least energy expenditure (wheel, plow) and of storing energy as fat. Thus, it could be reasonable to consider also the obesogenic environment as a result of the pressure of genetic factors. It is unlikely that interventions on populations aimed at inducing generic changes of the so called "life-style" may obtain relevant results since it would be necessary to radically change the environment we have built. What is reasonable to do is to investigate why individuals become obese and, once obesity is established, take the best care of them by changing their life-style and finding effective drugs and surgery techniques. Attention to risk factors is particularly relevant in childhood to effectively prevent adult obesity. More important is to prevent that obese develop pathological conditions leading to disabilities and, if these are already present, to slow down their worsening or induce their regression. According to the evolutionist medicine (Nesse et al. 2010) the cause of a disease under study should be searched in the story of the species. If this is true, which would be the story of the obese phenotype and the reasons of its expansion? The explosion of the obese phenotype may most probably origin from an interaction between genetic and environment factors with a different time pattern: environment changes developed in a shorter period not allowing genetic changes to select individuals capable of compensating for the aggressions of an environment that provides more calories with poor energetic expense. On the other hand, not all people living in an obesogenic environment become obese and it is logical to consider the impact of biological factors most probably deriving from genetics. The present problem of obesity may be better understood in relation to the evolution of mankind by considering the obesogenic environment and the genetic milieu.

1.2.1 The Obesogenic Environment

The obese phenotype is already present in the history of mankind in the Homo erectus age, about 500,000 years ago (Haslam and Rigby 2010), in the female sex, probably as symbol of fertility and abundance. Most likely, only women could become obese due to their sedentary task and breeding of children, whereas men had to go far for hunting. Thus, in the female sex one could already envisage the origins of the big two factors, the reduction of physical activity and increased food intake. Agriculture, the event that several anthropologists define the biggest mistake of the mankind developed about 15,000 years ago. This activity determined the aggregation of people in groups capable of providing stable sources of food deriving from both earth and animals. They were fatter than their ancestors, being used to consume food derived from milk and rich of saturated fats. The present obesogenic environment is only a degenerative aspect of our evolution and is one of the factors that have allowed our species to prevail over the other animals. Only more recently (Haslam and Rigby 2010), single patients with extreme obesity have been described as expression of a pathological conditions.

The mechanisms through which the big two factors lead to the development of obesity are still matter of debate. As an example, the time spent watching television seems to be associated with obesity in children (Robinson 2001), but, surprisingly, this is not linked to greater inactive time (Jackson et al. 2009). Rather, the contribution of watching TV to obesity may be due to either snack food consuming while watching, or to adverts prompting food intake after TV watching. Other studies however have found that increased time spent in passive leisure activities is a risk marker for obesity in the general population (Shuval et al. 2012). Recently, the importance of sugar rich beverages in US has been matter of debate (Clinical decision 2012). Furthermore, other factors such as reduction of sleeping hours or quitting smoke have emerged as possible contributors to the development of the obese phenotype (Keith et al. 2006). Chemicals agents defined as endocrine disruptors have been found to interfere with metabolic processes, thus favouring obesity. The organotin tributyltin, through high-affinity nuclear hormone receptor-mediated mechanism, affects adipocyte biology and promote long-term obesogenic changes. Other suspected obesogenic ligands have been identified. The excretion of Bisphenols, widely used in the plastic containers, has been reported to correlate with BMI (Grun 2010). Studies showing differences in gut bacteria of obese and lean subjects have appeared only during the last decade (Clarke et al. 2012). However, the association with obesity remains controversial because these differences are not always observed and it is also unclear whether they are secondary to or causative of obesity. It is widely acknowledged that stress levels are elevated in modern society and may contribute to weight gain. Isolating an effect of stress on obesity, however, is complicated because the manner in which we respond to stress seems to differ in relationship to social status (Moore and Cunningham 2012). In a recent editorial on the New England Journal of Medicine "The Weight of the Nation," the authors (Colleen et al. 2012) claim that "the obesity epidemic has been driven by structural changes in our environment, rather than by the reductionist view that the cause is poor decision making by individuals". The paper provides evidence that the obesity epidemic has been driven by a complex interaction of changing factors in several critical environments rather than by individual choices. However, only 18 % of Americans identify external factors as the main cause of childhood obesity, whereas 64 % identify personal factors (overeating, lack of exercise, and watching too much television) as the principal determinants. This view seems reasonable since the obesogenic environment is not an unforeseeable event of the modern age but it has been constructed over centuries of human evolution aiming at providing food at lesser effort. Early recognition of subjects at risk and prevention of progression from simple weight excess to obesity should be our goal, but, at the same time, this should not imply to avoid efforts to change the obesogenic environment and indeed some success has been obtained (Flegal et al. 2010).

Obesity is an individual problem; it is not a fault but everybody should not attribute to the society the reasons for his overweight and should face his own problem referring for help to physician and other reliable institutions. To loose fat and not only weight is a difficult task since individual genetic, epigenetic, psychological, and metabolic oppose to reverse obesity.

1.3 The Obese Individual

1.3.1 Genetics

The theory that environment is the only cause of obesity raises the obvious objection that not all the individuals living in an obesogenic environment accumulate fat. Even recognizing a substantial relevance to the obesogenic environment, the individual characteristics derived by parental influences cannot be disregarded. There is now compelling evidence that genetic factors, mainly acting through mechanisms controlling food intake, play a key role in determining body weight. A variable amount of fat has been a characteristic of the human species for thousands of years. Recently obesity-associated genes, as TMEM 18, were present already in species dissociated from Homo species 160 million years ago (Almen et al. 2010). Heritability studies show a marked genetic contribution and twin studies show an elevate concordance (66 %) rate in BMI between identical twins. BMI of adopted twins is associated with the BMI of the biological parents but not with that of the adoptive ones (Phan-Hug et al. 2012). Since obesity is a complex disorder with the contribution of non-genetic factors, estimates of heritability may be difficult although the evidence of obesity linked only to a single gene mutation show that Mendelian influences do exist. In the following paragraphs, we first review those forms of obesity where the clinical and metabolic phenotype is entirely due to a single gene mutation. These forms involve disruption of CNS function in particular in the hypothalamus, where signals coming from periphery are integrated and generate metabolic responses.

1.3.2 Monogenic Obesity

Monogenic forms of obesity are relatively infrequent but of great pathophysiological relevance since they have allowed to clarify several neuroendocrine mechanisms leading to weight gain (Phan-Hug et al. 2012).

Leptin is a peptide with hormonal activity produced by adipocytes in proportion to fat mass. Leptin signals to the hypothalamus the level of energy storage so that elevated levels of leptin induce decreased food intake and viceversa. Unexpectedly, for not fully explained reasons, obese patients have elevated circulating leptin levels due to a resistance to the hormone effects. Genetic leptin deficiency is a very rare cause of severe early-onset obesity. Patients with congenital leptin deficiency present with normal birth weight, severe early-onset hyperphagia from the first months of life, impaired satiety and, when food is denied, aggressive behaviour. Motor and cognitive development is normal. Pubertal delay, signs of hypothyroidism, reduced growth hormone secretion and immunodeficiency are also characteristic and point to neuroendocrine dysfunction. These patients respond very well to exogenous leptin administration.

Autosomal recessive loss-of-function mutations in the leptin receptor gene are rare. They have been first reported in three obese subjects from a Kabilian consanguineous family and subsequently in other five unrelated subjects. The presence of the mutant receptor causes high leptin serum levels but not appreciably different from those present in common obesity phenotypic similarities with the LEP-deficient subjects were noticed. In particular, LEPR deficient subjects exhibit rapid weight gain in the first few months of life, severe hyperphagia and several endocrine dysfunctions: hypogonadotropic hypogonadism, impaired somatotrophic or thyrotropic secretion. Body composition obviously shows a large amount of fat mass whereas energy expenditure is normal (Phan-Hug et al. 2012).

Pro-opiomelanocortin (POMC) is a precursor peptide whose products include ACTH and the melanocortin family. Autosomal recessive loss-of-function mutations in POMC, which are extremely rare, are associated with a clinical picture of early-onset obesity and ACTH-deficiency. Other symptoms may include hypo pigmentation of the skin, red hair and hyperinsulinemia. Heterozygous carriers of loss-of-function mutations have a mild phenotype and quite often obesity is present.

Autosomal recessive null mutations of pro-hormone convertase 1 are responsible for severe early-onset obesity, ACTH-deficiency, postprandial hypoglycaemia due to elevated proinsulin levels and hypogonadotropic hypogonadism due to GnRH-deficiency.

Melanocortin four Receptor mutations are the most frequent known monogenic forms of severe obesity being found in up to 6 % of children with early-onset obesity (Farooqi 2008). These mutations are characterized by either dominant or recessive mode of inheritance. The first is less frequent and more severe Patients bearing these mutations present with hyperphagia, severe hyperinsulinemia and accelerated linear growth. These findings may be related to the role of MC4R in the regulation of growth hormone resulting in increased GHRH levels. Mutations of MCR3 receptors have been found in patients with high metabolic efficiency but their role as cause of monogenic obesity is still unclear (Mencarelli et al. 2011).

Intellectual disabilities or significant learning difficulties are common features in childhood-onset obesity. This suggests that mutations in genes disrupting neuro development are often non-specific and may also affect circuitry involved in food intake regulation. Deletions of SIM1 and BDNF have been associated to early obesity. SIM1 encodes a transcription factor involved in the hypothalamic development. SIM1 expressing neurons are responsible for the production of neuropeptides such as oxytocin, vasopressin, CRH, TRH and somatostatin. Deletions of SIM1 have been reported in association with early-onset obesity, developmental delay and increased linear growth (Phan-Hug et al. 2012).

Brain-derived nerve growth factor is a neurotrophin playing an important role in regulating food intake. Mutations that cause haploinsufficiency of BDNF as well as some genetic variants, notably the BDNF p.Val66Met polymorphism, are also associated with the development of obese phenotype and hyperphagia (Rosas-Vargas et al. 2011).

In spite of the increasing knowledge on monogenic obesity, the majority of patients with early-onset and severe obesity remain without an etiological diagnosis as the causal gene has not been identified yet.

1.3.3 Genome-Wide Association Studies

A genome-wide association study (GWAS) is a case control study that compares the frequencies of common genetic variants in a group of affected individuals and in a group of healthy controls (Jeffrey et al. 2012). Thousands up to millions of single nucleotide polymorphisms (SNPs) are simultaneously genotyped to identify the SNPs with a significant higher frequency in cases than in controls. These SNPs are marker of a region of the human genome which may increase the patient susceptibility for the disease. In contrast to previous methodologies which specifically evaluated few genetic regions, a GWAS investigates the entire genome. Differently from the candidate-gene approach, this approach is therefore hypothesis-free. GWA studies identify SNPs and other DNA variants that are associated with a disease, but cannot, on their own, specify which are the causal genes. Genome-wide association studies are based on a discovery stage in which SNPs that show significant different allelic frequencies from controls are identified and a replication stage that validates the association observed in different populations (Loos 2012).

In the GWAS for obesity, 32 loci have been identified as associated with the obese phenotype (Jeffrey et al. 2012). Traits examined so far include BMI, waist circumference, waist-to-hip ratio (WHR), body fat percentage, and extreme and early-onset obesity. However, these loci have only small effects on obesity-susceptibility and explain just a fraction of the total variance. Thus, they are poorly predictive of obesity. The firstly identified locus, FTO, has the largest effect on obesity-susceptibility with each FTO risk allele increasing BMI by an average 0.39 kg/m^2 (equivalent to 100 g for a person 1.70 m tall) and obesity risk by 1.20 fold (Loos 2012). FTO was the most easily identified locus as not only its effect size is largest, but also the frequency of the BMI-increasing allele is high in white Europeans (i.e. 46 %). Out of the 32 loci associated with BM, the FTO locus explains most of the inter-individual variation.

1.3.4 Genetic and Evolution

To explain the genetic reasons for the high prevalence of obesity in modern age, different hypotheses have been formulated. The hypothesis still most popular is called thrifty genotype. It is based on the ability of some individuals to store energy as fat during periods of food abundance and to be more resistant during famine (Neel 1998). Since energetic stores are essential for survival and reproduction, the ability to store energy as fat has constituted a selective advantage during hundred of years. The thrifty genotype has allowed to compensate the variability of food availability with accumulation of fat. These selective mechanisms may have evolved more than 150,000–200,000 years ago. In the modern era, the transmission of this genetic panel in presence of a constant abundance of energetic food may have determined the diffusion of the obese phenotype. This attractive hypothesis has been challenged by mathematical models (Speakman 2008). If famines would have started at the origins of Homo species, to day we should be all obese. Considering shorter periods i.e., 12,000 years, mortality levels estimated during famines cannot give a selective strength capable of diffusing the thrifty genotype in nearly 30 % of population. It should be noted, however, that this high percentage refers to USA. In Italy, for example, only 10 % of the population has a BMI over 30 (D'Amicis et al. 2011).

Another potential factor to determine obesity increase has been proposed (Johnson et al. 2010). Fructose, unlike glucose, is able to increase triglyceride synthesis and fat deposition, an effect counteracted by Vitamin C. A mutation of L Gulono-oxydase occurred in the late eocene made humans unable to synthesize vitamin C. Since fructose was the most important nutrient in the fruits that were the main energetic source, this mutation could have favoured fat deposition. In our society fructose consumption is greatly increased whereas that of Vitamin C rich fruits is reduced thus favouring the effect of fructose on fat deposition.

A complementary hypothesis takes into consideration a selective pressure exerted by fertility rather than by vitality since fertility, in relatively short times, can bring about more rapid changes in species evolution. Actually, pituitary gonadal axis is extremely sensitive to signal on the nutritional status sent by fat stores through leptin. Thus, the thrifty genotype may have acted on the reproductive ability rather than on survival (Prentice et al. 2008).

Infectious diseases determine a relevant energetic stress due to the requirement of new protein synthesis as immunoglobulins and the necessity to face phlogosis and fever. In addition, lipolytic hormones such as cortisol, glucagon and catecholamines are released from fat tissue to provide energy. Leptin may represent the link between fat and immune functions as it has been demonstrated that leptin-deficient animals are more prone to infections. It is thus possible during famines people with more abundant fat tissue were more likely to survive favouring the transmission of genes associated with obesity (Prentice et al. 2008).

These hypotheses on the transmission of obesity are based on the concept that during the evolution of homo species a positive pressure on genes favouring the deposition of fat as energy storage, fertility or resistance to infections has played a major role. According to the alternative hypothesis of the drifty genotype, genes favouring obesity have been free to express as random drift when limiting conditions as predation have been compensated by protective social structure. In this situation, random mutations occur in the genes and the frequencies of these mutant alleles drift at random. These genes may then be called 'drifty genes', and, due to their influence, some of us are "unfortunate" in the genetic lottery and inherit a predisposition to conditions that in our modern society have such devastating effects" (Speakman et al. 2008).

1.4 Thrifty Phenotype: Epigenetic

Epigenetic has offered new perspectives to the understanding of the evolution of obese phenotype in the human species. At least some of the 'genetic' variation in obesity might actually reflect epigenetic effects that is, effects that are programmed by early-life experiences. Epigenetic in some way recall the Lamarck theories that adaptive characters may be transmitted within few generations and not during thousand of years by mean of selective mutations according with Darwin. Epigenetic may be defined (Dupont et al. 2009) as the study of changes in gene function that can be transmitted by meiosis or mitosis without affecting DNA sequence. DNA hypermethylation or small fragments of RNA called microRNA may be responsible for these epigenetic effects.

These effects were first observed by studies showing that low birth weight children had an increased risk for non-communicable diseases in later life (Hales et al. 1991). The effect of in utero nutrition on subsequent disease risk was shown by elegant follow-up studies of children conceived or born during periods of the Second World War. Risk of diabetes was again linked with low birth weight and famine exposure during the period in utero, however fatal undernutrition proved to be a less significant risk for obesity development in adult life than fatal overnutrition (Zhang et al. 2011). This suggests that, adult obesity might be partly attributed to factors acting on mothers or grand mother of the affected individuals when they were pregnant or breastfeeding. Actually, there are some evidences that the heritability of the acquired traits may be transmitted for more generations but the relevance of epigenetic mechanisms for the diffusion of obese phenotype still needs to be clarified. It may be of interest the study (Wang et al. 2010) showing differences in DNA methylation between lean and obese children.

1.5 Epidemiology

The most extensive report on the worldwide obesity epidemiology has appeared in The Lancet in 2011 (Finucane et al. 2011) and estimates trends for the mean BMI in adults 20 years and older in 199 countries and territories. Between 1980 and 2008, mean BMI worldwide increased by 0.4 kg/m^2 per decade. National BMI change for women ranged from non-significant decreases in 19 countries to increases of more than 2.0 kg/m^2 per decade in nine. Male BMI increased in all but eight countries and by more than 2 kg/m^2 per decade in Nauru and Cook Islands The USA had the highest BMI among the high-income countries. In 2008, an estimated 1.46 billion adults worldwide had BMI of 25 kg/m^2 or greater, of these 205 million men and 297 million women were obese. Globally, mean BMI has increased since 1980 but trends since 198, and mean population BMI in 2008 varied substantially between nations (Finucane et al. 2011).

Some recent studies have also suggested that USA obesity rates are continuing to increase. Data from the behavioural risk factor surveillance system (BRFSS) of the Centres for Disease Control and Prevention showed increases between 2007 and 2009 in the reported prevalence of obesity among adults (MMWR 2010). Based on these data it has been calculated that by 2050 almost all Americans will be overweight. Other reports, however, depict a less dramatic scenario. The National Health and Nutrition Examination Survey (NHANES) suggests that for men, the overall prevalence of obesity shows a significant linear trend over the 12-year period from 1999 through 2010. For women, within race/ethnicity groups, the data suggest slight increases that were statistically significant for non-Hispanic black and Mexican–American women but not significant for women overall. For both men and women, estimates for 2009–2010 did not differ significantly from estimates for 2003–2008. The authors conclude (Flegal et al. 2010) that the increases in the population prevalence of obesity previously observed may not continue at a similar rate, and in fact, the increases appear to be slowing or levelling off. In 2009–2010, the prevalence of obesity was 35.5 % among adult men and 35.8 % among adult women, with no significant change compared with 2003–2008. However, they did not find indication that the prevalence of obesity is declining in any group.

International comparisons of BMI and obesity could be biased by methodological differences among the studies but they seem to indicate that the phenomenon of slowing or levelling trends may not be limited to the United States. Indeed, a number of studies in other countries have suggested that trends previously observed in the prevalence of obesity may be slowing or not increasing. Data from the Health Survey for England (2009) showed that for men the prevalence of obesity was 22.2 % in 2005 and 22.1 % in 2009; comparable figures for women were 23.0 and 23.9 %. Reports from Sweden, Switzerland and Spain have also suggested a possible degree of levelling. Rokholm et al. (2010) reviewed evidence for a levelling off of trends in obesity since 1999 and found mixed results.

Since we know relatively little regarding the causes of the trends previously observed, the prediction of the future trends in obesity is uncertain. Several analyses have extrapolated future prevalence of obesity as function of time from these models but they assume that the causal factors for obesity will continue to rise with time and this assumption may be wrong. If the estimates from the NHANES seem most likely to reflect the actual prevalence of obesity in the United States, the question is why the prevalence of obesity has stabilized over the last ten at least years in US ?

An optimistic interpretation is that increasing recognition by community health authorities and government organizations of the adverse health effects of obesity may lead to interventions able to reduce the so called obesogenic environment. More likely, most of the people with strong genetic susceptibility to weight gain have already become obese while genetically lean people resist to the aggression of the toxic environment. Unfortunately, even if this is true, among the susceptible subjects there are an ever-increasing number of adults with extreme obesity (BMI $\geq$ 40) who represent the largest part of patients with disabling obesity. Indeed, although the rates of obesity have been stabilized also among children and adolescents the proportion of very heavy boys (i.e. BMI at or above the 97th percentile for their age), continue to increase (Ogden 2010).

Thus, it seems more relevant in the frame of this book to examine the data related to the so called morbid obesity potentially leading to disabling obesity. Actually, the concept of morbid obesity is poorly defined. It may be defined according to Dorland's Medical Dictionary for Health Consumers (2007) "a condition of weighing two or more times the ideal weight; so called because it is associated with many serious and life-threatening disorders". But the ideal weight is a rather philosophic concept without solid data to support it and this definition underlines more the presence of any comorbidity rather than obesity per se: not all the comorbidities of obesity may lead to disability. For these reasons to collect data on the epidemiology of disabling obesity as a single clinical condition is a difficult task.

The evidence that obesity increases mortality from all causes is well supported by a large study (Culver et al. 2010) on almost 1 million and half of individuals that evaluated subjects who never smoked and excluded of those with serious disease at baseline. Under these conditions the increase in mortality started for classes of BMI between 20 and 25 and had a sharp increase from 40–44.9. Although mortality will not necessarily imply a previous disabling pathology these data show that fat excess as derived by BMI may be a life threatening condition.

On the other hand, the problem of the so called "obesity paradox" has been raised in several papers. It has been observed that the prognosis for several diseases may be better in patients overweight or even with mild obesity suggesting that these subjects may be in more favorable conditions due to their energetic stores (Kastorini and Panagiotakos 2012). This assumption requires a more careful evaluation and a reliable analysis of the actual nutritional status of these patients. As reported in an extensive review (Kastorini and Panagiotakos 2012) the majority of the studies examining the role of overweight and obesity on survival have evaluated only BMI. However, Oreopoulos et al. (2011) suggested that higher lean body mass and lower fat mass, assessed with dual X-ray absorptiometry (DXA), were beneficial prognostic factors in heart failure patients, while the use of BMI, resulted in a misclassification of body fat in 41 % of the patients. Assessment of obesity with the use of waist circumference was associated with an obesity paradox in patients with heart failure, but not in renal failure or coronary heart disease patients (Kastorini and Panagiotakos 2012).

On the other hand, a consistent percentage of obese are metabolically healthy (Blurher 2012) and have a better prognosis for the development of cardiovascular disease but they probably are at risk for problems linked to fat mass excess per se such as social stigma, ostearthritis and possibly OSAS.

1.6 Comorbidities

Comorbidities are the prevalent causes of disability in obesity (Bray 2004). They can be divided roughly into three categories: (1) Conditions pathophysiologically linked to obesity; (2) Pathologic conditions where obesity may act as a risk factor; (3) Disease incidentally occurring in obese people but aggravated by the presence of the excess of body fat. Often, the distinction between (1) and (2) is difficult even for orthopaedic comorbidities, but it may be useful from a clinical point of view. Epidemiological data on obesity comorbidities can hardly be found and moreover not all obese can be defined as disabled and not all comorbidities can lead an obese individual to be classified as disable. On the other hand, obese individuals may simultaneously complain of more conditions that may be addictive in determining disability. For these reasons the presence or the level of disability should be evaluated on an individual basis with appropriate approaches (see Chap. 4). In the following paragraphs the main pathological conditions associated with obesity will be briefly mentioned taking in account that an extensive examen of them will be found in the dedicated chapters.

Obese are stigmatized for their fatness in the school, employment, health care, and elsewhere. Obese people referred to a weight management centre had serious abnormalities in health-related quality of life (Fontaine et al. 1996) in proportion with their BMI and more frequently in female. Intentional weight loss improves the quality of life.

Osteoarthritis is probably the most common obesity-associated pathology. In an old study (Felson et al. 1998), 33 % of 1,420 obese patients had radiographic evidence of knee osteoarthritis when evaluated over period of observation of about 30 years in the Framingham heart study. The link between obesity and subsequent osteoarthritis was obviously related to the degree of obesity. In more recent studies, however, the relevance of other metabolic and genetic (Elliot et al. 2012) factors has been underlined. Thus, it is possible that hormonal factors such as leptin or other adipokinin may contribute to the development of the osteoarthritis at the level of joints by mechanisms different from those linked to weight load. In a study on data collected in the health Survey for England and UK Back exercise and trial data (Lidstone et al. 2006) musculo-skeletal illnesses evaluated in a sample of almost 9,000 subjects were responsible for disability in 19 % of the individuals but the prevalence increased form 14.6 % in normal weight to 30.3 in obese with BMI greater than 35. Obesity often aggregates with obstructive sleep apnea syndrome (OSAS) due to upper airway obstruction caused by fat deposition around the neck, changes in upper airway muscle tone and changes in central mechanisms of breathing control. At variance with the general prevalence of 2–4 %, about 40 % of obese patients and above 90 % in those extremely obese

(BMI > 40 kg/m^2) suffer from OSAS (Dopp 2009). Further, in approximately 90 % of patients with obesity hypoventilation syndrome, the sleep-disordered breathing is represented by OSAS (Mokhlesi et al. 2008).

Obese may have a specific cardiomyopathy due to the hemodynamic changes linked to excess weight but data on its prevalence cannot be found in the literature (Poirier et al. 2006) The link between obesity and atherosclerotic vascular disease is well documented. Prospective studies that have reported follow-up data over more than two decades, such as Framingham Heart Study, the Manitoba Study, and the Harvard School of Public Health Nurses Study, have documented that obesity, mainly if abdominal, is an independent predictor of clinical CHD (Bray 2004). Also in the report from the Nurses' Health Study the risk for U.S. Women of developing coronary artery disease is increased 3.3-fold with a BMI greater than 29 kg/m^2 compared with that in women with a BMI below 21 kg/m^2 (Manson et al. 1995). Weight gain also strongly affects this risk at any initial BMI (Meigs et al. 1997).

Most hypertensive patients are overweight or obese. Hypertension is about six times more frequent in obese subjects than in lean men and women, in addition weight gain particularly in young patients is a relevant risk factor for hypertension Specific estimates for the prevalence of high blood pressure per age group and BMI group are reported by NHANES III.1 Among men, the prevalence of high blood pressure increases progressively with increasing BMI, from 15 % at a BMI < 25 kg/m^2 to 42 % at a BMI $\geq$ 30 kg/m^2. Women show a similar pattern with a prevalence of hypertension ranging from 15 % at a BMI < 25 kg/m^2 to 38 % at a BMI $\geq$ 30 kg/m^2. The increase in blood pressure is more pronounced in people with abdominal obesity. In the Swedish Obesity Study, hypertension was present at baseline in 44–51 % of the subjects. Key determinant of the weight-induced increases in blood pressure was a disproportionate increase in cardiac output related to an increase of sympathetic activity (Poirier et al. 2006).

Nonalcoholic fatty liver disease (NAFLD) is the term that describes several liver pathologies associated with obesity, including hepatomegaly, elevated liver enzymes, and histological pictures ranging from simple steatosis to fibrosis evolution to cirrhosis. Increased steatosis is characteristic of the livers of overweight individuals; the accumulation of lipid in the liver suggests that secretion of VLDL in response to hyperinsulinemia is inadequate to face the high rate of triglyceride turnover. In obese patients, the prevalences of steatosis, steatohepatitis, and cirrhosis has been reported to be approximately 75, 20, and 2 %, respectively (Bray 2004).

Cholelithiasis is the primary hepatobiliary pathology associated with overweight (Bray 2004). At BMI lower than 24, the incidence of clinically symptomatic gallstones was approximately 250/100,000 person-years of follow-up and gradually increased for BMI up to 30 and more sharply when BMI exceeded this limit. The explanation for this association is mainly due the increased cholesterol turnover related to total body fat.

Obesity may also affect the kidney. Glomerulopathy was significantly increased in pathological specimens compared with other forms of end-stage renal disease (Bray 2004).

Specific forms of cancer are significantly increased in overweight individuals (Bray 2004). Males face increased risk for neoplasms of the colon, rectum, and prostate. In women, cancer of the reproductive system and gallbladder is more common. One explanation for the increased risk of endometrial cancer in overweight women is the increased production of estrogens by adipose cells. This increased production is related to the degree of excess body fat that accounts for a major source of estrogen production in postmenopausal women. Breast cancer is not only related to total body fat, but may also have a more important relationship with central body fat. The increased visceral fat measured by computed tomography shows an important relationship with the risk of breast cancer.

Although dyslipidemia is not per se cause of disability, it may be important in the relationship with the increased risk of heart disease (Bray 2004). A positive correlation between BMI and triglycerides has been repeatedly demonstrated as well as an inverse relationship between HDL and BMI. Waist circumference alone as well as waist/hip ratio or sagittal diameter, accounted for most of the variance in triglycerides and HDL cholesterol. Other lipid abnormalities have been described in obesity but it should be noted that a relevant proportion of obese paradoxically in the highest levels of BMI have normal lipid profile probably in relation to the prevalence of gluteo femoral fat that can contain a large amount of triglycerides thus preventing lipotoxicity (Dixon and O'Brien 2006).

Type 2 diabetes mellitus is strongly associated with overweight (Bray 2004). The risk of type 2 diabetes mellitus increases with BMI levels, more pronounced visceral fat distribution as well as with the duration of weight excess According with the data of Nurses Health Study at a BMI of 35 kg/m^2, the relative risk increased 40-fold. Similar data are reported for the Health Professionals Follow-Up Study. At a BMI above 35 kg/m^2, the age-adjusted relative risk for diabetes in nurses increased to 60.9, or more than 6,000 %. Up to 65 % of cases of type 2 diabetes mellitus can be attributed to overweight. For the 11.7 million cases of diabetes, overweight may account for two-thirds of diabetic deaths (Bray 2004).

A variety of endocrine changes are associated with obesity (Bray 2004; Álvarez-Castro et al. 2011). Among these, the alterations of gonadal activity in both sexes have a clinical relevance in the comorbidities of obese individuals. Irregular menses and frequent anovulatory cycles are common, and the rate of fertility may be reduced. The presence of weight gain in polycystic ovary syndrome where insulin resistance plays a fundamental role is well known. Men also complain often of reduced libido associated with low testosterone levels and normal/low gonadotropins levels. A clinical and hormonal picture of functional secondary hypogonadism that often is difficult to differentiate from the organic forms.

GH secretion is reduced with increasing BMI but whether this may have a relevance in the pathophysiology of fat deposition is unclear.

Obesity is associated with increased cortisol production rate, which is compensated for by a higher cortisol clearance, resulting in plasma free cortisol levels that do not change when body weight increases (Álvarez-Castro et al. 2011). Aldosterone secretion is correlated to BMI and it can be involved in the pathophysiology of insulin resistance (Garg and Adler 2012).

Although obese subjects usually have a normal thyroid function, TSH and BMI are known to be positively correlated. In fact, many studies of children, adolescents, and adults have shown slightly increased TSH levels in obese individuals as compared to lean subjects (Reinehr 2010). By contrast, obesity increases susceptibility to autoimmune thyroid disease, with an emerging role of leptin as a peripheral determinant (Marzullo et al. 2010).

Other conditions such as gastroesophageal reflux, migraine, bronchial asthma are more frequent in obese patients but the pathophysiological meanings of these associations are still unclear.

1.7 Conclusions

Obesity could be easily defined as a percentage of fat tissue exceeding the limits of normality. But the limit of normal amount of fat mass is poorly defined. In the literature there are no studies on large populations set up to define the range of normality of this numeric variable. This is due to the fact that quantitative methods for measuring fat mass require sophisticated equipments i.e., DEXA or air plethysmography. The popular cut-off value of 30 for BMI does not imply an actual limit of fat mass but rather a limit over which the risk of cardiovascular disease sharply increases. This risk however is already present in overweight people when compared to individuals with normal values of BMI. Moreover, adipose tissue displays an elevated functional heterogeneity: thus, the crude amount of total fat mass may be misleading in the interpretation of risk of comorbidities even when assessed with sophisticated approach. Actually, the so called "healthy obese" or better metabolically healthy obese, in spite of elevated BMI values have a far better prognosis than individuals with altered risk factor profile. On the other hand, lean people with an obese metabolic profile have also been described. These inconsistencies can be explained by the well known evidence that visceral fat is more dangerous than gluteo-femoral fat as the latter that can even be protective for lipotoxicity i.e., the deposit of fat in liver and heart or on muscles. Thus, disabling obesity requires an individual assessment with methods capable of taking into consideration the possibility of multiple aggregated factors.

The discussion on the causes that have led a relevant percentage of people to become fat are interesting but probably usefulness. In a recent head to head appeared on BMJ (Wilding 2012; Frayling 2012) the two authors debated on whether the prevalance of obesity may be attributable to environment rather than to genetic. The conclusions however are very similar: the fight against obesity needs a control of the dense energy food market and the identification of patients who are genetically at risk for obesity. The most relevant problem is what to do with people needing medical and social assistance for their disabilities considering that the problem of disabling obesity will not be solved in a short time. Rather, we have to recognize that, whichever evolutionaristic hypothesis could be true, obesity is a characteristic of our species and the number of patients with problems, either medical or social, linked to their weight excess will remain in the most optimistic view, stable for a long time. Therefore, the problem is to prevent that obese individuals may become disabled and, if already disabled, to contrast the worsening of their disabilities. Recently, AACE (Jeffrey et al. 2012) stated that "To say that obesity is not a disease but rather a consequence of chosen lifestyle (i.e., overeating and/or inactivity) is equivalent to saying that lung cancer is also not a disease because it was brought about by volitional cigarette smoking. It is the strong contention of AACE that the view of obesity as a behavioral decision is debunked by biomedical evidence".

On the other hand dramatic perspectives of a future where 100 % of the people will be obese seem to have little sense according to the most recent epidemiological data. Moreover, we observe an increasing number of elderly obese suggesting that obesity not always causes a reduced life expectancy but increases the rate of disability with advancing age. Thus, our efforts should be directed more to contrast disabling obesity than obesity per se. For these reasons studies aimed at establishing reliable parameters to define not from a quantitative but from a pathophysiological point of view when obese individuals are at risk of developing disabilities are strongly needed.

References

Almen MS, Jacobsson JA, Shaik JH, Olszewski PK, Cedernaes J, Alsio J, Sreedharan S, Levine AS, Fredriksson R, Marcus C, Schiolth HB (2010) The obesity gene, TMEM18, is of ancient origin, found in majority of neuronal cells in all major brain regions and associated with obesity in severely obese children. BMC Med Genet 9(11):58

Ålvarez-Castro P, Sangiao-Alvarellos S, Brandon-Sandà I, Cordido F (2011) Endocrine function in obesity. Endocrinol Nutr. 8:422–832

Berrington de Gonzalez A, Hartge P, Cerhan JR, Flint AJ, Hannan L, MacInnis RJ, Moore SC, Tobias GS, Anton-Clarke S, Murphy E, Nilaweera K, Ross P, Shanahan F, O'Toole PW, Cotter PD (2012) The gut microbiota and its relationship to diet and obesity: New insights. Gut Microbes 3:186–202

Blurher M (2012) Are there still healthy obese patients? Curr Opin Endocrinol Diabetes Obes 19:341–1346

Bray G (2004) Medical consequences of obesity. J Clin Endocrinol Metab 89:2583–2589

Clarke SF, Murphy EF, Nilaweera K, Ross PR, Shanahan F, O'Toole PW, Cotter PD (2012) The gut micro biota and its relationship to diet and obesity: new insights. Gut Microbes 3:186–202

Colleen L, Sarah E, Gollust, Jeff N (2012) Are Americans Ready to Solve the Weight of the Nation? New Engl J Med 367

Culver H, Freeman LB, Beeson WL, Clipp SL, English DR, Folsom AR, Freedman DM, Giles G, Hakansson N, Henderson KD, Hoffman-Bolton J, Hoppin JA, Koenig KL, Lee IM, Linet MS, Park Y, Pocobelli G, Schatzkin A, Sesso HD, Weiderpass E, Willcox BJ, Wolk A, Zeleniuch-Jacquotte A, Willett WC, Thun MJ (2010) Body-mass index and mortality among 1.46 million white adults. New Engl N Engl J 363:2211–2219

D'Amicis A, Gargiulo L, Adamo D (2011) "Epidemiologia dell'obesità in Italia".7° Rapporto sull'obesità in Italia, Obesità e genetica: oltre lo stile di vita, Il Pensiero Scientifico Editore, Roma

Del Parigi A, Pannacciulli N, Le DN, Tataranni P (2005) In pursuit of neural risk factors for weight gain in humans. Neurobiol Aging 26(Suppl 1):50–155

Dixon J, O'Brien P (2006) A disparity between conventional lipid and insulin resistance markers at body mass index levels greater than 34 kg/m^2. Int J Ob 256:793–797

Dopp J (2009) Sleep-disordered breathing an obesity: Pathophysiology, complications, and treatment. Nutr Clin Pract 24:675–687

Dupont C, Armant R, Brenner CA (2009) Epigenetics: Definition, Mechanisms and Clinical Perspective. Semin Reprod Med 27:351–357

Elliott KS, Chapman K, Day-Williams A, Panoutsopoulou K, Southam L, LindgrenCM; the GIANT consortium, Arden N, Aslam N, Birrell F, Carluke I, Carr A, Deloukas P, Doherty M, Loughlin J, McCaskie A, Ollier WE, Rai A, Ralston S, Reed MR, Spector TD, Valdes AM, Wallis GA, Wilkinson M The arcOGEN consortium (2012) Evaluation of the genetic overlap between osteoarthritis with body mass index and height using genome-wide association scan data. Ann Rheum Dis. Sep 6. [Epub ahead of print] PubMed PMID: 22956599

Farooqi IS (2008) Monogenic human obesity. Front Horm Res. 36:1–11

Felson DT, Anderson JJ, Naimark A, Walker AM, Meenan RF (1998) Obesity and knee osteoarthritis. The Framingham Study. Ann Intern Med. 109:18–24

Finucane MM, Stevens GA, Cowan MJ, Danaei G, Lin JK, Paciorek CJ, Singh GM, Gutierrez HR, Lu Y, Bahalim AN, Farzadfar F, Riley LM, Ezzati M (2011) National, regional, and global trends in body-mass index since 1980: systematic analysis of health examination surveys and epidemiological studies with 960 country-years and 91 million participants. Lancet 377:557–567

Flegal K, Carroll M, Ogden C et al (2010) Prevalence and Trends in Obesity among U S Adults. JAMA 303:235–241

Fontaine KR, Cheskin LJ, Barofsky I (1996) Health-related quality of life in obese persons seeking treatment. J Fam Pract 43:265

Frayling TM (2012) Are the causes of obesity primarily environmental? No. BMJ 345:e5844 2012

Garg R, Adler GK (2012) Role of mineralocorticoid receptor in insulin resistance. Curr Opin Endocrinol Diabetes Obes 19:168–175

Gomez-Ambrosi J, Silva C, Galofre J, Escalada J, Santos S, Millàn D, Vila N, Ibanez P, Gil M, Valentì F, Rotellar F, Ramirez B, Salvador J, Fruhbeck G (2012) Body mass index classification misses subjects with increased cardiometabolic ris factors related to elevated adiposity. Int J Obesity 36: 286–94

Grun F (2010) Current Opinion in Endocrinology, Diabetes and Obesity. Obesogens 17:453–459

Hales CN, Barker DJ, Clark PM, Cox LJ, Fall C, Osmond C, Winter PD (1991) Fetal and infant growth and impaired glucose tolerance at age 64. BMJ 303:1019–1922

Haslam D, Rigby N (2010) A long look at obesity. Lancet 376:85–86

Jackson D, Stewaart J, Djafarian K (2009) Increased television viewing is associated with elevated body fatness but not with lower total energy expenditure in children. Am J Clin Nutr 89:1031–1036

Jeffrey IM, Alan JG, Yehuda H, Timothy Garvey W (2012) 'American Association of Clinical Endocrinologists' position statement on obesity and obesity medicine. Endocrine Practice 18 No. 5 September/October

Joly-Amado A, Denis RG, Castel J, Lacombe A, Cansell C, Rouch C, Kassis N, Dairou J, Cani PD, Ventura- Clapier R, Prola A, Flamment M, Foufelle F, Magnan C, Luquet (2012) Hypothalamic AgRP-neurons control peripheral substrate utilization and nutrient partitioning. EMBO J. 18 doi: 10.1038

Johnson RJ, Andrews P, Benner SA, Oliver W (2010) Theodore E. Woodward award. The evolution of obesity: insights from the mid-Miocene. Trans Am Clin Climatol Assoc. 121:295–305; discussion 305–308

Keith S, Redden DT, Katzmarzyk PT, Boggiano MM, Hanlon EC, Benca RM, Ruden D, Pietrobelli A, Barger JL, Fontaine KR et al (2006) Putative contributors to the secular increase in obesity: Exploring the roads less travelled. Int J Obes 30:1585–1594

Health Survey for England, 2009: trend tables. National Health Service Information Center. http://www.ic.nhs.uk/statistics-and-data-collections/health-and-lifestyles-related-surveys/health-surveyfor England/health-survey-for-England–2009-trend-tables

Kastorini CM, Panagiotakos DB (2012) The obesity paradox: methodological considerations based on epidemiological and clinical evidence–new insights. Maturitas. 72:220–224

Krebs JR (2009) The gourmet ape: evolution and human food preferences. Am J Clin Nutr 90:707S–711S

Le DS, Chen K, Del Parigi A, Salbe AD, Reiman EM, Krakoff J (2006) Less activation of the left dorsolateral prefrontal cortex in response to a meal: a feature of obesity. Am J Clin Nutr 84:725–731

Lidstone J, Ellsa L, Finna P, Whittaker V, Wilkinson C, Summerbella C (2006) Independent associations between weight status and disability in adults: Results from the health survey for England. Public Health 10:412–417

Loos RJ (2012) Genetic determinants of common obesity and their value in prediction. Best Pract Res Clin Endocrinol Metab 26:211–226

Lutkenhaus G http://www.geraldlutkenhaus.com

Manolopoulos KN, Karpe F, Frayn KN (2010) Gluteofemoral body fat as a determinant of metabolic health. Int J Obes 34:949–959

Manson JE, Willett WC, Stampfer MJ, Colditz GA, Hunter DJ, Hankinson E, Hennekens CH, Speizer FE (1995) Body weight and mortality among women. N Engl J Med 333:677–685

Marzullo P, Minocci A, Tagliaferri MA, Guzzaloni G, Di Blasio A, De Medici C, Aimaretti G, Liuzzi A (2010) Investigations of thyroid hormones and antibodies obesity: leptin levels are associated with thyroid autoimmunity independent of bioanthropometric, hormonal, and weight-related determinants. J Clin Endocrinol Metab 95:3965–3972

Meigs JB, D'Agostino RB Sr, Wilson PW, Wilson PW, Cupples LA, Nathan DM, Singer DE (1997) Risk variable clustering in the insulin resistance syndrome. Diabetes: The Framingham Offspring Study 46:1594–1600

Mencarelli M, Dubern B, Alili R, Maestrini S, Benajiba L, Tagliaferri M, Galan P, Rinaldi M, Simon C, Tounian P, Hercberg S, Liuzzi A, Di Blasio AM, Clement K (2011) Rare melanocortin-3 receptor mutations with in vitro functional consequences are associated with human obesity. Hum Mol Genet 20:392–399

MMWR (2010) Vital signs: state-specific obesity prevalence among adults—United States, 2009. Morb Mortal Wkly Rep 59:951–955

Mokhlesi B, Kryger MH, Grunstein RR (2008) Assessment and management of patients with obesity hypoventilation syndrome. Proc Am Thorac Soc 15:218–225

Moore CJ, Cunningham SA (2012) Social position, psychological stress, and obesity: a systematic review. J Acad Nutr Diet 112:518–526

Neel JV (1998) The thrifty genotype. Nut Rev 1999(57):S2–S9

Nesse RM, Bergstromb CT, Ellison PT et al (2010) Making evolutionary biology a basic science for medicine. PNAS 107:1800–1807

Ogden CL, Carroll MD, Curtin LR, Lamb MM, Flegal KM (2010) Prevalence of high body mass index in US children and adolescents, 2007–2008. JAMA 303:242–249

Oreopoulos A, Fonarow GC, Ezekowitz JA, McAlister FA, Sharma AM, Kalantar-Zadeh K, Norris CM, Johnson JA, Padwal RS (2011) Do anthropometric indices accurately reflect directly measured body composition in men and women with chronic heart failure? Congest Heart Fail. 17:90–92

Phan-Hug F, Beckmann JS, Jacquemont S (2012) Genetic testing in patients with obesity. Best Pract Res Clin Endocrinol Metab. 26:133–143

Poirier P, Giles TD, Bray GA, Hong Y, Stern JS, Pi-Sunyer FX, Eckel RH (2006) American Heart Association; Obesity Committee of the Council on Nutrition, Physical activity, and metabolism. Obesity and cardiovascular disease: pathophysiology, evaluation, and effect of weight loss: an update of the 1997 American Heart Association Scientific Statement on Obesity and Heart Disease from the Obesity Committee of the Council on Nutrition, Physical Activity, and Metabolism. Circulation 113:898–918

Prentice AM, Hennig BJ, Fulford AJ (2008) Evolutionary origins of the obesity epidemic: natural selection of thrifty genes or genetic drift following predation release? Int J Ob. 32:1607–1610

Reinehr T (2010) Obesity and thyroid function. Mol Cell Endocrinol 316:165–171

Clinical decision: regulation of Sugar-Sweetened Beverages (2012) New Engl J Med 367:1464–1466
Robinson TN (2001) Television viewing and childhood obesity. Pediatr Clin North Am 48:1017–1025
Rokholm B, Baker JL, Sorensen TI (2010) The levelling off of the obesity epidemic since the year 1999: a review of evidence and perspectives. Obes Rev 11:835–846
Rosas-Vargas H, Martinez-Ezquerro JD, Bienvenu T (2011) Brain-derived neurotrophic factor, food intake regulation, and obesity. Arch Med Res 42:482–494
Speakman JR (2008) Thrifty genes for obesity, an attractive but flawed idea, and an alternative perspective: the 'drifty gene' hypothesis. Int J Ob. 32:1611–1617
Shuval K, Leonard T, Murdoch J, Caughy M, Kohl H, Skinner C (2012) Sedentary behaviours and obesity in a low income, ethnic minority population. J Phys Act Health 2 Feb 29 Epub ahead of print
Swinburn BA, Egger G, Raza F (1999) Dissecting obesogenic environments: the development and application of a framework for identifying and prioritizing environmental interventions for obesity. Prev Med 29:563–570
Wang X, Zhu H, Snieder H, Su S, Munn D, Harshfield G, Maria BL, Dong Y, Treiber F, Gutin B, Shi H (2010) Obesity related methylation changes in DNA of peripheral blood leukocytes. BMC Med 8:87
Wilding J (2012) Are the causes of obesity primarily environmental? Yes. BMJ 11(345):e5843
Yanovski SZ, Yanovski JA (2011) Obesity prevalence in the United States -Up, Down or sideway. New Engl J Med 364:987–989
Zhang S, Rattanatray L, Morrison JL, Nicholas LM, Lie S, McMillen IC (2011) Maternal obesity and the early origins of childhood obesity: weighing up the benefits and costs of maternal weight loss in the periconceptional period for the offspring. Exp Diabetes Vol 2011, Article ID 585749, 10 pages

Chapter 2
Physiological Bases of Physical Limitations During Exercise

Claudio L. Lafortuna

Abstract During the execution of physical activity, obese individuals suffer from considerable functional limitations which are principally related to the excess of their body mass, more than to other morbid conditions co-occuring with obesity. Important changes in body composition are associated with the rise in body adiposity, so that obese persons, and women in particular, result with a lower amount of fat-free mass and leg muscle volume per unit body mass, but with absolute larger quantities. Therefore, motor performance of obese individuals during anaerobic activity, which includes the execution of short and intense efforts, is largely reduced by the imbalance between the size of available skeletal muscle and the disproportionate accumulation of fat tissue, although the capacity of strength and absolute leg power output production are increased. Differently, activities relying upon aerobic metabolism, such as bicycle ergometer exercise or walking, are limited in obese subjects principally due to the greater metabolic energy required to move the heavier body, or single body segments involved in movements, which may ultimately exceed the limits of the aerobic capacity. The physiological mechanisms underlying these functional limitations during exercise in obesity should be considered when devising protocols of physical activity and rehabilitation aiming to cure the body mass excess.

2.1 Engines, Fuel, and Performance

In much the same way as cars do, humans move in the space, perform mechanical work, consume fuel, and achieve a power performance. It is remarkable that Dorothy M. Needham, one of the most eminent biochemists of the past century, titled *Machina Carnis* [i.e. muscle machine] her comprehensive book considering the nature of the biochemical processes fuelling the muscle engine and the complex regulation of its energy supply (Needham 1971).

C. L. Lafortuna (✉)
Istituto di Bioimmagini e Fisiologia Morecolare del Consiglio Nazionale delle Ricerche, Segrate, Milan, Italy
e-mail: Claudio.lafortuna@cnr.it

P. Capodaglio et al. (eds.), *Disabling Obesity*, DOI: 10.1007/978-3-642-35972-9_2,

Indeed, muscles accomplish the essential task to produce force and motion taking energy from the chemical fuel originally coming from foods, not so different from the combustion engines of ordinary cars. To this end, the energy contained in chemical bonds of the high-energy phosphate substance adenosine triphosphate (ATP), considered as the ultimate energy source for muscle contraction, is made available as free energy by chemical splitting and is used to produce mechanical work, while part is degraded as heat. For a detailed account of the biochemical pathways supporting muscle energetics, the readers are referred to the excellent review by Pietro E. di Prampero (di Prampero 1981), which links with clearness the energy processes at cellular level with physiological phenomena detectable at whole organism level in the exercising individual.

In a concise formulation, all energy required for muscle shortening is provided by the breakdown of ATP at cellular level. Nevertheless, energy supply provided by ATP stores within the muscle is relatively limited, and the process of ATP refurnishing takes place concomitant with utilization.

When exercise has an intensity requiring an amount of energy compatible with the kinetics of the refurnishing process, ATP is restored by means of oxidative mechanisms consuming oxygen and utilizing the body stores of carbohydrates and fats as substrate, and exercise can be maintained for a long duration (several minutes or longer) in a steady-state condition (aerobic exercise). Differently, when exercise has an intensity which outbalances the capability of the mechanisms for energy restoration, it is supported for a limited period (several seconds) by the ATP sources at muscular level (anaerobic exercise).

The great deal of physiological responses intervening in the course of aerobic exercise substantially represent the functional adjustment of the elements belonging to the oxygen transport pathway from ambient air to mitochondria to the increased oxygen flow required by exercising muscles. In conditions of metabolic steady-state during exercise at constant intensity, the pulmonary inflow of oxygen measured by assessing the respiratory gas exchanges with the technique of indirect calorimetry does correspond to the oxygen consumed at cellular level. Thus, due to the stoichiometric nature of the relation between utilized oxygen and ATP yield, the amount of oxygen consumed during aerobic submaximal exercise corresponds to the actual energy required to sustain that physical activity. In fact, depending on the oxidized substrate, 1 L of oxygen consumed has an energy equivalent ranging from 4.7 kcal (or 19.6 kJ), when lipid oxidation is the sole source of energy, to 5.1 kcal (or 21.1 kJ), when carbohydrates are the sole source of energy.

Therefore, during exercise when muscles actively shorten producing mechanical work under aerobic conditions, the rate of oxygen consumption (VO2) increases progressively as exercise intensity rises. Figure 2.1 illustrates the relation between VO2 and exercise intensity for a young man running on a treadmill. Data shown in the figure are taken from a historical record obtained by Rodolfo Margaria in Pavia in the late 1930s (Margaria 1938), but they are fairly representative of the physiological responses associated with the energetics of exercise which can also be investigated with modern technologies.

Figure 2.1 represents the common knowledge of how VO2 relates substantially linearly to exercise intensity during a typical incremental test (region between

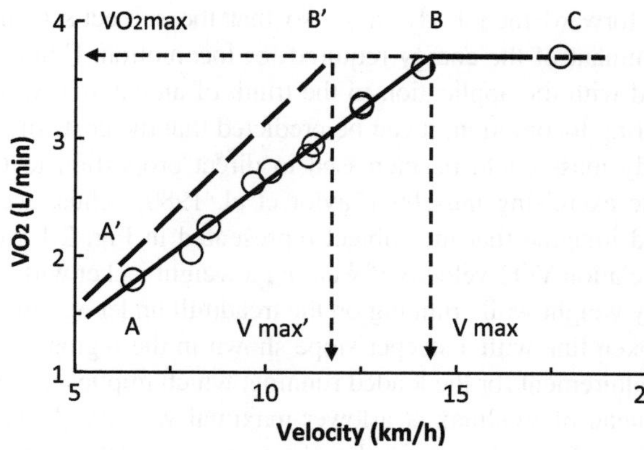

Fig. 2.1 The relationship between the rate of oxygen consumption (VO2) and running velocity on treadmill. The single data points are obtained in a young man exercising at steady-state conditions and enable the determination of the individual's maximal aerobic capacity (VO2max) and the maximal aerobic performance (Vmax). The prediction of exercise energetics performed while wearing an added weight is represented by the broken line, which also indicates maximal aerobic performance during loaded running (V′max). See text for details. (Data from Margaria 1938)

points A and B) and ignores the upward deviation in VO2 occurring during prolonged exercise at high workloads. Above point B, the relation levels off (region between points B and C) and the plateau in VO2 represents the individual's maximal capacity to transfer oxygen from the ambient air to the mitochondria (VO2max), either determined by metabolic oxidative capacity limitation in muscle mitochondria or by oxygen supply limitation at some step in the oxygen transport pathway, depending on the ongoing physiological scenario. For an integrative view of the factors limiting maximal oxygen transfer, the readers may enjoy a recent comprehensive review by Wagner (2006).

To a first approximation, point B also indicates the maximal exercise performance (Vmax) which could be sustained aerobically. Below point B, in conditions corresponding to submaximal exercise, VO2 increases with exercise intensity and in steady-state conditions exactly matches the amount of oxygen required by exercising muscles to perform that kind and intensity of mechanical work. The incline of the line for the relation between VO2 and exercise intensity is dependent on the energy required for that specific activity (i.e. running on treadmill, for the case depicted in Fig. 2.1). For individuals with similar body mass, structure, and shape, moving their body segments according to the same biomechanical paradigm and to the same physiological coordination pattern of skeletal muscles involvement, the energy cost of a given physical activity is expected to be similar (Taylor et al. 1982), so that VO2 required for running at a given submaximal velocity is the same for different individuals, independent of their VO2max.

Nevertheless, the overall energy required for exercise performance is also strongly influenced by the mass of the body or the body segments involved in motion during the specific exercise activity under consideration. During terrestrial locomotion, humans have to perform at each step a given amount of mechanical work to propel forward their body mass, so that the subject's own weight is an important determinant of the energy required for locomotion. From several experiments performed with the application to the trunk of an external weight in healthy lean subjects during locomotion, it can be predicted that the costs of supporting and accelerating body mass could be increased in direct proportion to the extra mass supported by the exercising muscles (Taylor et al. 1980). Thus, according to this model, we could imagine that the subject represented in Fig. 2.1 would display a left shift in the relation VO2-velocity if wearing a weight jacket with a load of about 15 % of his body weight while running on the treadmill under aerobic conditions. In Fig. 2.1, the broken line with a steeper slope shown in the region A′–B′ indicates a higher energy requirement for the loaded running, which implies the consequence of attaining the plateau of VO2max at a lower maximal velocity (Vmax′). Therefore, it appears that the individual's maximal performance in aerobic conditions depends upon both his maximal aerobic capacity and the energy cost of exercise.

It is relevant to understand the interplay between the variables such as the individual's aerobic capacity, the metabolic energy required for exercise, and the maximal aerobic performance since in obesity these reference parameters undergo changes which may seriously interfere with the individual's capability to sustain the execution of specific physical exercise and, in certain cases, even of simple movements associated with everyday life activities.

2.2 Muscle Engine and Muscle Functioning: Implications for Maximal Strength, Power Output, and Anaerobic Performance

2.2.1 Body Composition

The most self-evident structural body change in obesity is an enormous global increase in fat tissue accumulation, which greatly affects body composition in obese people, with important consequences for motor activity.

The main parameters characterizing body composition according to a two-compartment model (see Ellis 2000, for a comprehensive review of models and methods for the estimate of body composition) in a typical sample of young seriously obese men and women (BMI > 35 kg/m^2) are shown in Table 2.1 and compared with those coming from a sample of normal weight individuals of the same age and body frame. Apart from the striking difference in the amount of fat tissue between the two groups, it is interesting to note the trend of fat-free mass (FFM). It can be appreciated that, in normal weight individuals, FFM constitutes the largest share of body mass (although with significant differences between men and women), and this percent is to a great extent reduced in obese individuals and even outbalanced by fat mass. By contrast, when considered in absolute terms of mass units, FFM is actually significantly increased in obese men and women in comparison with their respective normal weight counterpart.

Table 2.1 Average values (±SD, in parentheses) of anthropometric and body composition parameters in two healthy groups of normal weight (NW) and obese (OB) young men and women, matched for age and height

	NW ($n = 22$)[a]		OB ($n = 22$)[b]	
	Males	Females	Males	Females
Age (years)	27.3 (4.3)	27.9 (3.8)	27.2 (3.2)	27.9 (3.6)
Height (m)	1.75 (0.05)	1.64 (0.06)	1.75 (0.07)	1.64 (0.05)
Body mass (kg)	68.0 (6.6)	57.8 (7.1)	130.6[c] (12.3)	113.7[c] (12.1)
Body mass index (kg/m²)	22.2 (1.7)	21.6 (2.0)	42.6[c] (3.3)	42.1[c] (2.9)
Fat mass (kg)	7.3 (3.7)	16.6 (5.1)	59.2[c] (6.0)	58.9[c] (7.3)
Fat-free mass (kg)	60.7 (4.7)	44.2 (3.8)	71.4[c] (8.0)	54.9[c] (5.8)
Fat-free mass (%)	89.6 (4.7)	77.0 (6.5)	54.6[c] (2.7)	48.3[c] (2.5)

[a]Unpublished data from the Laboratory of Biomechanics at Istituto di Bioimmagini e Fisiologia Molecolare del CNR, Segrate (MI); [b]data selected from the cohort published in Lafortuna et al. (2012) to match the NW group; [c]significantly different from the NW counterpart ($p < 0.001$, Student's t test for unpaired data)

Bearing in mind that FFM is an indicator, though rather crude, of skeletal muscle mass, it can be easily realized that these very simple relations have relevant repercussions on the performance of the different motor activities powered by the muscle action. In fact, during brisk movements entailing body displacement which require a considerable power output under anaerobic conditions (such as stepping up a ramp of stairs or performing short sprints) the determinant of actual performance is the ratio between the muscle mass and total body mass, which dictates the power output per unit body mass (Lafortuna et al. 2002).

Differently, during static (or quasi static) actions with a relevant development of strength, the performance is mainly determined by the absolute size of involved muscles (Maughan et al. 1983; Fukunaga et al. 2001b).

2.2.2 *Skeletal Muscle Size*

In spite of a relative abundance of studies which report body composition trends in relation with body adiposity and provide evidence of a definite increase of FFM as a function of BMI (Norgan 1994; Schutz et al. 2002; Lafortuna et al. 2004, 2005), investigations addressing the effects of obesity on muscle characteristics are surprisingly scanty.

Indeed, skeletal muscle can be characterized by quantitative (size) and qualitative (composition) structural attributes which concur in determining functional performance. A recent study, conducted on a sample of men and women with different body adiposity, investigated these structural characteristics in lower limb muscles with computed tomography (CT) and image analysis (Lafortuna et al. Influence of body adiposity on structural characteristics of skeletal muscle in men and women. unpublished data). The volume of muscle, after accounting for limb length and age, resulted to increase significantly with the degree of body adiposity in both genders, although at a lower rate in women, a trend similar to that observed for FFM. In fact, from these results it can be established that overall leg muscle volume increases on average by about 2.4 kg every 5 BMI units in man and about 1.3 kg in females, when variations due to age are accounted for. This muscle enlargement can be viewed as an adaptive response to increased body mass which acts as a chronic training load during everyday motor activity. Indeed skeletal muscle is an adaptable tissue actively responding to mechanical loading, in addition to growth factors and nutrients, and is finely regulated by a complex and not completely understood cellular signalling network ultimately controlling protein synthesis and sarcomere addition [for a recent review of muscle responses to mechanical loading of exercise the reader may (see Miyazaki and Esser 2009)]. Unfortunately, at present no study has been performed to detect the involvement of any regulatory process specifically controlling the muscle response to the obesity-related mechanical overload.

Otherwise, when the volume of muscle in lower limbs is expressed per unit body mass, which may be considered as an indicator of leg muscle amount available for the displacement of each kilogram of the whole body during the movements relying upon leg action, adiposity plays a negative role, particularly in women.

2.2.3 Muscle Structure: Function Relationships

Thus, most importantly, these variations of muscle size due to the effect of body adiposity closely influence the individual's functional capabilities for movements, both in relation to its absolute quantity or relative to body mass, according to the relevant function considered. In Fig. 2.2 leg muscle volume normalized by limb length (panel *a*), representing the average muscle cross-sectional area (CSA) through the whole limb, or normalized by body mass (panel *b*), representing the limb muscle available per unit body mass, is plotted against BMI in men and women over a range of body adiposity. In the lower panels of the figure, the functional correlates of these parameters obtained in obese individuals are plotted against BMI. Panel *c* shows the trend of peak force expressed during the push phase of a vertical jump performed under conditions of maximal effort, whereas average power output per unit body mass developed during a Margaria stair climbing test (Margaria et al. 1966), which correlates directly with the real performance such as the ascending velocity, is illustrated in panel *d*. Overall, the figure shows how motor functions parallel the structural changes intervening at different degrees of obesity. A similar relationship between body composition and muscle performance has been observed also among obese children and adolescent in whom FFM was a positive

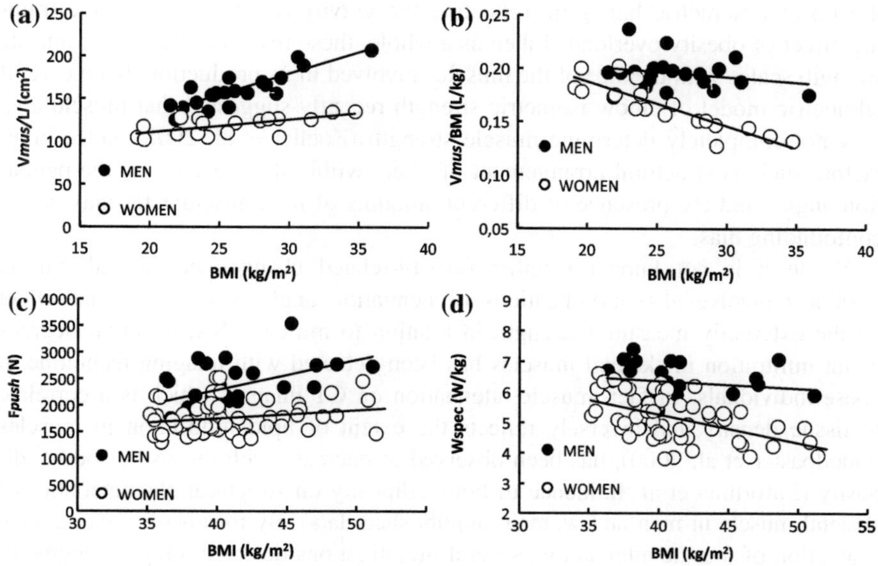

Fig. 2.2 The effect of body adiposity on skeletal muscle structure and performance in men and women. The volume of leg muscle normalized per limb length (Vmus/Ll) and per body mass (Vmus/BM) is plotted as a function of BMI (panels *a* and *b*). Peak force expressed during the push phase (Ppush) of a vertical jump performed under conditions of maximal effort (panel *c*) and average mass-specific maximal power output (Wspec) during a stair climbing test (panel *d*) are plotted against BMI. *Closed circles* males; *open circles* females. Regression lines through the data are shown as continuous lines. (Data in panels *a* and *b* from Lafortuna et al. Influence of body adiposity on structural characteristics of skeletal muscle in men and women. unpublished data; in panels *c* and *d* from Lafortuna et al. 2005)

linear correlate of absolute lower limb power output, independent of the different body composition ensuing between boys and girls at puberty (Sartorio et al. 2006).

Contrasting a group of extremely obese young men with a counterpart of lean individuals matched by age and stature, Maffiuletti et al. (2007) found that obese subjects display significantly more elevated absolute but lower relative (per unit body mass) muscle torque and power output during isometric and isokinetic knee extensions. Pescatello et al. (2007) reported similar results concerning peak isometric and isotonic elbow strength in a quite large sample of overweight/obese and normal weight individuals, detecting also a significantly larger CSA of biceps muscle in overweight/obese group. Such trends in absolute and relative strength for effect of obesity have been found also among male adolescents performing isometric knee extensions (Abdelmoula et al. 2012).

Interestingly, in obese subjects the stronger muscle groups during isotonic exercise, in comparison with normal weight persons, were those mainly involved in movements against the force of gravity (Lafortuna et al. 2004), as similarly reported by Hulens et al. (2001a, b) who observed higher strength in obese women than in the lean counterpart during isokinetic knee extension, but not during knee flexion and isometric handgrip, reflecting the gravity-related nature of the training effect of obesity overload. Taken as a whole, these results indicate that muscle strength scales with the size of the muscles involved in its production, but a careful allometric model of elbow isometric strength recently suggested that muscle area may not completely determine muscle strength (Zoeller et al. 2008), since other factors such as structural arrangement of fibers within the muscle (i.e. the pennation angle) and the presence of different amounts of intramuscular fat may act as confounding bias.

While it is not known whether obesity-related changes in skeletal muscle attributes involve also modifications of pennation angle, with significant effect on the externally measured strength in relation to muscle CSA, different degrees of fat infiltration in skeletal muscles has been detected with imaging technique in obese individuals. In fact, muscle attenuation on CT images, which is a correlate of tissue density and inversely reflects the extent of lipid infiltration in muscles (Goodpaster et al. 2000), has been observed to decrease with the rise of body adiposity (Lafortuna et al. Influence of body adiposity on structural characteristics of skeletal muscle in men and women. unpublished data). By means of a quantitative evaluation of muscle attenuation, several investigations detected a higher degree of such infiltration within muscles in obesity and older age (Kelley et al. 1991; Ryan and Nicklas 1999) and found it to be associated with reduced motor performances (Hilton et al. 2008; Goodpaster et al. 2001; Visser et al. 2005), besides an increased metabolic risk (Kelley et al. 1991; Goodpaster et al. 1997). Hence, although muscle size is known to be the major predictor of force and power production in the general context (Maughan et al. 1983; Fukunaga et al. 2001b), qualitative attributes of muscle composition such as fat infiltration may be important in determining the association between muscle mass and its function, and, especially in obesity during older age, may be of value to characterize the functional limitation of motor capabilities.

Thus, mainly due to a larger muscle mass, muscle strength and power output in absolute terms have been reported to be substantially higher in obese people under different experimental conditions, although their motor performance during anaerobic tasks requiring brief and intense efforts results considerably reduced. This functional condition is summarized in Fig. 2.3, which compares the value of peak vertical force attained during the push phase of a vertical jump performed under maximal effort by lean and obese young men and shows the relationship resulting between BMI and the vertical displacement of the center of body mass (an indicator of the anaerobic performance) during the same experimental conditions. In spite of the remarkably higher strength developed during this maneuver, the actual performance of obese individuals is fairly lower and decreases with their degree of adiposity.

Finally, a last issue concerning the different effect exerted by body adiposity on muscle size in the two genders should be addressed, in view of the clinical relevance and the potential impact on rehabilitation strategies. As reported above, men and women undergo different changes in body composition along with the increase in body adiposity (see also panel *a* in Fig. 2.2). Due to the lower increase of muscle mass with rising adiposity, the muscle volume per unit body mass (structural correlate of anaerobic performance) decreases considerably in women as a function of

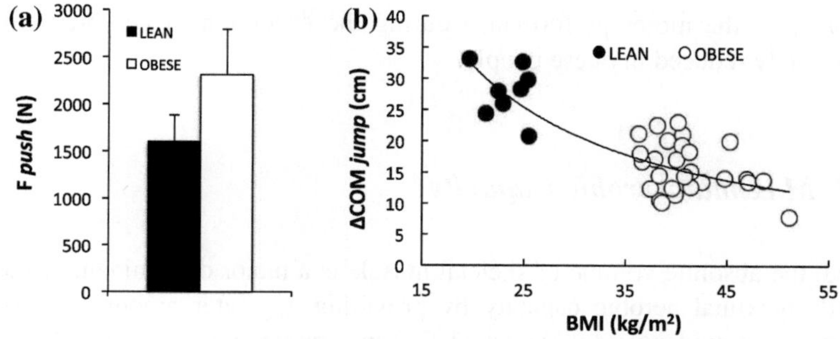

Fig. 2.3 Strength and performance during maximal anaerobic effort in obese and lean young men. Average peak force (±SD) expressed during the push phase (Ppush) of a vertical jump (panel *a*) and effect of BMI on the vertical displacement of body center of mass (ΔCOM) attained during the same test (panel *b*). *Closed symbols* lean subjects; *open symbols* obese subjects. (Data from Lafortuna et al. 2002)

BMI (see panel *b* in Fig. 2.2), with important practical implications. Similar trends have also been reported in the variation of FFM/BM with the degree of obesity (Lafortuna et al. 2005). This raises the hypothesis that different mechanisms may intervene in men and women to regulate muscle size as a function of body adiposity. In fact, it has been observed that even a prolonged heavy strength training stimulus induces a significantly greater increase of muscle volume in males than in females (Ivey et al. 2000). Besides a possible role of the higher levels of masculine testosterone, several gender-related differences in the regulation of growth hormone, as differences in body composition and fat distribution (Roelen et al. 1997), and in leptin plasma concentration (Llopis et al. 1998), may participate in determining such a gender-dependent response of muscle mass to the increase of body adiposity among obese individuals. Moreover, due to the loss of muscle size with advancing age, the concomitance of obesity and older age in females appears to involve a particularly unfavorable condition which exposes them to functional derangements of dynamic performance. Thus, as also evidenced with functional testing in a clinical context (Lafortuna et al. 2004), older women with high degree of obesity are prone more than men to suffer from major reductions in anaerobic leg power output induced by obesity, which may interfere with everyday locomotor abilities and even ultimately lead more rapidly toward thresholds for disabled mobility.

2.3 The Energy Expenditure for Body Movements: Implications for Aerobic Performance

In spite of the definite increase in the size of muscle engine under the effect of body adiposity, which permits the production of higher strength and absolute power, besides a diminished anaerobic performance, as discussed in the previous section, also the motor performance during the execution of aerobic activity is considerably reduced in obese people.

2.3.1 Maximal Aerobic Capacity

Indeed, the absolute volume of skeletal muscle is a major determinant of the individual's maximal aerobic capacity by providing a greater amount of oxidative machinery, as evidenced by Davies (1973) in a pioneering study on African subjects with a different condition of nutrition and involvement in physical activity. More recently, Tolfrey et al. (2006) showed with an allometric approach in boys and men that lower limb muscle mass is a better predictor of VO2max than body mass or FFM. Nevertheless, a significant positive linear correlation has been detected for the relationship between VO2max or VO2 at anaerobic threshold and FFM in study groups including both lean and obese young women (Lafortuna et al. 2006) or adolescent obese girls (Lafortuna et al. 2009). As a consequence, among obese adults and adolescents, a higher aerobic capacity in absolute terms, although remarkably reduced when considered per unit body mass, is often reported (Lafortuna et al. 2006, 2009; Hulens et al. 2001a). Goran et al. (2000) investigated a sample of subjects composed of a group of children in a spectrum of different body compositions and a group of overweight and normal women, and found that the amount of body fat has no direct influence on maximal aerobic capacity, which in turn depends mainly on FFM, absolute VO2max resulting significantly higher in obese than in lean subjects, without any significant difference when expressing VO2max by unit FFM.

Nevertheless, as remarked for anaerobic performance, also during tasks with an aerobic profile, the exercise capacity is remarkably reduced in obesity (Hulens et al. 2001a). In fact, the execution of most modalities of exercise require more mechanical work and demand more energy for obese than for lean individuals, in proportions dictated by the mass of the body parts involved and the biomechanical paradigm employed for the specific movements. Hence, while the greater VO2 attained by obese subjects at maximal exercise may be considered as the consequence of a larger FFM and muscle mass (as discussed above), the greater values displayed during submaximal work are related to higher energy requirements of the specific motor activity.

2.3.2 Bicycle Exercise

During exercise on bicycle ergometers, the same mechanical external work is performed by obese and normal-weight individuals, since the subjects use mainly their legs to generate a rotary motion and overcome an imposed load acting at the level of the crank resistance. Notwithstanding this mechanical setting, the energy cost of bicycle ergometer exercise has been found to be increased by about 15–20 % for effect of body mass in the two genders at different ages by several investigators (Anton-Kuchly et al. 1984; Berry et al. 1993; Hulens et al. 2001a; Salvadori et al. 1999; Lafortuna et al. 2006, 2008, 2009).

Anton-Kuchly et al. (1984) estimated that about 60–70 % of such a difference could be attributed to the extra work of moving heavier legs, while the remaining 20–30 % could be explained by the postural activity required to stabilize the larger mass of trunk while pedaling on the ergometer. In fact, Lafortuna et al. (2008) assessed that the net energy required to move the legs was dependent on the degree of body adiposity, and that, when this amount of energy is accounted for, the energy cost of exercise is independent of BMI at all workloads and no difference is detectable between obese and lean individuals. Also, using a model based on pedal frequency, leg mass and workload devised by Francescato et al. (1995) to predict the energy expended for cycling, Lafortuna et al. (2006) obtained evidence that the obesity-related higher energy cost of cycling is largely due to the effect of the higher mass of legs.

In principle, also other factors such as higher oxygen cost of breathing (Kress et al. 1999) and increased proportion of glycolytic IIb type muscle fibers (Kriketos et al. 1997), which have been demonstrated to be substantially less efficient than type I fibers during cycling (Coyle et al. 1992), may contribute to the increased cost of exercise. Nonetheless, the intrinsic muscle efficiency during cycling, as represented by *work efficiency* calculated by the changes of work performed and the energy expended above the condition of free pedaling without load (Berry et al. 1993), has been reported to be similar among obese and lean people (Lafortuna et al. 2008), confirming that the main determinant of higher energy cost of cycling may be better due to the higher energy entailed in limb movements and body stabilization than to intrinsic differences in muscle performance. In Fig. 2.4 is summarized the key findings concerning the energetics of bicycle ergometer exercise in obese and lean individuals. The figure shows the net metabolic energy required for pedaling at different workloads in the two groups (panel *a*), the energy required for the movements of the legs without any mechanical load (panel *b*), and the mechanical efficiency of cycling after accounting for the energy of leg movements (panel *c*).

2.3.3 Walking

Different from the setting of bicycle exercise, during walking, mechanical work is performed involving a large number of body segments to cyclically raise and accelerate the body center of mass at each step, so that the subject's own weight results to be an important load. Therefore, obese people expend much more metabolic energy during walking than non-obese individuals (Melanson et al. 2003, Browning and Kram 2005; Browning et al. 2006; Lafortuna et al. 2008), with values of energy expenditure 1.8 to 2.3-fold higher than those required for

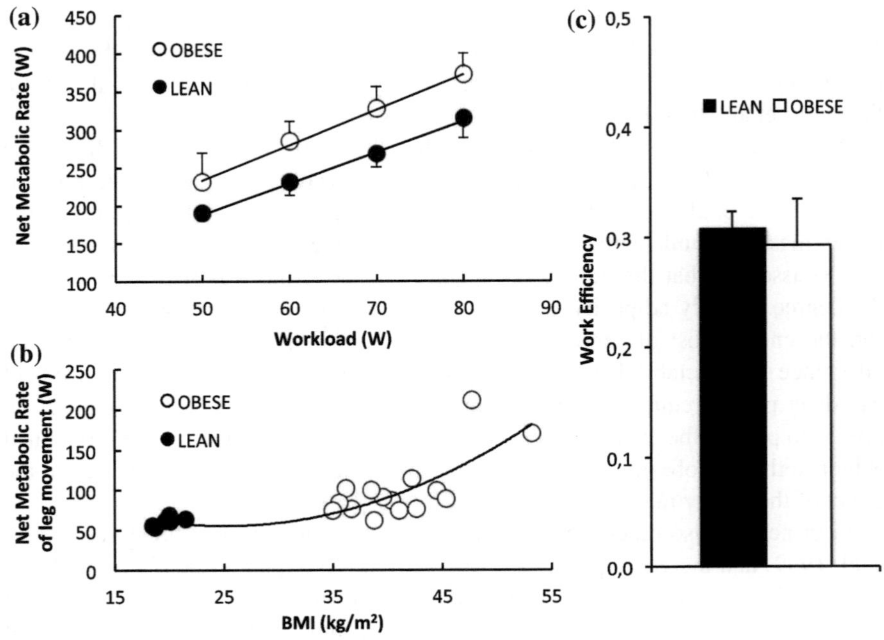

Fig. 2.4 The energetics of bicycle ergometer exercise in obese and lean women. The net metabolic rate required for cycling is plotted against the mechanical workload in panel *a* for the two groups, while the net metabolic rate pertaining to the leg movements without mechanical load is plotted against BMI for each individual subject in panel *b*. The average pedaling *work efficiency*, which accounts for leg movements without load, is shown in panel *c* for obese and lean subjects. *Closed symbols* lean subjects; *open symbols* obese subjects. Vertical bars denote 1 SD. Regression lines through the data are shown as continuous lines. (Data from Lafortuna et al. 2008)

lean individuals, depending on body mass difference, walking speed, and ground incline during locomotion. When the net metabolic rate (gross metabolic cost of standing) is expressed per unit body mass, the difference is considerably reduced but not suppressed. The same trend is also detectable in the energy cost per distance during walking (i.e., the metabolic energy per kilogram body mass to walk a given distance) which has been found to be 5–15 % higher for effect of obesity, as observed by different investigators in obese adults and adolescents (Browning and Kram 2005; Browning et al. 2006; Lafortuna et al. 2008; Peyrot et al. 2009). Interestingly, Browning et al. (2006) found that mass specific metabolic rate and energy cost per distance are related to body composition (body fat %) but not to body mass distribution (thigh mass to body mass ratio).

Figure 2.5 illustrates in panel *a* the large difference in net metabolic energy expended by obese and lean subjects for walking at different velocities, showing in (panel *b*) also the energy per unit body mass and distance, while (panel *c*) presents the relationship between mass-specific metabolic rate expended for walking at 1 m/s and body composition for single individuals with a different degree of body adiposity.

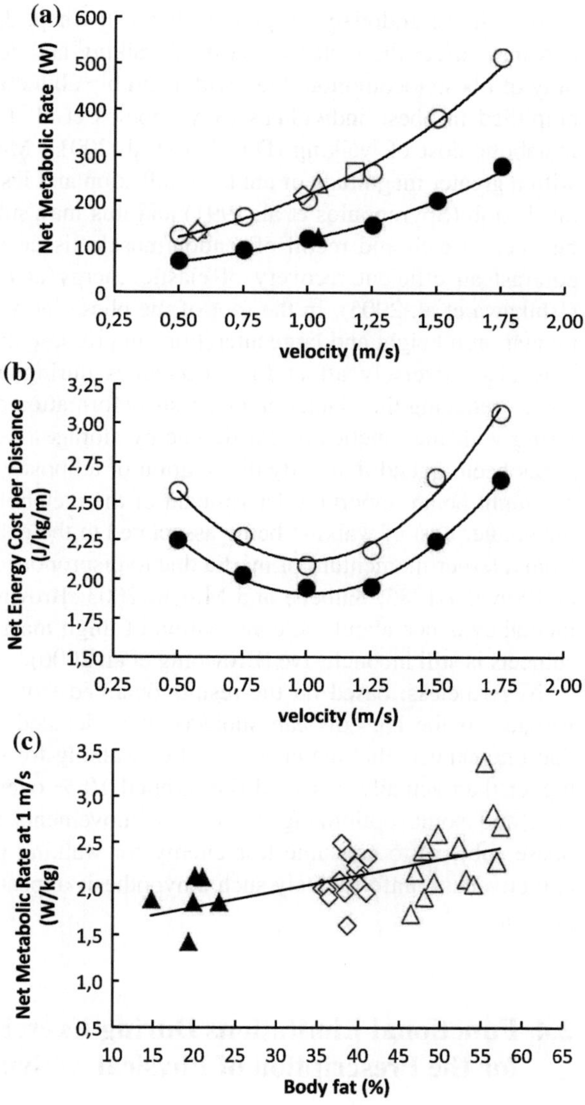

Fig. 2.5 The energetics of walking in obese and lean women and girls. Net metabolic rate in absolute terms and expressed per body mass and per distance (cost per distance) are plotted against walking velocity (panels *a* and *b*). Mass-specific net metabolic rate at the velocity of 1 m/s is plotted as a function of body composition (body fat %) in single individuals ranging from normal weight to obesity classes II and III (panel *c*). Note that mass-specific net metabolic rate at 1 m/s in W/kg numerically corresponds to the net cost per distance in J/kg/m, at the same velocity. Closed symbols: lean subjects; open symbols: obese subjects. Regression lines through the data are also shown in panel *c*. [*Circles* adult women from Browning and Kram 2005; *triangles* adult women from Lafortuna et al. (2008); *squares* adult women from Mattsson et al. (1997); *diamonds* adolescent girls from Lafortuna et al. (2010)]

Thus, due to the higher metabolic energy expended by obese individuals, walking may be an exhausting task, requiring a considerable fraction of the individual's aerobic maximal capacity. Young obese women walking at 1.3 m/s on 4 % incline have been reported to exercise at about 75 % of their maximal capacity, while the lean counterpart did at about 50 % (Lafortuna et al. 2008). It is conceivable that such a task may be scarcely sustained by older and less fit individuals, as confirmed by the findings of Mattsson et al. (1997).

Overall, these findings evidence that total body mass is a primary determinant of metabolic energy required for walking, but obesity plays an unfavorable role also due to other factors, besides weight excess.

As will be addressed in greater detail in Chap. 3, several biomechanical factors may affect the metabolic cost of walking and could explain the lower economy of obese locomotion. Step width and hip abduction have been observed to be amplified in obese individuals (Spyropoulos et al. 1991), which require a larger metabolic cost of walking (Donelan et al. 2001). Moreover, obese subjects walk with a greater magnitude of ankle dorsiflexion and lesser magnitude of ankle plantar flexion (Spyropoulos et al. 1991) and this may substantially alter the interplay between stretch and recoil of tendon/muscle tissue (Fukunaga et al. 2001a) and contrast an efficient recovery of elastic energy at the end of the contact phase (Ishikawa et al. 2005). In the foot of the obese individual, a significantly reduced plantar arch height and large alterations in pressure distribution (Hills et al. 2001), may also adversely affect foot mechanics during walking gait (Messier et al. 1994), reducing the plantar aponeurosis deformation during foot contact and interfering with the kinetics of elastic energy storage (Ker et al. 1987). Furthermore, it has been argued that body distribution of adiposity rather than body weight per se, might be an important determinant of the metabolic cost of obese locomotion, the greater cost of walking being associated to the extra mechanical work deriving from a larger momentum of inertia due to disproportionately heavier limbs (Myers and Steudel 1985; Saibene and Minetti 2003; Browning et al. 2007), but experimental evidence about the contribution of thigh mass to body mass ratio in obese subjects is still inconclusive (Browning et al. 2006).

Nevertheless, based on the results obtained from experiments with additional weights to the legs of lean subjects, it is debated (Browning and Kram 2009) that one expects that the energy cost of walking for obese people would be fairly higher than actually observed (i.e. around 10 % over that of non-obese individuals), and some optimizing strategy in movements might be involved enabling obese subjects to consume less energy for walking than expected. In spite of its attractiveness, unfortunately such a hypothesis appears substantially speculative at present.

2.4 Functional Limitations During Exercise and Impact for the Prescription of Physical Activity

While the execution of short and brisk movements of chiefly anaerobic character is principally limited in obese people by the imbalance between the size of skeletal muscle and the disproportionate accumulation of fat tissue, the aerobic performance is mainly reduced due to the greater metabolic energy required to move the heavier body, or single body segments.

Thus, a major priority of exercise prescription is to focus on activities which contribute to muscular fitness and preserve muscle mass. This goal seems important also in view of the effects of caloric restriction (a powerful tool against weight excess concomitantly used with physical activity) which result in reductions in both fat and fat-free mass. A net loss of muscle tissue would have negative outcomes on both anaerobic and aerobic functions, especially in women who are inherently endowed with a lower amount of skeletal muscle relative to body mass.

During aerobic activity such as ergometer cycling or walking, the relative aerobic effort at a given exercise intensity (cycling workload or walking velocity) is higher for obese individuals compared with the lean counterpart and during relatively severe exercise the energy requirements may approach or even exceed the limits of the individual's aerobic capacity (Salvadori et al. 1999; Browning and Kram 2005; Lafortuna et al. 2006; Lafortuna et al. 2008). Therefore, also in view of a considerable degree of sedentariness and poor aerobic conditioning frequently observed in obese individuals, the employment of protocols aiming at improving the cardiovascular endurance should be encouraged with a scheme appropriate for obese people.

Finally, the comparison of the physiological responses to cycling and walking is very interesting for the formulation of exercise protocols suitable for obese individuals.

Based on the fact that a smaller muscle mass is used in cycling exercise to attain the same metabolic energy expenditure of walking (Hermansen and Saltin 1969), with a greater metabolic stress and energy requirement per unit of contracting muscle (Koyal et al. 1976), it has been evidenced in obese adults and adolescents (Lafortuna et al. 2008, 2010) that walking is a convenient mode of exercise, compared to cycling. Walking permits in fact to attain any given energy expenditure at a comparatively lower average heart rate (or in a shorter time), with lower lactic acid blood concentration and higher fat oxidation.

In practice, by some reckoning from these data comparing the metabolic responses of the two modalities, to obtain the energy expenditure of 250 kcal with a bout of 25 min activity, cycling should be performed at an intensity requiring an average heart rate of about 160 b/min with 3 g of lipids oxidation, whereas it will be sufficient to walk at intensity requiring an average heart rate of about 130 b/min, with over 11 g of lipid oxidation.

Under this perspective, for obese individuals, inherently limited in their work capacity, it is very attractive to devise forms of physical activity enabling the attainment of a considerable energy expenditure by preference promoting substantial fat oxidation with the lower subjective perception of effort and exercise intensity, which could ultimately allow a better tolerance and adherence to physical activity protocols.

References

Abdelmoula A, Martin V, Bouchant A, Walrand S, Lavet C, Taillardat M, Maffiuletti NA, Boisseau N, Duché P, Ratel S (2012) Knee extension strength in obese and nonobese male adolescents. Appl Physiol Nutr Metab 37:269–275

Anton-Kuchly B, Roger P, Varene P (1984) Determinants of increased energy cost of submaximal exercise in obese subjects. J Appl Physiol 56:18–23

Berry MJ, Storsteen JA, Woodard CM (1993) Effects of body mass on exercise efficiency and VO2 during steady-state cycling. Med Sci Sports Exerc 25:1031–1037

Browning RC, Kram R (2005) Energetic cost and preferred speed of walking in obese versus normal weight women. Obes Res 13:891–899

Browning RC, Kram R (2009) Pound for pound: Working out how obesity influences the energetics of walking. J Appl Physiol 106:1755–1756

Browning RC, Baker EA, Herron JA, Kram R (2006) Effects of obesity and sex on the energetic cost and preferred speed of walking. J Appl Physiol 100:390–398

Browning RC, Modica JR, Kram R, Goswami A (2007) he effects of adding mass to the legs on the energetics and biomechanics of walking. Med Sci Sports Exerc 39:515–525

Coyle EF, Sidossis LS, Horowitz JF, Beltz JD (1992) Cycling efficiency is related to the percentage of type I muscle fibers. Med Sci Sports Exerc 24:782–788

Davies CT (1973) The contribution of leg (muscle plus bone) volume to maximum aerobic power output: the effects of anaemia, malnutrition and physical activity. J Physiol 231:108P–109P

Donelan JM, Kram R, Kuo AD (2001) Mechanical and metabolic determinants of the preferred step width in human walking. Proc Biol Sci 268:1985–1992

Ellis KJ (2000) Human body composition: in vivo methods. Physiol Rev 80:649–680

Francescato MP, Girardis M, di Prampero PE (1995) Oxygen cost of internal work during cycling. Eur J Appl Physiol Occup Physiol 72:51–57

Fukunaga T, Kubo K, Kawakami Y, Fukashiro S, Kanehisa H, Maganaris CN (2001a) In vivo behaviour of human muscle tendon during walking. Proc Biol Sci 268:229–233

Fukunaga T, Miyatani M, Tachi M, Kouzaki M, Kawakami Y, Kanehisa H (2001b) Muscle volume is a major determinant of joint torque in humans. Acta Physiol Scand 172:249–255

Goodpaster BH, Thaete FL, Simoneau JA, Kelley DE (1997) Subcutaneous abdominal fat and thigh muscle composition predict insulin sensitivity independently of visceral fat. Diabetes 46:1579–1585

Goodpaster BH, Kelley DE, Thaete FL, He J, Ross R (2000) Skeletal muscle attenuation determined by computed tomography is associated with skeletal muscle lipid content. J Appl Physiol 89:104–110

Goodpaster BH, Carlson CL, Visser M, Kelley DE, Scherzinger A, Harris TB, Stamm E, Newman AB (2001) Attenuation of skeletal muscle and strength in the elderly: The Health ABC Study. J Appl Physiol 90:2157–2165

Goran M, Fields DA, Hunter GR, Herd SL, Weinsier RL (2000) Total body fat does not influence maximal aerobic capacity. Int J Obes Relat Metab Disord 24:841–848

Hermansen L, Saltin B (1969) Oxygen uptake during maximal treadmill and bicycle exercise. J Appl Physiol 26:31–37

Hills AP, Hennig EM, McDonald M, Bar-Or O (2001) Plantar pressure differences between obese and non-obese adults: a biomechanical analysis. Int J Obes Relat Metab Disord 25:1674–1679

Hilton TN, Tuttle LJ, Bohnert KL, Mueller MJ, Sinacore DR (2008) Excessive adipose tissue infiltration in skeletal muscle in individuals with obesity, diabetes mellitus, and peripheral neuropathy: association with performance and function. Phys Ther 88:1336–1344

Hulens M, Vansant G, Lysens R, Claessens AL, Muls E (2001a) Exercise capacity in lean versus obese women. Scand J Med Sci Sports 11:305–309

Hulens M, Vansant G, Lysens R, Claessens AL, Muls E, Brumagne S (2001b) Study of differences in peripheral muscle strength of lean versus obese women: an allometric approach. Int J Obes Relat Metab Disord 25:676–681

Ishikawa M, Komi PV, Grey MJ, Lepola V, Bruggemann GP (2005) Muscle-tendon interaction and elastic energy usage in human walking. J Appl Physiol 99:603–608

Ivey FM, Roth SM, Ferrell RE, Tracy BL, Lemmer JT, Hurlbut DE, Martel GF, Siegel EL, Fozard JL, Jeffrey Metter E, Fleg JL, Hurley BF (2000) Effects of age, gender, and myostatin genotype on the hypertrophic response to heavy resistance strength training. J Gerontol A Biol Sci Med Sci 55: M641–648

Kelley DE, Slasky BS, Janosky J (1991) Skeletal muscle density: effects of obesity and non-insulin-dependent diabetes mellitus. Am J Clin Nutr 54:509–515

Ker RF, Bennett MB, Bibby SR, Kester RC, Alexander RM (1987) The spring in the arch of the human foot. Nature 325:147–149

Koyal SN, Whipp BJ, Huntsman D, Bray GA, Wasserman K (1976) Ventilatory responses to the metabolic acidosis of treadmill and cycle ergometry. J Appl Physiol 40:864–867

Kress JP, Pohlman AS, Alverdy J, Hall JB (1999) The impact of morbid obesity on oxygen cost of breathing (VO2RESP) at rest. Am J Respir Crit Care Med 160:883–886

Kriketos AD, Baur LA, O'Connor J, Carey D, King S, Caterson ID, Storlien LH (1997) Muscle fibre type composition in infant and adult populations and relationships with obesity. Int J Obes Relat Metab Disord 21:796–801

Lafortuna CL, Fumagalli E, Vangeli V, Sartorio A (2002) Lower limb alactic anaerobic power output assessed with different techniques in morbid obesity. J Endocrinol Invest 25:134–141

Lafortuna CL, Agosti F, Marinone PG, Marazzi N, Sartorio A (2004) The relationship between body composition and muscle power output in men and women with obesity. J Endocrinol Invest 27:854–861

Lafortuna CL, Maffiuletti NA, Agosti F, Sartorio A (2005) Gender variations of body composition, muscle strength and power output in morbid obesity. Int J Obes (Lond) 29:833–841

Lafortuna CL, Proietti M, Agosti F, Sartorio A (2006) The energy cost of cycling in young obese women. Eur J Appl Physiol 97:16–25

Lafortuna CL, Agosti F, Galli R, Busti C, Lazzer S, Sartorio A (2008) The energetic and cardiovascular response to treadmill walking and cycle ergometer exercise in obese women. Eur J Appl Physiol 103:707–717

Lafortuna CL, Agosti F, Busti C, Galli R, Sartorio A (2009) The energy cost of cycling and aerobic performance of obese adolescent girls. J Endocrinol Invest 32:647–652

Lafortuna CL, Lazzer S, Agosti F, Busti C, Galli R, Mazzilli G, Sartorio A (2010) Metabolic responses to submaximal treadmill walking and cycle ergometer pedalling in obese adolescents. Scand J Med Sci Sports 20:630–637

Lafortuna CL, Agosti F, De Col A, Pera F, Adorni F, Sartorio A (2012) Prevalence of the metabolic syndrome and its components among obese men and women in Italy. Obes Facts 5:127–137

Llopis MA, Granada ML, Cuatrecasas G, Formiguera X, Sánchez-Planell L, Sanmartí A, Alastrué A, Rull M, Corominas A, Foz M (1998) Growth hormone-binding protein directly depends on serum leptin levels in adults with different nutritional status. J Clin Endocrinol Meta 83:2006–2011

Maffiuletti NA, Jubeau M, Munzinger U, Bizzini M, Agosti F, De Col A, Lafortuna CL, Sartorio A (2007) Differences in quadriceps muscle strength and fatigue between lean and obese subjects. Eur J Appl Physiol 101:51–59

Margaria R (1938) Sulla fisiologia, e specialmente sul consumo energetico, della marcia e della corsa a varie velocità ed inclinazioni del terreno. Atti Accademia Nazionale dei Lincei. Memorie 7: 299-368

Margaria R, Aghemo P, Rovelli E (1966) Measurement of muscular power (anaerobic) in man. J Appl Physiol 21:1662–1664

Mattsson E, Larsson UE, Rössner S (1997) Is walking for exercise too exhausting for obese women? Int J Obes Relat Metab Disord 21:380–386

Maughan RJ, Watson JS, Weir J (1983) Strength and cross-sectional area of human skeletal muscle. J Physiol 338:37–49

Melanson EL, Bell ML, Knoll JR, Coelho LB, Donahoo WT, Peters JC, Hill JO (2003) Body mass index and sex influence the energy cost of walking at self-selected speeds (Abstract). Med Sci Sports Exerc 35:S183

Messier SP, Davies AB, Moore DT, Davis SE, Pack RJ, Kazmar SC (1994) Severe obesity: effects on foot mechanics during walking. Foot Ankle Int 15:29–34

Miyazaki M, Esser KA (2009) Cellular mechanisms regulating protein synthesis and skeletal muscle hypertrophy in animals. J Appl Physiol 106:1367–1373

Myers MJ, Steudel K (1985) Effect of limb mass and its distribution on the energetic cost of running. J Exp Biol 116:363–373

Needham DM (1971) Machina Cranis. The Biochemistry of Muscular Contraction in its Historical Development. Cambridge University Press, Cambridge

Norgan NG (1994) Population differences in body composition in relation to the body mass index. Eur J Clin Nutr 48(Suppl 3):S10–S25

Pescatello LS, Kelsey BK, Price TB, Seip RL, Angelopoulos TJ, Clarkson PM, Gordon PM, Moyna NM, Visich PS, Zoeller RF, Gordish-Dressman HA, Bilbie SM, Thompson PD, Hoffman EP (2007) The muscle strength and size response to upper arm, unilateral resistance training among adults who are overweight and obese. J Strength Cond Res 21:307–313

Peyrot N, Thivel D, Isacco L, Morin JB, Duche P, Belli A (2009) Do gait mechanical parameters explain the higher metabolic cost of walking in obese adolescents? J Appl Physiol 106:1763–1770

di Prampero PE (1981) Energetics of muscular exercise. Rev Physiol Biochem Pharmacol 89:143–222

Roelen CA, Koppeschaar HP, de Vries WR, Snel YE, Doerga ME, Zelissen PM, Thijssen JH, Blankenstein MA (1997) Visceral adipose tissue is associated with circulating high affinity growth hormone-binding protein. J Clin Endocrinol Metab 82:760–764

Ryan AS, Nicklas BJ (1999) Age-related changes in fat deposition in mid-thigh muscle in women: relationships with metabolic cardiovascular disease risk factors. Int J Obes Relat Metab Disord 23:126–132

Saibene F, Minetti AE (2003) Biomechanical and physiological aspects of legged locomotion in humans. Eur J Appl Physiol 88:297–316

Salvadori A, Fanari P, Fontana M, Buontempi L, Saezza A, Baudo S, Miserocchi G, Longhini E (1999) Oxygen uptake and cardiac performance in obese and normal subjects during exercise. Respiration 66:25–33

Sartorio A, Agosti F, De Col A, Lafortuna CL (2006) Age- and gender-related variations of leg power output and body composition in severely obese children and adolescents. J Endocrinol Invest 29:48–54

Schutz Y, Kyle UU, Pichard C (2002) Fat-free mass index and fat mass index percentiles in Caucasians aged 18–98 years. Int J Obes Relat Metab Disord 26:953–960

Spyropoulos P, Pisciotta JC, Pavlou KN, Cairns MA, Simon SR (1991) Biomechanical gait analysis in obese men. Arch Phys Med Rehabil 72:1065–1070

Taylor CR, Heglund NC, McMahon TA, Looney TR (1980) Energetic cost of generating muscular force during running. J Exp Biol 86:9–18

Taylor CR, Heglund NC, Maloiy GM (1982) Energetics and mechanics of terrestrial locomotion. I. Metabolic energy consumption as a function of speed and body size in birds and mammals. J Exp Biol 97:1–21

Visser M, Goodpaster BH, Kritchevsky SB, Newman AB, Nevitt M, Rubin SM, Simonsick EM, Harris TB (2005) Muscle mass, muscle strength, and muscle fat infiltration as predictors of incident mobility limitations in well-functioning older persons. J Gerontol A Biol Sci Med Sci 60:324–333

Wagner PD (2006) The oxygen transport system: integration of functions. In: Tipton CM (ed) ACSM's Advanced Exercise Physiology. Lippincott-Williams and Wilkins, Philadelphia

Zoeller RF, Ryan ED, Gordish-Dressman H, Price TB, Seip RL, Angelopoulos TJ, Moyna NM, Gordon PM, Thompson PD, Hoffman EP (2008) Allometric scaling of isometric biceps strength in adult females and the effect of body mass index. Eur J Appl Physiol 104:701–710

Chapter 3
Biomechanics of Basic Activities

Veronica Cimolin, Paolo Capodaglio and Manuela Galli

Abstract As excess weight is able to influence the biomechanics of several activities of daily living (walking, standing up, bending, and other movements) causing functional limitations, and possibly predisposing to injury, the quantitative analysis of movement appears a very important tool for the definition of the functional profile in the obese population to then plan appropriate rehabilitation interventions. Three-dimensional motion analysis is a powerful tool for quantitative assessment of movement in patients with movement disorders, including obese patients. In this chapter the applications of quantitative movement analysis in obese individuals are illustrated and reviewed and the experience of a motion analysis laboratory, which routinely applies the quantitative evaluation in obese patients for an improvement of rehabilitation and therapeutic managements, is presented. After a brief overview of the equipment commonly present in a motion analysis laboratory, some experimental setups for the acquisition of movement clinically important for obese individuals (walking, posture, sit to stand, and trunk movement), their application and limitations are described.

V. Cimolin (✉) · M. Galli
Bioengineering Department, Politecnico di Milano, Milano, Italy
e-mail: veronica.cimolin@biomed.polimi.it

M. Galli
e-mail: manuela.galli@biomed.polimi.it

V. Cimolin
Research Lab in Biomechanics and Rehabilitation, San Giuseppe Hospital, Istituto Auxologico Italiano IRCCS, Piancavallo, Verbania, Italy

P. Capodaglio
Rehabilitation Unit and Research Lab in Biomechanics and Rehabilitation, San Giuseppe Hospital, Istituto Auxologico Italiano IRCCS, Piancavallo, Verbania, Italy
e-mail: p.capodaglio@auxologico.it

M. Galli
IRCCS "San Raffaele Pisana" Tosinvest Sanità, Rome, Italy

P. Capodaglio et al. (eds.), *Disabling Obesity*, DOI: 10.1007/978-3-642-35972-9_3,

3.1 Introduction

Three-dimensional motion analysis has turned out to be a powerful tool for a quantitative assessment of the movement. Thanks to several important features—it is noninvasive, allows to repeat the examination more times within a short period of time, provides quantitative and three-dimensional data as for kinematics (trajectories, velocity, accelerations, angles), kinetics (forces, joint moments, joint powers), and quantitative evaluation of muscle activity (electromyography)—quantitative evaluation of motion represents a fundamental instrument in human movement analysis. Assessment of the characteristics of a patient's movement and of significant deviations from normality can be extremely useful in clinical settings, both in the diagnosis of diseases with implications of one of the above systems, and in the planning and monitoring of specific rehabilitative treatments. In clinical realities the low number of reliable, valid, and objective measures to monitor treatment outcomes is one of the most crucial problems in the management of pathological subjects. Clinical evaluations, such as passive range of motion, measures of muscle tone, and even video recording, which are frequently used by clinicians, have all shown poor reproducibility between observers and sessions, even when tested under standardized conditions. From these limitations, it's evident the need to introduce a method of instrumental assessment that is able to supply the clinician with quantitative, three-dimensional information relating both to kinematic and kinetic aspects of motion and to the pattern of muscle activation during the motor task. For this reason this method is largely applied in clinics to assess quantitatively patients with different pathologies with movement disorders, both in children and in adults.

In the rehabilitation of patients with obesity quantitative motion analysis is of increasing importance. Obesity has, in fact, a profound effect on disability and quality of life (Bray 2004). As the prevalence of obesity is increasing at an alarming rate worldwide, obesity-related disabilities will eventually become a serious threat to national health systems. The excessive amount of fat modifies the body geometry by adding passive mass to different regions and causing impairment in skeletal statics and dynamics. Excess weight is able to influence the biomechanics of several activities of daily living, such as walking, standing up, bending, and other movements (Sibella et al. 2003; Saibene and Minetti 2003; Vismara et al. 2007; de souza et al. 2005a, b), causing functional limitations, and possibly predisposing to injury (Wearing et al. 2006). Investigating quantitatively these capacities appears necessary to define the functional profile in the obese population to then plan appropriate rehabilitation interventions.

The aim of this chapter is to illustrate and review the applications of quantitative movement analysis in obese individuals and to show the experience of the motion analysis laboratory at Istituto Auxologico Italiano IRCCS, San Giuseppe Hospital, Piancavallo (VB), Italy, which routinely applies the quantitative evaluation in obese patients for an improvement of rehabilitation and therapeutic managements.

3.2 Equipment

The technology of Istituto Auxologico Italiano IRCCS, San Giuseppe Hospital, laboratory used to perform quantitative analysis of movement is characterized by the following technical aspects:

- it is three-dimensional;
- it is noninvasive;
- it is able to provide extremely precise quantitative information;
- it is able to perform multifactorial integrated analyses, and thus to acquire contemporaneously kinematic data (e.g., motion trajectories), kinetic data (e.g., ground reaction forces) and data relating to muscle activation (electromyographic data);
- it is easy to use;
- it is relatively inexpensive.

In particular, the equipment employed for the patients evaluation is composed of:

- 6-camera optoelectronic system (Vicon, UK): its cameras (Fig. 3.1a) measure the three-dimensional coordinates of reflective markers (Fig. 3.1b) positioned at specific points of reference on the patient's body. From the three-dimensional coordinates of the markers it is possible to calculate angles of flex-extension, abd-adduction, and intra-extrarotation of the main joints, speed, accelerations… and thus to know in detail the kinematics of the body segments on which the markers have been positioned (Davis et al. 1991).
- Two piezoelectric force plates (Kistler, CH): they measure the system of ground reaction forces (Fig. 3.2). From the system of ground reaction forces and the kinematic parameters, which have been acquired using the optoelectronic systems, the moments and powers of the different joints can be calculated.
- One 8-channel telemetric electromyographic unit (Noraxon, USA): it records, via surface electrodes, the responses of muscles to electrical stimulation—i.e., electrical signals generated by muscle contraction (Fig. 3.3).
- Two-camera system for the video recording of patients: they allow the clinician to observe a patient's motor action from a qualitative point of view.

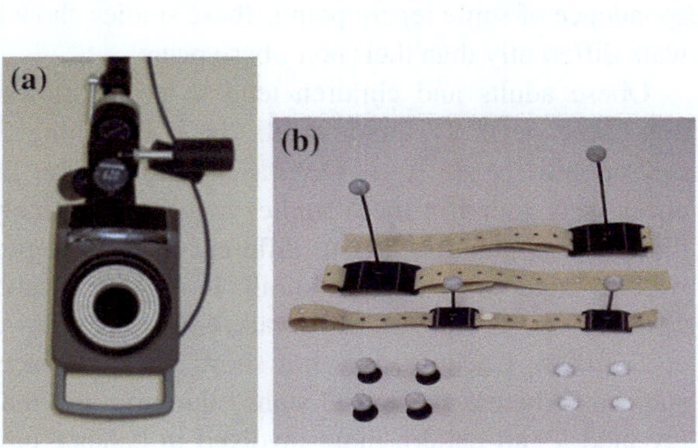

Fig. 3.1 Optoelectronic camera system (**a**); markers (**b**)

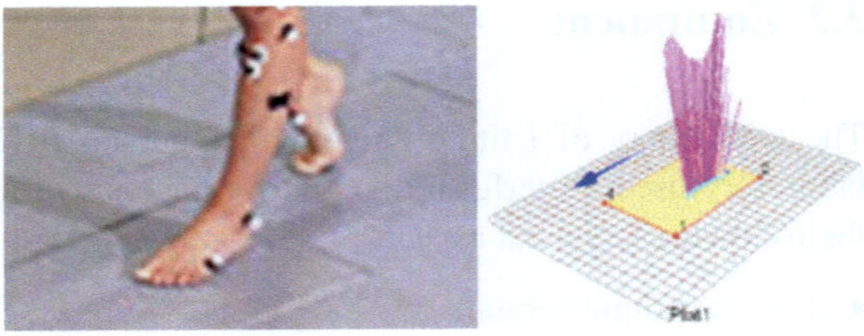

Fig. 3.2 On the *left*, a force plate; on the *right*, the vectogram during foot contact with the force plate during walking is represented

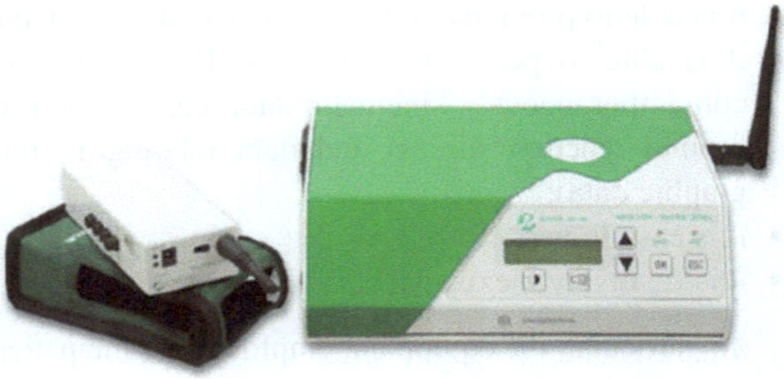

Fig. 3.3 Electromyographic unit

3.3 Quantitative Analysis of Walking (Gait Analysis)

Locomotor activity, walking in particular, is the most popular form of physical activity for weight management (Jakicic et al. 2003), presumably because it is easy to do and requires considerable metabolic energy. While studies that investigate how obesity affects the physiological responses to locomotion are essential and ongoing (Browning 2006; Lafortuna et al. 2008), biomechanical studies are also critical. Quantifying how obesity affects the biomechanics of gait provides important insights into the relationship between metabolic and mechanical energetics, mechanical loading (e.g., joint loads), and the associated risk of musculoskeletal injury and/or pathology. Our understanding of how obesity affects gait biomechanics in adults and children is increasing, and there is now a body of literature and several reviews (Shultz et al. 2011; Runhaar et al. 2011). While potentially limited by motion capture accuracy due to the presence of fat in correspondence of some repere points, these studies show that obese adults and children walk differently than their non-obese peers.

Obese adults and children tend to prefer slower walking speeds than their non-obese peers (Vismara et al. 2007; Browning 2005; Melanson et al. 2003; Spyropoulos et al. 1991; DeVita and Hortobagyi 2003). However, some differences were found in these studies and they are likely related to the variability in the obese population and to different methodologies in data collection, such as walking outdoor or on a treadmill. In children, walking speed is dependent upon age, but obese children still walk ~10–15 % slower than non-obese (McGraw et al. 2000). Cadence does not show any difference between obese and healthy subjects, whereas prolonged stance duration and reduced single support duration revealed a gait pattern more involved in balance control for obese patients, with a longer period of double support (both feet on the ground) (Spyropoulos et al. 1991; DeVita and Hortobagyi 2003; Lai et al. 2008), and an increased step width (Browning and Kram 2007; Ko et al. 2010). Collectively, these adaptations are thought to arise from anthropometric characteristics (e.g., increased thigh diameter) as well as the need to maintain balance during gait.

Movements of the lower extremities are affected by obesity, with differences being reported in the three planes of movement at hip, knee, and ankle joint angles. However, there is no clear consensus on kinematic adaptations associated with adult or pediatric obesity. While some investigations report that obese adults and children walk with a more erect sagittal plane posture compared to non-obese individuals (e.g., less flexed lower extremity joints), (DeVita and Hortobagyi 2003; Gushue et al. 2005), others do not report such a finding (Spyropoulos et al. 1991; Lai et al. 2008; Browning and Kram 2007; Hortobágyi et al. 2011). One reason for the conflicting results is that walking speed has not been controlled in some studies, even though walking speed influences kinematics (Lelas et al. 2003). Thus, the differences in kinematics (less flexion) may be as much due to speed as obesity. Vismara et al. (2007) found ROMs at hip, knee, and ankle on sagittal plane of young obese individuals similar to those of healthy participants. These results support that obesity does not determine major and immediate changes in the learned motor strategy in young adult obese patients. Many obese patients older than those recruited for this study often show articular problems and pathological gait pattern (Syed and Davis 2000; Messier et al. 2005) that could be due to the progressive effect of excessive joint loads over the years. Then, the effect of obesity on joint biomechanics could be not immediate, but progressive. Not surprisingly, frontal plane walking kinematics may also be affected by obesity, but again, consensus is lacking and the differences between obese and non-obese are typically small. Hip abduction has been reported to be greater in obese versus non-obese adults and children (Spyropoulus et al. 1991; Ko et al. 2010), but these results are not reported in all studies (Lai et.al. 2008; McMillan et al. 2010). Differences in frontal plane hip angles in obese individuals could result from the need for increased leg circumduction due to thigh dimensions or an increase in pelvic tilt due to stance leg hip abductor weakness. Foot eversion is greater in obese versus non-obese adults (Lai et.al. 2008; Messier 1994), but not in children (Shultz et al. 2011; McMillan et al. 2010). Very few studies have reported transverse plane angles in obese adults and the primary finding is that obese individuals walk with extra-rotated feet (Vismara et al. 2007; Messier 1994; de Souza et al. 2005a, b).

As for gait kinetics, as a primary determinant of walking kinetics is body mass, obesity has a significant effect on ground reaction forces (GRFs) and lower extremity kinetics (joint reaction forces and net muscle moments).. While the peak of sagittal plane moments normalized by body mass of hip and knee reveal similar values to controls, ankle moments are reduced in obese versus non-obese individuals (Lai et al. 2008; Browning and Kram 2007; Ko et al. 2010; McMillan et al. 2010). This suggests that obese individuals may rely on more proximal joints to perform the work required to walk. A few studies have quantified frontal plane moments in healthy obese individuals, particularly at the knee due to the relationship between this moment (internal abduction or external adduction) and medial compartment loading (Shelburne et al. 2008). Particularly troubling is the finding that obese children may also experience these greater adduction moments (Shultz 2009; Gushue et al. 2005), although this has not been reported in adolescents (McMillan et al. 2010). Greater adduction moments in children may result in the development of altered frontal plane alignment, further increasing the risk of osteoarthritis (OA) development as adults (Taylor et al. 2006). Given the greater loads experienced by the lower extremity joints of obese individuals, the link between obesity and OA is thought to be due, in part, to mechanical loading (Griffin and Guilak 2005; Andriacchi et al. 2009) and varus limb alignment (i.e., bowlegged) (Felson et al. 2004). Few reports of lower extremity gait mechanics in obese individuals with OA have been conducted. Most notable is the quantification of ADAPT (Arthritis, Diet, and Activity Promotion Trial) outcomes in terms of gait biomechanics and knee joint forces before and after weight loss (Messier et al. 2005). The participants were found to walk slower and have associated changes in lower extremity kinematics (Aaboe et al. 2011). In addition, these individuals walk with a greater external knee adduction moment (Andriacchi and Mundermann 2006).

Obesity is often associated with low back pain (LBP) (Fanuele et al. 2002; Hangai et al. 2008). Some studies have shown a correlation between obesity and functional impairment of the spine secondary to weakness and stiffness of the lumbar muscles, leading to LBP and disability (Larsson 2004). In the literature, the analysis of gait pattern in obese subjects affected by LBP has received scant attention. Some studies have been performed using different techniques (clinical assessment, accelerometers, 3D movement analysis) and under different experimental conditions (activities of daily living, treadmill, and ground walking) (Spenkelink et al. 2002; Lamoth 2006a, b). Spenkelink et al. (2002) investigated differences between LBP patients and healthy subjects in daily living activities and the day-to-day variability using an ambulant monitoring system based on accelerometers. In this study LBP patients showed lower activity levels with lower gait velocity and standing time and higher resting time than control group. Lamoth et al. (2006a, b) examined the capacity of LBP versus healthy subjects to adapt their gait pattern to changes in velocity during treadmill walking. Their analysis was conducted in terms of trunk and pelvis rotations and erector spinae muscular activity to evaluate the trunk-pelvic coordination. They found that LBP patients had difficulties in modulating trunk-pelvis coordination flexibility, especially on the transversal plane, and erector spinae activity following velocity perturbations. In a more recent study, the same authors (Lamoth et al. 2008) examined the relationship between attention and gait by varying conditions of attentional load during treadmill walking. In line with the previous study, they confirmed the reduced motility of the upper body in LBP patients during gait, which was accentuated during an attention demanding task.

To our knowledge, the only study which quantified the gait pattern of obese women with and without LBP by using 3D-Gait Analysis was conducted by Vismara et al. (2010). They demonstrated that the presence of LBP induces some further alterations of the gait pattern compared to obese subjects without LBP.

The coexistence of obesity and LBP seems therefore to affect gait more severely than obesity alone. The presence of LBP leads to reduced stability during gait, as assessed by the spatial–temporal parameters, and a less physiological knee and ankle strategy. Their hip pattern, which was also present and more flexed in presence of LBP, may represent a strategy to favor body advancement. It can be speculated that pain relief may represent the major underlying cause of this strategy: in order to reduce traction on the sciatic nerve, subjects limit their ankle dorsal flexion. It has to be reminded, however, that subjects affected by sciatica were excluded. These results seem to stratify the gait patterns of obese with LBP from those without and from normal-weight subjects. Women without LBP, in fact, showed a gait pattern in-between those of women with LBP and normal weight subjects.

3.4 Postural Analysis

Reduced balance affects a variety of daily and occupational tasks, particularly those performed with the upper limbs from a standing posture (Chau et al. 2004). Adipose tissue accumulation and body mass increases can be a major factor contributing to the occurrence of falls, which explains why obese persons appear to be at greater risk than normal weight subjects under daily postural stresses and perturbations (Finkelstein et al. 2007). There is now evidence that body weight is a strong predictor of postural instability (Hue et al. 2007), with obesity-associated postural perturbations appearing in adolescence (McGraw et al. 2000). In particular, obesity has been associated with greater forward displacement of the CoP during dynamic standing balance activities (Berrigan et al. 2006). Excessive body weight affects posture linearly with the increase of BMI (Gilleard and Smith 2007), akin to what occurs in the later stages of pregnancy (Teasdale et al. 2007): the center of gravity shifts forward, the lumbar lordosis increases together with the pelvic forward tilt, the dorsal kyphosis and a secondary cervical lordosis become more pronounced (Rodacki et al. 2005; Syed and Davis 2000). Frequently, internal rotation of the hips, knee valgism, and flat feet coexist and the feet tend to splay apart during standing to optimize the center of gravity and stability. Discomfort and reduced tolerance of fixed postures are also consequences of a redundant mass. Pain has been shown to affect posture (Rodacki et al. 2005) and reduced sensory integration has been hypothesized for the poor balance (Syed and Davis 2000). As for the presence of higher oscillations in obese individuals respect to controls, two hypotheses are proposed by the literature: (a) the reduction of plantar sensitivity due to the hyper activation of the plantar mechanoreceptors for the continuous pressure of supporting the large mass; (b) the presence of high mechanical request in obese subjects due to a whole body center of mass further away from the axis of rotation causing a greater gravitational torque (Hue et al. 2007).

In addition, body mass distribution usually shows gender differences (gynoid and android shape), even if android fat distribution is also observed in females, particularly in postmenopausal women: thus, whether shape induces possible gender-specific consequences on balance is still controversial. Few dated studies on normal weight persons reported no gender differences in sway area during standing and in perturbed conditions (Maki et al. 1996; Wolfson et al. 1994), although elderly women appear to have poorer balance and functional performances (Capodaglio et al. 2007). In 2003, Gravante et al. (2003) studied gender differences in obese and healthy subjects using a baropodometric platform. They reported no gender-related differences in foot pressure, foot contact surface, and center of pressure (CoP) mean location. They also found that the CoP mean location was not influenced by weight. More recent studies have investigated obese male subjects. Hue et al. (2007) reported the lower postural stability of obese men, assessed by the antero–posterior (AP) and medio–lateral (ML) displacement of the CoP, than their lean counterparts, although other authors showed that postural stability improved in severely obese men after weight loss (Teasdale et al. 2007) and specific balance training (Maffiuletti et al. 2005). In general, the increased body mass seems to produce AP instability in both genders and ML destabilization only in males (Menegoni et al. 2009). Two possible hypotheses could explain these results, both related to the mass distribution. The fat mass is usually concentrated in the thorax–abdominal region in males (android shape), while it is usually around the hips and the upper portion of legs in females (gynoid shape) (Clark 2004). In both genders the increased body mass contributes to an increased ankle torque (AP destabilization), but the android shape involves a greater amount of mass/load over the hips, which could account for the increased ML CoP excursion. The second hypothesis, partially overlapping with the previous one, is related to the effect of a different mass distribution on the center of mass. The mass distribution alters the center of mass position, which is higher in the android than in gynoid shape. Furthermore, the fact that the males are usually heavier than females should be taken into account. In summary, men have to cope with heavier body weights and higher center of mass, possibly leading to an increased loading/unloading mechanism in ML direction.

3.5 Quantitative Analysis of Trunk Kinematics

The morphology of the spine accounts for its own mechanical behavior and changes in morphology often have consequences for biomechanical performance. As a consequence, the analysis of spinal mobility can play a central role in understanding the relationship between function and clinical conditions. Obese patients often report musculoskeletal disorders with a high incidence and prevalence of nonspecific and specific low back pain which may influence the trunk movements. Methods for the quantitative analysis of trunk movements in vivo need to take in account noninvasiveness together with accuracy and clinical feasibility. Some techniques based on optoelectronic systems have been proposed but most studies are focused on specific segments (for example lumbar and/or pelvis) and considering different movements (during walking, posturo-locomotor tests, forward and lateral flexion) (Lamoth 2006a, b; Schön-Ohlsson et al. 2005; Sibella et al. 2003). To our knowledge, only Menegoni et al. (2008) described and quantified the functional mobility of the spine during flexion, lateral bending and rotation in obese women. They estimated potential errors introduced by different skin movement artifacts affecting obese and normal weight participants, without detecting any marker movements from the bony landmarks during rotation and bending. Only during forward flexion the markers located on the spine and pelvis dislocated from the bony landmarks; the errors were similar in healthy and obese participants and only the error concerning the angle relative to the lumbar segment was greater in obese. Intra-subject and inter-subject variability was found relatively limited with a standard deviation lower than 6°. The investigation of the relationship between obesity and spinal mobility and its clinical consequences appears that body weight influences standing posture in the sagittal plane: obese subjects are characterized by a forward flexed trunk and anteversion of the pelvis. Obese individuals seem to compensate for the forward translation of the center of mass with an increased pelvic tilt. Body weight is also a constraint for thoracic movement, demonstrated by the angle related to kyphosis and the thoracic movement in forward flexion and by the limited range of motion in lateral bending. In addition, dorsal stiffness with normal lumbar ROM appears to be a feature of obese subjects. These strategies could play an important role in the onset of nonspecific and even more of specific low-back pain, which is known to affect many obese patients. In this field, few studies demonstrate a correlation between obesity and functional impairment of the spine secondary to weakness and stiffness of the lumbar muscles, possibly leading to LBP and disability (Larsson 2004); moreover, there is a lack of quantitative data on spinal mobility in obese subjects who already suffer from LBP. Vismara et al. (2010) investigated the relationship between obesity and LBP quantitatively during flexion and lateral bending. They showed biomechanical differences in spinal mobility between healthy and obese subjects under static and dynamic conditions, with more pronounced differences when comparing obese patients with to those without LBP. They suggested that obesity may modify spinal posture and function favoring the onset of LBP. Significant differences were found at lumbar and pelvic level among groups. They confirmed previous results (Menegoni et al. 2008) showing an increased anterior pelvic tilt while maintaining a normal lumbar lordosis under static conditions in obese. The increased anterior pelvic tilt induces a greater flexion of the sacroiliac joints, and therefore a higher torque on the L5-S1 joint and disks. This possibly increases the shear forces at this level and overload the disk, thus increasing the risk of disk degeneration (Hangai et al. 2008; Like et al. 2005). In line with Gilleard and Smith (2007), they observed an increased lumbar lordosis in obese patients with LBP which may well represent a pain-related strategy. Abdominal circumference and gravity may influence the lumbar lordosis and its mobility during forward flexion or lateral bending. All these factors could impair the dynamic function of some muscles, in particular the erector spinal muscles, so that their counteraction to the anterior shear forces on the spine could be jeopardized (McGill et al. 2000). During forward flexion, they observed that thoracic

ROM was significantly lower in obese and significantly lower in LBP as compared to controls, while lumbar ROM remained similar among the three groups. Due to thoracic stiffness, forward flexion in obese and particularly in LBP appears to be performed mainly by the lumbar spine, which is most frequently involved in pain syndromes. Thoracic stiffness with normal lumbar ROM appears to be a feature of obesity and it appears plausible that it might play a role in the onset of LBP in obese patients. According to these results, lateral bending is performed in a qualitatively different modality when LBP is present and it appears the most meaningful clinical test for detecting lower spinal impairments and monitor functional consequences of obesity.

3.6 Quantitative Analysis of Sit-to-Stand

Sit-to-stand movement (STS) is one of the most common movements in daily activities and is an important functional task that may become difficult for patients. Rising from the seated position is a complex activity: it requires an adequate postural control during the motor transfer from a stable 3-points base, the sitting position, to a 2-points base, the standing position, and needing adequate torques to be developed about each body joint. Difficulty or inability to stand up is common in individuals with a variety of motor disabilities and STS evaluation is often considered in clinical evaluation scales of different pathologies. Studies on STS have been conducted mainly in healthy subjects (Janseen et al. 2002; Lindemann et al. 2003) and no quantitative analyses have been conducted in obese patients even if STS has been a topic for many studies (Coglin et al. 1994; Doorenbosch et al. 1994). In these patients, weight conditions and muscular weakness and low back pain change motion task strategy and, consequently, the torque distributions of hip, knee and ankle joints. STS is likely to be a very good model to assess the ability of obese patients to properly execute daily living tasks. To our knowledge only Galli et al. (2000) and Sibella et al. (2003) developed a biomechanical model for a quantitative description of STS motion strategy in obese subjects compared to normal individuals. Kinematic analysis showed that while normal individuals rise from the chair by flexing the trunk forward and keeping the feet in their initial position, on the contrary, obese subjects rise from the chair by limiting the forward trunk flexion and moving the feet backwards from the initial position. In the healthy subjects, from the first to the last trial a small decrease in forward trunk flexion is of note, even though the flexion degree remains always sensibly higher than that of the obese group. Vice versa, no changes in forward trunk flexion values are visible for the obese group. The kinetic analysis confirms and explains kinematic results, as the developed model allowed to compute torques at the different joints. To high degrees of trunk flexion correspond a high hip joint moment which, in general, leads to a high low back loading and a minimization of knee joint torque. Obese subjects tend to minimize trunk flexion: this kinematic strategy brings a minimization to hip joint torque (and therefore a minimization of low back loading), but it maximizes the moment at knee joint. During the trials execution, while for the control group the criteria imposed at the beginning of the trial are conserved trial-by-trial, the obese group, trial-by-trial, changed the initial strategy because of fatigue into the one used by the controls and characterized by a forward flexion that minimizes knee joint load. Doing so they are able to stand up but they overload the vertebral column, worsening their lower back pain. At the beginning of STS task, when obese patients were asked to stand up, they tried to protect the vertebral column. When fatigue increased, during the execution of multiple STS, the protection of vertebral column was secondary with respect to the aim of standing up and this why they changed their rising strategy.

3.7 Conclusions

In this chapter an overview of the role of 3D motion analysis to quantify functional tasks useful to assess the functional limitation of obese subjects from a biomechanical point of view is presented. These studies show how motion analysis could represent a useful tool to quantify the performance of these subjects giving crucial information that clinical evaluations and video recording are not able to provide. However, although there is increasing evidence that obesity affects lower extremity joint kinematics and kinetics and trunk movements, these data must be viewed with a clear understanding of the challenges associated with quantifying motions of a skeletal system covered by extensive soft tissue. Almost all studies reporting lower extremity kinematics/kinetics in obese individuals during gait and STS have used standard gait marker set (Davis et al. 1991) which could be influenced by inaccurate marker placement and soft tissue artifact (Baker 2006; Della Croce et al. 2005). Of particular concern especially in obese subjects is the reliance on anterior superior iliac spine (ASIS) and/or greater trochanter markers to establish the pelvis and hip joint centers leading to inaccurate estimates of joint centers and errors in the resultant kinematics/kinetics (hip and knee in particular) (Stagni et al. 2000). However, we know that some parameters are not so much affected by these errors (i.e. ranges of motion) (Kirtley et al. 2002) and so they could be considered reliable measures in these patients. Future studies that report kinematic data using obese subjects should need to clearly identify marker placement and skeletal model development procedures and how errors regarding marker placement were addressed. A combination of using dual-energy X-ray absorptiometry images for determining inter-ASIS distance and estimating segment inertial parameters (Chambers et.al. 2010), a sacral marker cluster and digitized pelvic anatomical landmarks (Segal et al. 2009) are suggested to improve the accuracy of marker-based motion capture.

As for trunk kinematics the proposed experimental setup can represent a non-invasive clinically useful technique for functional investigation in various spinal conditions and evaluation of effectiveness in rehabilitation. Potential biases of spinal motion measurement with surface markers are soft tissue artifacts which can affect the measure especially during forward flexion. However, these artifacts seemed to produce a similar systematic error between the obese and the healthy. A limit of this approach is that it is unable to estimate the error introduced by the thickness of the skin, a layer interposed between the marker and the bone, which hinders the correct localization of the bony landmarks.

Despite the evidenced considerations, the use of motion analysis appears necessary to define the functional profile in the obese population and to have indications for planning appropriate rehabilitation interventions.

References

Aaboe J, Bliddal H, Messier SP et al (2011) Effects of an intensive weight loss program on knee joint loading in obese adults with knee osteoarthritis. Osteoarthritis Cartilage 19(7):822–828
Andriacchi TP, Koo S, Scanlan SF (2009) Gait mechanics influence healthy cartilage morphology and osteoarthritis of the knee. J Bone Joint Surg Am 91(Suppl 1):95–101
Andriacchi TP, Mundermann A (2006) The role of ambulatory mechanics in the initiation and progression of knee osteoarthritis. Curr Opin Rheumatol 18(5):514–518
Baker R (2006) Gait analysis methods in rehabilitation. J Neuroeng Rehabil 3:4
Berrigan F, Simoneau M, Tremblay A, Hue O, Teasdale N (2006) Influence of obesity on accurate and rapid arm movement performed from a standing posture. Int J Obes (Lond) 30:1750–1757
Bray GA (2004) Medical consequences of obesity. J Clin Endocrinol Metab 89:2583–2589
Browning RC, Kram R (2007) Effects of obesity on the biomechanics of walking at different speeds. Med Sci Sports Exerc 39(9):1632–1641
Capodaglio P, Capodaglio EM, Facioli M, Saibene F (2007) Long-term strength training for community-dwelling people over 75: impact on muscle function, functional ability and life style. Eur J Appl Physiol 100:535–542
Chambers AJ, Sukits AL, McCrory JL, Cham R (2010) The effect of obesity and gender on body segment parameters in older adults. Clin Biomech (Bristol, Avon) 25(2):131–136
Chau N, Gauchard GC, Siegfried C, Benamghar L, Dangelzer JL, Français M et al (2004) Relationships of job, age, and life conditions with the causes and severity of occupational injuries in construction workers. Int Arch Occup Environ Health 77(1):60–66
Clark KN (2004) Balance and strength training for obese individuals. ACSM'S Health Fitness J 8:14–20
Coglin S et al (1994) Transfer strategy used to rise from a chair in normal and low-back pain subjects. Clin Biomech 9:85–92
Davis RB, Ounpuu S, Tyburski D, Gage JR (1991) A gait analysis data collection and reduction technique. Hum Mov Sci 10:575–587
de Souza SA, Faintuch J, Valezi AC, Sant'Anna AF, Gama-Rodrigues JJ, de Batista Fonseca IC, de Melo RD (2005a) Postural changes in morbidly obese patients. Obes Surg 15(7):1013–1016
de Souza SA, Faintuch J, Valezi AC, Sant' Anna AF, Gama-Rodrigues JJ, de Batista Fonseca IC, Souza RB, Senhorini RC (2005b) Gait kinematic analysis in morbidly obese patients. Obes Surg 15(9):1238–1242
Della Croce U, Leardini A, Chiari L, Cappozzo A (2005) Human movement analysis using stereophotogrammetry: part 4: assessment of anatomical landmark misplacement and its effects on joint kinematics. Gait Posture 21(2):226–237
DeVita P, Hortobagyi T (2003) Obesity is not associated with increase knee joint torque and power during level walking. J Biomech 36:1355–1362
Doorenbosch C, Harlaar J, Roebroeck ME, Lankhorst GJ (1994) Two strategies of transferring from sit-to-stand; the activation of monoarticular and biarticular muscles. J Biomech 27(11):1299–1307

Fanuele JC, Abdu WA, Hanscom B, Weinstein JN (2002) Association between obesity and functional status in patients with spine disease. Spine (Phila Pa 1976) 27(3):306–312

Felson DT, Goggins J, Niu J et al (2004) The effect of body weight on progression of knee osteoarthritis is dependent on alignment. Arthritis Rheum 50(12):3904–3909

Finkelstein EA, Chen H, Prabhu M, Trogdon JG, Corso PS (2007) The relationship between obesity and injuries among U.S. adults. Am J Health Promot 21:460–468

Galli M, Crivellini M, Sibella F, Montesano A, Bertocco P, Parisio C (2000) Sit-to-stand movement analysis in obese subjects. Int J Obes Relat Metab Disord 24(11):1488–1492

Gilleard W, Smith T (2007) Effect of obesity on posture and hip joint moments during a standing task, and trunk forward flexion motion. Int J Obes 31:267–271

Gravante G, Russo G, Pomara F, Ridola C (2003) Comparison of ground reaction forces between obese and control young adults during quiet standing on a baropodometric platform. Clin Biomech (Bristol, Avon) 18:780–782

Griffin TM, Guilak F (2005) The role of mechanical loading in the onset and progression of osteoarthritis. Exerc Sport Sci Rev 33(4):195–200

Gushue DL, Houck J, Lerner AL (2005) Effects of childhood obesity on three-dimensional knee joint biomechanics during walking. J Pediatr Orthop 25(6):763–768

Hangai M, Kaneoka K, Kuno S, Hinotsu S, Sakane M, Mamizuka N, Sakai S, Ochiai N (2008) Factors associated with lumbar intervertebral disc degeneration in the elderly. Spine J 8(5):732–740

Hill JO, Peters JC (1998) Environmental contributions to the obesity epidemic. Science 280(5368):1371–1374

Hills AP, Parker AW (1991) Gait characteristics of obese children. Arch Phys Med Rehabil 72:403–407

Hortobágyi T, Herring C, Pories WJ et al (2011) Massive weight loss induced mechanical plasticity in obese gait. J Appl Physiol 111(5):1391–1399

Hue O, Simoneau M, Marcotte J et al (2007) Body weight is a strong predictor of postural stability. Gait Posture 26:32–38

Jakicic JM, Marcus BH, Gallagher KI, Napolitano M, Lang W (2003) Effect of exercise duration and intensity on weight loss in overweight, sedentary women: a randomized trial. JAMA 290(10):1323–1330

Janssen WG, Bussmann HB, Stam HJ (2002) Determinants of the sit-to-stand movements: a review. Phys Ther 82(9):866–879

Kirtley C (2002) Sensitivity of the modified Helen Hayes model to marker placement errors. In: Seventh international symposium on the 3-D analysis of human movement, Newcastle, UK, 10–12 July 2002

Ko S-u, Stenholm S, Ferrucci L (2010) Characteristic gait patterns in older adults with obesity–results from the Baltimore longitudinal study of aging. J Biomech 43(6):1104–1110

Lafortuna CL, Agosti F, Galli R et al (2008) The energetic and cardiovascular response to treadmill walking and cycle ergometer exercise in obese women. Eur J Appl Physiol 103(6):707–717

Lai PP, Leung AK, Li AN, Zhang M (2008) Three-dimensional gait analysis of obese adults. Clin Biomech (Bristol, Avon) 23(1):S2–S6

Lamoth CJ, Daffertshofer A, Meijer OG, Beek PJ (2006a) How do persons with chronic low back pain speed up and slow down? Trunk-pelvis coordination and lumbar erector spinae activity during gait. Gait Posture 23(2):230–239

Lamoth CJ, Meijer OG, Daffertshofer A, Wuisman PI, Beek PJ (2006b) Effects of chronic low back pain on trunk coordination and back muscle activity during walking: changes in motor control. Eur Spine J 15(1):23–40

Lamoth CJ, Stins JF, Pont M, Kerckhoff F, Beek PJ (2008) Effects of attention on the control of locomotion in individuals with chronic low back pain. J Neuroeng Rehabil 5:13

Larsson UE (2004) Influence of weight loss on pain, perceived disability and observed functional limitations in obese women. Int J Obes Relat Metab Disord 28(2):269–277

Lelas JL, Merriman GJ, Riley PO, Kerrigan DC (2003) Predicting peak kinematic and kinetic parameters from gait speed. Gait Posture 17(2):106–112

Like M, Solovieva S, Lamminen A, Luoma K, Leino-Arjas P, Luukkonen R et al (2005) Disc degeneration of the lumbar spine in relation to overweight. Int J Obes (Lond) 29(8):903–908
Lindemann U, Claus H, Stuber M, Augat P, Muche R, Nikolaus T et al (2003) Measuring power during the sit-to-stand transfer. Eur J Appl Physiol 89:466–470
Maffiuletti NA, Agosti F, Proietti M et al (2005) Postural instability of extremely obese individuals improves after a body weight reduction program entailing specific balance training. J Endocrinol Invest 28:2–7
Maki BE, McIlroy WE, Perry SD (1996) Influence of lateral destabilization on compensatory stepping responses. J Biomech 29:343–353
McGill SM, Hughson RL, Parks K (2000) Changes in lumbar lordosis modify the role of the extensor muscles. Clin Biomech 15(10):777–780
McGraw B, McClenaghan BA, Williams HG, Dickerson J, Ward DS (2000) Gait and postural stability in obese and nonobese prepubertal boys. Arch Phys Med Rehabil 81:484–489
McMillan AG, Pulver AM, Collier DN, Williams DS (2010) Sagittal and frontal plane joint mechanics throughout the stance phase of walking in adolescents who are obese. Gait Posture 32(2):263–268
Melanson EL, Bell ML, Knoll JR (2003) Body mass index and sex influence the energy cost of walking at self-selected speeds. Med Sci Sports 35(5 suppl):S183
Menegoni F, Vismara L, Capodaglio P, Crivellini M, Galli M (2008) Kinematics of trunk movements: protocol design and application in obese females. J Appl Biomater Biomech 6(3):178–185
Menegoni F, Galli M, Tacchini E, Vismara L, Cavigioli M, Capodaglio P (2009) Gender-specific effect of obesity on balance. Obesity (Silver Spring) 17(10):1951–1956
Messier SP, Gutekunst DJ, Davis C, DeVita P (2005) Weight loss reduces knee-joint loads in overweight and obese older adults with knee osteoarthritis. Arthritis Rheum 52(7):2026–2032
Rodacki AL, Fowler NE, Provensi CL, Rodacki C de L, Dezan VH (2005) Body mass as a factor in stature change. Clin Biomech (Bristol, Avon) 2(8):799–805
Runhaar J, Koes BW, Clockaerts S, Bierma-Zeinstra SMA (2011) A systematic review on changed biomechanics of lower extremities in obese individuals: a possible role in development of osteoarthritis. Obes Rev 12(12):1071–1082
Saibene F, Minetti AE (2003) Biomechanical and physiological aspects of legged locomotion in humans. Eur J Appl Physiol 88(4–5):297–316 (review)
Segal NA, Yack HJ, Khole P (2009) Weight, rather than obesity distribution, explains peak external knee adduction moment during level gait. Am J Phys Med Rehabi 88(3):180–188
Shelburne KB, Torry MR, Steadman JR, Pandy MG (2008) Effects of foot orthoses and valgus bracing on the knee adduction moment and medial joint load during gait. Clin Biomech (Bristol, Avon) 23(6):814–821
Schön-Ohlsson CU, Willén JA, Johnels BE (2005) Sensory motor learning in patients with chronic low back pain: a prospective pilot study using optoelectronic movement analysis. Spine (Phila Pa 1976) 30(17):E509–E516
Shultz SP, Browning RC, Schutz Y et al (2011) Childhood obesity and walking: guidelines and challenges. Int J Pediatr Obes 6(5–6):332–341
Sibella F, Galli M, Romei M, Montesano A, Crivellini M (2003) Biomechanical analysis of sit-to-stand movement in normal and obese subjects. Clin Biomech 18(8):745–750
Spenkelink CD, Hutten MM, Hermens HJ, Greitemann BO (2002) Assessment of activities of daily living with an ambulatory monitoring system: a comparative study in patients with chronic low back pain and nonsymptomatic controls. Clin Rehabil 16(1):16–26
Spyropoulos P, Pisciotta JC, Pavlou KN et al (1991) Biomechanical gait analysis in obese men. Arch Phys Med Rehabil 72(13):1065–1070
Stagni R, Leardini A, Cappozzo A et al (2000) Effects of hip joint centre mislocation on gait analysis results. J Biomech 33(11):1479–1487
Syed IY, Davis BL (2000) Obesity and osteoarthritis of the knee: hypotheses concerning the relationship between ground reaction forces and quadriceps fatigue in long-duration walking. Med Hypotheses 54(2):182–185

Taylor ED, Theim KR, Mirch MC et al (2006) Orthopedic complications of overweight in children and adolescents. Pediatrics 117(6):2167–2174

Teasdale N, Hue O, Marcotte J et al (2007) Reducing weight increases postural stability in obese and morbid obese men. Int J Obes (Lond) 31:153–160

Vismara L, Menegoni F, Zaina F, Galli M, Negrini S, Capodaglio P (2010) Effect of obesity and low back pain on spinal mobility: a cross sectional study in women. J Neuroeng Rehabil 7:3

Vismara L, Romei M, Galli M, Montesano A, Baccalaro G, Crivellini M, Grugni G (2007) Clinical implications of gait analysis in the rehabilitation of adult patients with "Prader-Willi" syndrome: a cross-sectional comparative study ("Prader-Willi" syndrome versus matched obese patients and healthy subjects). J Neuroeng Rehabil 10(4):14

Wearing SC, Hennig EM, Byrne NM, Steele JR, Hills AP (2006) The biomechanics of restricted movement in adult obesity. Obes Rev 7:13–24

Wolfson L, Whipple R, Derby CA, Amerman P, Nashner L (1994) Gender differences in the balance of healthy elderly as demonstrated by dynamic posturography. J Gerontol 49:M160–M167

Chapter 4
The Obesity-Related Disability

Amelia Brunani, Paolo Capodaglio, Matilde Leonardi and Alberto Raggi

Abstract Obesity, particularly with early onset and severe degree, imposes function limitations in different areas. We need appropriate assessment tools to identify and evaluate the obesity-related disability. Generally, tests of physical and cardiorespiratory performance (FIM, 6MWT, dyspnoea score, sub-maximal exercise test), score of daily life activities (ADL o IADL), and quality of life (SF36, IWQL) are used. However, these instruments have not been specifically developed for obese patients and cover only a part of the complex disability that is associated with this condition, thus underestimating the disability risk profile. In the last years, a reference classification of disability (ICF) proposed by WHO has been used also in obesity. Areas such as mobility, self-care, handling stress, and social relationships appear hindered. Recently, an obesity-specific disability questionnaire (TSD-OC) has been validated against other standard tools and linked with ICF. A correct identification and evaluation of functional, psychological, and social limitations/disability secondary to obesity might promote long-term successful rehabilitation intervention to reduce the impact of this disease.

A. Brunani (✉)
Rehabilitation Unit, Istituto Auxologico Italiano IRCCS, Piancavallo, Verbania, Italy
e-mail: brunani@auxologico.it

P. Capodaglio
Rehabilitation Unit and Research Lab in Biomechanics and Rehabilitation,
Istituto Auxologico Italiano IRCCS, Piancavallo, Verbania, Italy
e-mail: p.capodaglio@auxologico.it

M. Leonardi
Head Neurology, Public Health, Disability Unit, Scientific Director Coma Research Centre,
Director Italian WHO-FIC Collaborating Centre Research Branch, Foundation IRCCS
Istituto Neurologico Carlo Besta, Milan, Italy
e-mail: leonardi@istituto-besta.it

A. Raggi
Public Health and Disability Unit, Foundation IRCCS Istituto Neurologico Carlo Besta,
Milan, Italy
e-mail: alberto.raggi@istituto-besta.it

P. Capodaglio et al. (eds.), *Disabling Obesity*, DOI: 10.1007/978-3-642-35972-9_4,

4.1 Introduction

Obesity has become a global public health concern. The World Health Organization (WHO) estimated that there were 400 million obese adults (as measured by body mass index (BMI)) worldwide in 2005 and expected to be 700 million in 2025 (WHO 2006). Studies using data from the National health and Nutrition Examination Survey (NHANES) reported an increase in prevalence of obesity from 11.1 % in the 1970s to 19.3 % in the early 2000s (Ford et al. 2009). The most recent data from NHANES report obesity prevalence approximately of 32 % for men and 36 % for women (Ford et al. 2011). The difference between men and women is not statistically significant. Using data from the Europe Action on Secondary and Primary Intervention through Intervention to Reduce Events (EUROASPIRE) surveys in hospital-based patients, the average prevalence increased from 25 % in study I to 38 % in study III (Kotseva et al. 2009). Limited numbers of studies were available on the epidemiology of obesity in other continents. Comparison between regions around the world indicate a wide variation in prevalence of obesity; this is primary due to the different cut-offs considering in defining obesity but probably also to race/ethnic and cultural effects. Despite these regional differences, over time the prevalence of obesity has increased worldwide (WHO 2011). The projections for the future are not better if the life expectancy from NHANES showed a reduction by almost 1 year for an 18-year-old person, assuming trends in smoking continue to decrease and obesity to increase. The overall prevalence of obesity for adults in the USA will be 45 % by the year 2020 (Stewart et al. 2009). The major health consequences associated with obesity include hypertension, dyslipidemia, diabetes mellitus, metabolic syndrome, and cardiovascular diseases (Haslam and James 2005). As a result, the global burden of disease attributable to obesity amounted to 2.6 million deaths: for these reasons, obesity ranked seventh in terms of mortality and tenth in terms of burden of disease due to leading global risk factors (Ezzati et al. 2002). In the last years the major obesity-related cause of death, cardiovascular mortality, has been spectacularly declining, partially by improved therapies and cardiovascular risk management. Also, if primary prevention efforts reduced the burden of disease, obesity could become more disabling over time. In a previous analysis (Alley and Chang 2007), data from NHANES 1988–1994 and NHANES 1999–2004 were used to examine change in disability by BMI. In the NHANES two types of disability indicators were collected: functional limitations, which refer to restrictions in basic movement ability, and limitations in activities of daily living (ADL), which represent the most severe disabilities and reflect an individual's ability to live independently. The study shows that despite recent improvements in cardiovascular health among obese adults, it appears that obesity-associated disability did not decrease during the years and, on the contrary, the disparity between obese and normal-weight individuals has increased over time. The odds of functional impairment between NHANES 1988–1994 and NHANES 1999–2004 increased by 43 % among obese participants. Furthermore, while the odds of ADL disability declined significantly over the 1990s for non-obese older individuals, it did not change among obese older individuals. If disability is increased but not mortality, the number of obese survivors expand morbidity, increasing life years lived with disability, care dependence, and health-care costs (Andreyeva et al. 2004; Reynolds et al. 2005). In a recent study (Reuser et al. 2009), mild obesity was associated with increased life duration with ADL disability of 2.0 years in men and 3.2 years in women. For females, even overweight increases disabled LE by 2.1 years and shortens life free from ADL disability by 1.5 years. Several studies report that obesity is an independent risk factor for developing disability over time; the relationship between weight or BMI and disability has a "U" or a "J" shape, meaning that not only obesity but underweight have increased risk of disability (Al Snith et al. 2007).

The disability risk associated with obesity may be related to cumulative exposure, because obesity affects disability risk not only through cardiovascular disease burden, but also through several chronic diseases. It is demonstrated that obesity itself, independent of the presence of metabolic syndrome, is a risk factor for mobility limitation among obese older adults (Stenholm et al. 2010). Excess body weight can cause biomechanical stress on the lower extremity joints leading to pain, osteoarthritis, reduced physical activity, and impaired muscle strength, all of which can predispose an individual to mobility limitation (Ling et al. 2003). In addition, in older obese persons, the lower extremity muscle strength (Stenholm et al. 2009) or cardiorespiratory fitness may be inadequate to perform weight-bearing activities without difficulties. The pro-inflammatory state caused by obesity has been linked to age-related muscle loss or sarcopenia that contribute independently to disability (Zamboni et al. 2005). Additionally, weight loss has been reported as a risk factor for adverse events in some older adults including fractures, falls, and mortality (Locher et al. 2007).

Available evidence indicates that the threshold for which mobility limitations appears to develop is between 30 and 35 kg/m^2; BMI is a significant contributor to lower body functional impairment in women ($R2 = 0.050$ to the regression model, $P = 0.032$) (Apovian et al. 2002). Some evidence shows this relationship occurs in women but not men (Chen et al. 2002). A model for the development of mobility disability into old age was proposed (Vincent et al. 2010). We can start with the consideration that aging is related with a progressive loss of muscle mass and increase of fat mass (Schultz et al. 2002). Storage of adipose tissue between organs and in the viscera increases the overall body mass the obese individual needs to move (Peyrot et al. 2009); accumulation of fat within muscle tissue creates marbled (Zamboni et al. 2007), lower quality muscle capable of generating less strength and power for mobility tasks (Marks 2007). Poor lifestyle choices coupled with accelerated hip/knee/ankle joint degeneration cause the older individual to be less able to perform mobility-related tasks. Joint pain worsens with increased weight gain over time. Not only these but also confounding factors such as family and job situation, educational level, sleep problems, or being a smoker could influence the functional limitation. Disablement is the end result of this process.

4.2 The Obesity-Related Disability

The definitions of disability are different and such differences have an impact on the way in which disability is assessed in different studies. A first definition of disability is based on a medical model: disability is defined in terms of disease's consequence, and therefore problems that can be experienced by patients are viewed as a direct expression of the disease. In the field of obesity, examples of this are the definitions of disability in terms of problems with mobility, joint pain, or breathing problems. A second definition of disability is based on a social model: disability is defined in terms of the social disadvantages a person (or a group of persons) suffers from. Problems with employment are the most outstanding examples of such a model, with studies demonstrating that obesity is associated with reduced employment (Tunceli et al. 2006) and use of disability benefits (Neovius et al. 2008). However, both the approaches fail in getting a comprehensive picture of the "disability topic" in obesity. The reasons for this are different and, likely, difficult to comprehensively report. A first reason is the presence, in obese subjects, of several comorbidities including, but not limited to, diabetes, joint pain, hypertension, cardiovascular diseases, metabolic syndrome and psychiatric conditions such as schizophrenia, PTSD, bipolar disorder, anxiety disorders, eating disorders, as well as personality disorders (Chwastiak et al. 2011; Carpiniello et al. 2009). Therefore, obesity might underline very different conditions, and if disability is a direct consequence of a health condition, the descriptions that can be drawn may vary as well. A second reason is the exposure to some risk factors that can be, at the same time, both the consequence and the determinant: the consequence of this is that the outcome and the intervention sometimes might overlap. For example, it is known that reduced mobility is a consequence of obesity. Excessive body weight can cause biomechanical stress on the lower extremity joints leading to pain or osteoarthritis, which in turn determine reduced physical activity, which in turn determine reduced muscle strength: all the aforementioned issues can predispose to mobility limitation (Ling et al. 2003). At the same time, however, change in lifestyle is recommended in the treatment of obesity, and physical activity is one of the most common recommendations.

In sum, the main problems of the medical and social models of disability are that they do not provide precise and linear directions between the cause (the disease and the social environment) and disability as an outcome. This is particularly critical for obesity which is complicated by several comorbidities and by the exposure to several risk factors.

A complete evaluation and description of obese patients' disability features should therefore try to put together these different elements that are peculiar of the complexity of this condition. Therefore, it should take into account both the level of person's bodily impairments, that of limitations and restrictions in performing activities and in participation, and the effect that facilitating or hindering environmental factors act in performing single activities. The International Classification of Functioning, Disability, and Health (ICF) (WHO 2001) has been proposed by

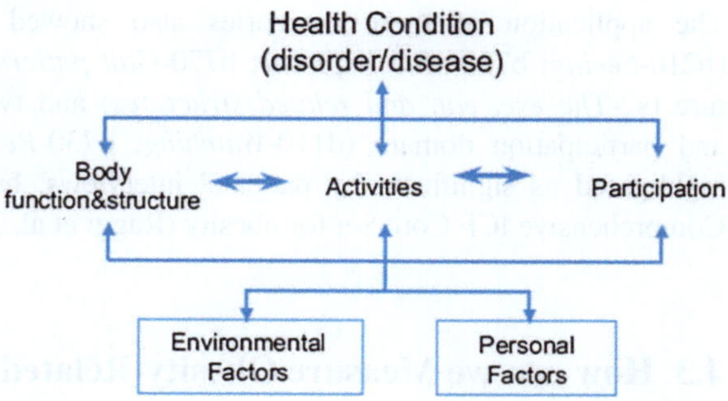

Fig. 4.1 Interactions between the components of the WHO International Classification of Functioning, Disability, and Health (ICF)

the WHO to comprehensively describe limitations in functioning and relevant environmental factors according to a bio-psycho-social model. It provides a framework to integrate professional assessment and the individual's point of view: the utilization of ICF categories for drawing patients' functional profiles can therefore be envisaged to facilitate the description and classification of all aspects of functioning and health in individuals, independently of specific assessment tools.

The ICF relies on this bio-psycho-social model (Fig. 4.1). The ICF contains lists of so-called ICF categories organized into the same components as the biopsychosocial model. Personal factors are not included in the present version of the ICF.

The ICF is very comprehensive but its practical use in disability evaluation presents some problems due to its size and multi-factorial nature; for this reason was developed the ICF Core Sets, a list of categories that are relevant to most patients with a specific health condition. The ICF Core Set for obesity was developed by Stucki et al. (2004) to define the typical spectrum of problems in functioning of patients with obesity. According to the large number of body organs involved in obesity, the selection of *body structures* included almost all chapters of this component. Limitations and restrictions in *activities and participation* are represented in almost as many categories (28 categories) as the *body-functions* (30 categories) component. The areas that are covered represent key issues for patients with obesity, including mobility, self care, and major life areas such as education, work, and employment. Especially the two categories, *walking* (d450) and *moving around* (d455), including *climbing*, were ranked highest. Similarly, the category d570 (*looking after one's health*) comprising managing diet and fitness received highest ranks and *Handling stress and other psychological demands* (d240) is also represented. It is significant that 33 categories representing 30 % of the categories are in the *environmental factors*.

In a previous study we defined an extended list of ICF categories for application to obese patients, by mapping a clinical protocol to the ICF (Brunani et al. 2010). A total of 166 ICF categories were identified: of them, 43 were reported as problematic by at least 20 % of obese patients (Raggi et al. 2009), and a correlation analysis showed a strong effect of body functions' impairments and a limited one of environmental factors towards difficulties in daily activities (Raggi et al. 2010).

The application of these categories also showed that three body functions (b210-*Seeing*; b730-*Muscle power*; b770-*Gait pattern functions*), one body structure (s2-*The eye, ear, and related structures*) and two categories from activities and participation domain (d110-*Watching*; d730-*Relating with strangers*) were highlighted as significant by patients' interviews, but were not included in the Comprehensive ICF Core Set for obesity (Raggi et al. 2009).

4.3 How can we Measure Obesity-Related Disability?

Unfortunately, the tests available do not meet the requirements of complete evaluation because we need to distinguish functional limitations, in physical action used in daily life, from disability that is restriction also in social life. Tests such as the Barthel index, the Katz ADL index, the functional independence measures (FIM), and the Performance ADL Test assess very basic ADL (feeding, continence, need of human assistance); others, such as the Sickness Impact Profile (SIP), the Instrumental ADL Scale, and SF36 also assess mental and/or social functions. We lack obesity-specific and detailed instruments to give us enough information and understanding of the physical difficulties and problems in everyday life that the obese experience.

For these reasons, in clinical practice we need to use different instruments of assessment in relation to the area that we want to investigate.

4.3.1 Functional Scales

Walking, more specifically gait, is an essential component of daily living. Walking is a very different activity for individuals with a BMI of 40 or more (Class III obesity); when compared with the lower weight cohort, this Class III group had decreased velocity and wider distance (Ling et al. 2012).

Previous studies have shown that the 6-min walking test (6MWT) can be used to evaluate "global" function in obesity that registered a deficit in distance walked and in work of walking when compared with normal-weight subjects (Larsson and Reynisdottir 2008). Enright (2003) confirms that people with a BMI more than 30 kg/m^2 walked a substantially shorter distance compared with a control group of people of normal weight. Hulens et al. (2003) reported that BMI is the most important predictor of walking distance and this is significantly greater in males than females, with an average difference of approximately 57 m. Furthermore, 75 % of variance in walking distance could be explain by BMI, peak VO2, quadriceps muscle strength, age, and hours of TV watching or sport participation. For this reason predicted values for age and gender frequently estimated using the formula from Gibbons et al. (2001), are now under revision to consider also weight and height (Chetta et al. 2006; Capodaglio et al. 2012) for validation in morbid obesity. Reference values for 6MWT obtained from an adult obese population have been recently proposed (Capodaglio et al. 2012). The demonstration that weight reduction improved walking speed and maximum oxygen uptake (VO2max) supports a direct link between weight and mobility (Larsson and Mattsson 2003). This relationship is likely to be mediated by respiratory function as research has shown a strong relationship between respiratory function and walking capacity ($r = 0.5$–0.8 between VO2max and 6MWT) (Wise and Brown 2005). A two-month weight-loss program including energy restriction and exercise training improved body composition, in particular maximum lipid-oxidation, and cardiorespiratory tolerance, measured by decrease of HR at rest, reduction in dyspnoea scores without modification of O2 saturation and VO2max during test calculated according to the American College of Sports Medicine guideline (Dumortier et al. 2003; Elloumi et al. 2011).Walking capacity and VO2max has been found to relate to HRQL in obese people: 6MWT showed good correlation (Spearman's ρ: 0.39–0.49; $P < 0.01$) with the physical function of SF-36 (Kolotkin et al. 2011).

Breathlessness and dyspnoea are common symptoms in people with obesity (Gibson 1998; Sin et al. 2002). Assessments of breathlessness may include sensation (intensity, sensory quality, or unpleasantness) and/or the behavioral/emotional consequences of the sensation (respiratory-related functional impairment, disability, or quality of life). The Modified Borg Scale (perceived exertion-intensity), the Medical Research Council (MRC) Scale, and Baseline Dyspnoea Index (BDI); (both assess respiratory-related functional impairment) were, respectively, the most frequently reported instruments. Few instruments had been tested for reliability and validity in people with increased adiposity. Visual Analogue Scale, Modified Borg Scale, descriptors of sensory quality, MRC, and BDI can be recommended as instruments based on their psychometric properties (Gerlach et al. 2012) (Table 4.1).

4.3.2 ADL and IADL

Among daily life activities, a range of different tasks such as showering, dressing, indoor walking, cooking, family activities, driving, but also work, heavy lifting, and sport activities are generally considered. The obese perceive disability to a much greater extent ($P < 0.003$) than normal-weight subjects (Larsson and Mattsson 2001). The main problems are related to occupational work in strenuous positions, sports, walking outdoors, climbing stairs, moderate housework requiring squatting, stooping, or lifting. Rising from sofas, pedicure, and stress incontinence were problematic. The correlation between perceived disability and functional limitations appears fairly good (rs. 0.56) although variable (range 0.14–0.61).

Results from a recent meta-analysis (Backholer et al. 2012) of nine selected studies (four longitudinal and eight cross-sectional cohorts) revealed a graded increase in the risk of ADL limitations from normal weight to overweight, obesity

Table 4.1 Functional scales

Test	Definition	Variables measured	References
6-min walking test (6MWT)	6-min walking "as far as you can"	6-min walk distance (6MWD) Blood oxygen saturation Heart frequency Perception of dyspnoea	American Thoracic Society (ATS) (2002)
Modified Borg Scale (MBS)	Vertical 0 to 10+ item scale with words describing degrees of perceived exaction anchored to numbers	Patient's subjective state of dyspnea at rest	Borg (1982)
Medical Research Council Scale (MRC)	Five steps that describe the entire range of respiratory disability. All the questions relate to everyday activities and are easily understood by patients	Respiratory disability from none (grade 1) to complete incapacity (grade 59). It does not quantify breathlessness itself	Fletcher (1952)
Baseline Dyspnea Index (BDI)	Rating of severity of dyspnea at rest. There are three different categories: functional impairment, magnitude of taste, magnitude of effort	Dyspnea related in five grades from severe (grade 0) to 4 (not impaired) for each category and the sum forms a final score (range 0–12)	Mahler et al. (1984)

class I and obesity class II or more. The increased risk of ADL limitation ranged from 1.04 (95 % CI, 1.00–1.08) for overweight, 1.16 (95 % CI, 1.11–1.21) for obese class I, and 1.76 (95 % CI, 1.28–2.41) for obesity class II or higher. The magnitude of this relationship was stronger among women than men; however, this gender difference was not statistically significant for any BMI category. Meta-analyses of longitudinal studies revealed a similar graded relationship; however, the magnitude of this relationship was slightly greater for all BMI categories, at 1.11 (95 % CI, 1.06–1.17) for overweight and 1.21 (1.05–1.40) for obese class I. Data from the Atherosclerosis Risk in Communities study (Houston et al. 2005) showed that the population-attributable fraction percentages (PAF %) of ADL and IADL impairment due to obesity ranged from 6 % in African-American men to 38 % in white women and 8 % in white men to 39 % in African-American women, respectively. A positive association was found between weight status in young adulthood and weight gain from young adulthood to middle age and functional limitations and disability in late adulthood. It is postulated that obesity early in life and weight gain may lead to joint wear and tear, reduced exercise capacity, and increased risk of chronic disease such as cardiovascular disease, diabetes, and arthritis earlier in life, thus resulting in functional limitations and disability. Most chronic diseases were themselves related to disabilities. Previous studies (Chen et al. 2002; Guallar-Castillón et al. 2007) showed that adjusting for selected chronic conditions did not change or slightly attenuated the associations between obesity and functional disabilities (Chen and Guo 2008). This suggests that obesity may relate to disability independently or alternatively, obesity may relate to frailty syndrome, which in turn may lead to further development of disability in elderly people (Hirsch et al. 2006) (Table 4.2).

4.3.3 Functional Independence Measure

The FIM scale includes 18 items, each of them is scored from 1 to 7 based on level of independence, where 1 represents total dependence and 7 complete independence. Possible scores range from 18 to 126. The 13 physical items can be scored separately from the 5 cognitive items (Hamilton et al. 1987). The association between the FIM, the gold standard for measuring outcomes in rehabilitation and BMI, has been recently investigated in unfit individuals with medical complications who are undergoing intensive rehabilitation: they showed higher gains in FIM scores, mostly accounted for by the motor subscale, as compared with normal weight patients (Jain et al. 2008). Vincent and Vincent (2008) showed that a high BMI does not prevent FIM gains during inpatient rehabilitation after total knee replacement. However, these gains were achieved less efficiently and at a higher cost than when BMI is normal. Two small studies also reported no association between BMI and FIM scores in post-acute stroke and joint arthroplasty rehabilitation patients (Vincent et al. 2006). Since post-acute rehabilitation individuals may encompass a range of diagnoses, it is likely that the BMI–FIM association

Table 4.2 Daily activities

Activities of daily living (ADL)
Definition: all daily self-care activities within an individual's place of residence, in outdoor environments, or both. We include: personal hygiene, dressing and undressing, self feeding, functional transfers (getting into and out of bed or wheelchair, getting onto or out of toilet), bowel and bladder management, ambulation (walking with or without use of an assistive device—walker, cane, or crutches—or using a wheelchair)
Evaluation: the ability or inability to perform ADLs as a measurement of the functional status of a person
Tools: Katz scale, Bristol scale
References: Katz et al. (1963), Bucks et al. (1996)
Instrumental activities of daily living (IADL)
Definition: instrumental activities of daily living are not necessary for fundamental functioning, but they let an individual live independently in a community. We include: housework, taking medications as prescribed, managing money, shopping for groceries or clothing, using the telephone or other form of communication, using technology (as applicable), transportation within the community
Evaluation: the ability or inability to perform IADLs as a measurement of the functional status of a person
Tools: Lawton IADL scale
References: Lawton and Brod (1969)
Functional independence measure (FIM)
Definition: assesses physical and cognitive disability. It is useful in clinical settings of rehabilitation
Evaluation: The scale includes 18 items, each of them is scored from 1 to 7 based on level of independence, where 1 represents total dependence and 7 complete independence. Possible scores range from 18 to 126. The 13 physical items can be scored separately from the 5 cognitive items
References: Hamilton et al. (1987)

may vary within subgroups. The studies investigate whether a high BMI might actually impede the rehabilitation process after surgery or stroke and the gains in physical function (Table 4.2).

4.3.4 Health-Related Quality of Life

Obesity is associated with impairment of health-related quality of life (HRQL) in psychological, social, and physical domains (Karlsson et al. 2007; Herpertz et al. 2003). Factors reported to be associated with greater impairment of quality of life among treatment seeking obese patients include female sex (Kolotkin et al. 2001), higher BMI (Kolotkin et al. 2002), binge eating disorder, and psychopathology (Marchesini et al. 2003; Rieger et al. 2005). The metabolic syndrome and HRQOL were more strongly associated in obesity-specific QOL than in generalized QOL (Han et al. 2009). Different questionnaires were developed such as the Medical Outcome Survey Short-Form 36 (SF-36) and the Psychological General Well-Being (PGWB). In 2002, the United States Task Force on Developing Obesity

Outcomes and Learning Standards (TOOLS) recommended, for the obese patient, the use of SF-36 or its 12-items version (SF-12). The SF-36 is frequently used to describe the health status and physical ability of people with numerous impairments who are receiving physical therapy (McHorney et al. 1993). This questionnaire is a generic measure of HR-QOL, with high content and external validity. The questionnaire contains 36 questions grouped into 8 scales, and these scales are further clustered into the physical component summary (PCS) and mental component summary (MCS). The PCS includes scales to measure physical functioning, role limitation—physical, bodily pain, general health. The MCS includes scales to measure role limitation—emotional, vitality, mental health, and social functioning. Other obesity-specific measures have been developed: Obesity-related problem scale (OP) was constructed to measure the impact of obesity on psychosocial functioning in everyday life (Karlsson et al. 2003); Obesity-related coping (OCQ) and obesity-related distress questionnaires (ODQ) define the impact of coping strategies and obesity-related distress (Ryden et al. 1999, 2001); Impact of weight on quality of life questionnaire (IWQOL) is a self-reported 74-item measure to assess the effect of weight along eight domains of functioning (Kolotkin et al. 2002); Obesity adjustment survey-short form (OAS-SF) was constructed to assess the psychological distress of individuals who are morbidly obese (Butler et al. 1999); short specific quality of life scale (OSQOL) is a self-reported 11-item questionnaire evaluating four dimensions: physical state, vitality, desire to do things, relations with other people, and psychological state (Le Pen et al. 1998); Obesity-related well-being questionnaire (ORWELL 97) represents the intensity and the subjective relevance of physical and psychosocial distress (Mannucci et al. 1999); BAROS, designed to measure outcomes in the surgical treatment of morbid obesity, is a self-reported seven-item instrument, evaluating three main areas; weight loss, improvement of medical conditions, and quality of life (Oria and Moorehead 1998); BQL was developed to measure quality of life in relation to weight, weight-related comorbidity and surgery-related gastrointestinal symptoms (Weiner et al. 2005); Obesity and weight-loss quality-of-life (OWLQOL) instrument intend to be used together with the weight-related symptom measure (WRSM), self-report measure for the presence and bothersome ness of symptoms associated with obesity (Niero et al. 2002). We recently evaluated the relation between HRQoL and disability: subjects with higher disability and HRQoL decrement were older and had higher BMI. Women were more likely to present moderate disability and reduction in HRQoL, while men more likely presented mild disability and HRQoL reduction (Sirtori et al. 2012).

Changes in generic and weight-specific quality of life scores are associated with weight loss. Wadden and Phelan (2002) reported that people who were obese had decreased physical functioning scores, as measured by the SF-36, and that weight loss improved HR-QOL scores, specifically in physical functioning. In particular, the percent changes in weight during a clinical trial of weight loss for 24 months were significantly associated with percent changes in the physical and mental component scores of the SF-36. For every 5 % decrease in weight, the mental component score increased by 3.21 % ($p = 0.003$), whereas the physical component score increased by 1.43 % ($p = 0.003$). For every 5 % decrease in weight, the physical functioning subscale score increased by 2.30 % ($p = 0.002$), the role physical subscale score increased by 4.55 % ($p = 0.001$), the general health subscale score increased by 3.38 % ($p = 0.001$), the vitality subscale score increased by 9.38 % ($p = 0.001$), the role—emotional subscale score increased by 2.83 % ($p = 0.024$), and the mental health subscale score increased by 2.78 % ($p = 0.009$) (Mindi et al. 2011). A review conducted by Teixeira et al. (2005) indicates that general measures of HRQoL, such as the SF-36, are not likely predictors of weight loss success, but weight-specific measures may have some predictive value.

A battery of generic and condition-specific measures was used for HRQL assessment in Swedish obese subjects (SOS) Quality of Life Survey (Sullivan et al. 2001), a controlled longitudinal trial of the health effects of weight loss before and at 0.5, 1, 2, 3, 4, 6, 8, and 10 years after bariatric surgery. The current health scale (CH), short version of the mood adjective check list (MACL), hospital anxiety and depression scale (HAD), social interaction (SI) category from the SIP, OP show improvements and deteriorations in HRQL that were associated with the magnitude of weight loss or regain, except regarding anxiety. Peak improvements in the surgical group were observed during the first year of weight loss, whereas the weight regain phase (mainly between 1- and 6-year follow-up) was accompanied by a gradual decline in HRQL. Long-term results of the study suggest that a maintained weight loss of about 10 % is sufficient for positive long-term effects on HRQL (Karlsson et al. 2007) (Table 4.3).

4.3.4.1 The Obesity-Related Disability Questionnaire

In 2009, the Italian Society of Obesity (SIO) developed a new scale for assessing disability correlated to obesity in adult populations, the short-form questionnaire the Obesity-related Disability test (*Test SIO Disabilita` Obesita` Correlata*, TSD-OC) (Donini et al. 2010). It is composed of 36 items divided into seven sections (pain: 5 items; stiffness: 2 items; ADL and indoor mobility: 7 items; housework: 7 items; outdoor activities: 5 items; occupational activities: 4 items and social life: 6 items); each item was evaluated on a 0–10 visual analogue scale (VAS), where 10 indicates the highest level of disability and 0 no difficulties in performing the task. The 'disability scores' as the sum of each item's raw score divided by the maximum possible score, expressed as a percentage according to the following linear transformation: (raw score/max score) × 100. In a validation study, significant low to moderate correlations between TSD-OC, SF-36 scores, 6MWT, and grip strength were observed and a total of 26 ICF categories were linked, mostly related to the area of mobility (Donini et al. 2011). In a group of 355 obese patients we defined the disability levels (low, medium, and high) and we found that after a short-term, intensive multidisciplinary rehabilitation program registering a 5 % BMI reduction, TSD-OC is sensitive to change in improving functional status with higher functional improvement for younger individuals with higher level of obesity (Precilios et al. 2012).

Table 4.3 Hearth-related quality of life (HRQoL) evaluation

Test	Evaluation	Measures	References
Medical Outcome Survey-Short Form 36 (SF-36)	The health status and physical ability	36 questions for 8 scales further clustered into the physical component summary (PCS) and mental component summary (MCS). The PCS includes scales to measure physical functioning, role limitation-physical, bodily pain, general health. The MCS includes scales to measure role limitation-emotional, vitality, mental health, and social functioning	Ware et al. (1993)
Psychological general well-being (PGWB)	The psychological distress	22 questions are arranged in 6 affective states: anxiety, depressed mood, positive well-being self-control, general health, and vitality	Dupuy (1984)
Obesity-related problem scale (OP)	The impact of obesity on psychosocial functioning in everyday life	8 items on a four-point scale. Self-reported in a broad range of social activities: private gatherings at home or at a friend's or relative's home, going to restaurants, participation in community activities (courses, etc.), holidays away from home, trying on and buying clothes, bathing in public places, and intimate relations	Karlsson et al. (2003)
Obesity-related coping (OCQ) and obesity-related distress questionnaires (ODQ)	The impact of coping strategies and obesity-related distress	16 items grouped into three factors: social trust, fighting spirit and wishful thinking. 13 items grouped into two factors labeled intrusion and helplessness	Rydén et al. (2001)

(Continued)

Table 4.3 (Continued)

Test	Evaluation	Measures	References
Impact of weight on quality of life questionnaire (IWQoL) and IWQOL-Lite	The effect of weight on functioning	74 and 31. The short version consists of five scales: physical function, self esteem, sexual life, public distress, work.	Kolotkin et al. (1995) Kolotkin and Crosby (2002)
Obesity adjustment survey-short form (OAS-SF)	The psychological distress of individuals who are morbidly obese	20 items	Butler et al. (1999)
Short specific quality of life scale (OSQOL)	Aspects of quality of life affected by weight	11 items into four dimensions: physical state, vitality, desire to do things, relations with other people, and psychological state	Le Pen et al. (1998)
Obesity-related well-being questionnaire (ORWELL 97)	The intensity and the subjective relevance of physical and psychosocial distress	18 items in 4 domains: somatic complaints, physical functioning, emotional functioning, and social functioning	Mannucci et al. (1999)
Bariatric analysis and reporting outcome system (BAROS)	The outcomes in the surgical treatment of morbid obesity	7 items in three main areas: weight loss, improvement of medical conditions, and quality of life	Oria and Moorehead (1998)
Bariatric quality of life (BQL)	The quality of life in relation to weight, weight-related comorbidity, and surgery-related gastrointestinal symptoms	19 items in the following domains: psychological well-being, social functioning, physical functioning, and problems and symptoms related to obesity surgery and obesity-related comorbidity	Weiner et al. (2005)
Obesity and weight-loss quality-of-life (OWLQoL)	A person's global evaluation of position in life related to weight, weight loss, and weight loss treatment	17 items in 7-point scale that ranges from 0 ("not at all") to 6 ("a very great deal")	Niero et al. (2002)

(Continued)

Table 4.3 (Continued)

Test	Evaluation	Measures	References
Weight-related symptom measure (WRSM)	The presence and bothersomeness of symptoms associated with obesity	20-item in 7-point scale. Responses are "yes" or "no", in the previous 4 weeks, that indicate the degree of the symptom	See OWLQoL
Mood adjective check list (MACL)	Three bipolar dimensions of mood: pleasantness/unpleasantness (e.g., satisfied, optimistic/depressed, resigned), activation/deactivation (e.g., alert, active/passive, apathetic) and calmness/tension (e.g., relaxed/tense, distressed)	38 adjectives in a 4-point scale with two acceptance and two rejection categories	Sjo¨berg et al. (1979)
Current health scale (CH)	General health perceptions	9 general statements on perceived current health	Davies et al. (1988)
Hospital anxiety and depression scale (HAD)	To detect mood disorders: for screening of anxiety and depression disorders in somatically ill patients	14 items in a 4-point scale, which are summed to separate scores on anxiety and depression	Zigmond and Snaith (1983)
Social interaction (SI) category from the sickness impact profile (SIP)	To assess health-related dysfunction in social life	20 statements on quality and quantity of social interaction within the family, among friends and in the community	Bergner et al. (1981)

4.3.5 Work Disability

Although occupational disability is a great burden to the individual and extremely costly to the society in countries with an established welfare system, only recently the role of obesity as a risk factor for occupational disability has been addressed (Capodaglio et al. 2010). Obesity increases the risk of work disability mainly due to osteoarthritis and other obesity-related health consequences, including coronary heart disease, limiting work ability, and eventually leading to work disability (Claessen et al. 2009, 2012). In a previous study (Scott et al. 2006), increasing obesity was associated with decreasing odds of workforce participation, with Classes I, II, and III obesity having odds ratios (95 % confidence interval) of 0.94 (0.89–0.99), 0.85 (0.77–0.94), and 0.66 (0.57–0.78), respectively. This association appears to be independent of associated comorbidity and sociodemographic factors. Recently, fast-food meals were associated with decreased productivity in men ($P = 0.038$, independent of BMI). The results suggest that obesogenic dietary behaviors and higher BMI are associated with decreased QOL and productivity of different degrees in women and men (Cash et al. 2011). A preliminary investigation on the impact of obesity-related disability among different job categories has been recently published (Bogni et al. 2011) and tools already used for evaluating mental and physical areas of concern during occupational tasks (The Job Content Questionnaire by Karasek) are currently under investigation in obese workers (Brunani et al. unpublished data).

4.4 Conclusions

The identification of critical areas and specific disabilities in obese patients might facilitate the introduction of more effective person-centered measures, either in the long term, such as lifestyle modifications, or in the short–medium term, such as rehabilitative interventions. All together this could promote better health outcomes. Furthermore, specific tools designed for assessing obesity-related disability can more clearly identify factors that are responsible for the onset and maintenance of obesity disability in order to develop targeted rehabilitation programs as well as public health programs to reduce the impact and social cost of obesity.

The multidisciplinary approach, supported by ICF bio-psycho-social model, is the first-choice strategy for evaluating the obesity-related disability. Using specific disability measures for obesity requires in fact a multidisciplinary approach, since disability is multifaceted, ranging from motor and mobility to energetic and cardiorespiratory disabilities. The TSD-OC appears a tool able to measure the aspects of disability described by obese subjects: in particular, the TSD-OC proved to be significantly correlated with functional (grip strength and distance walked in 6 min) and QoL parameters (SF-36). The TSD-OC showed better correlations with the SF-36 items related to physical component, although a significant correlation was also found with the items describing the mental health component (Donini et al. 2011), suggesting that disability in obesity is not only a physical issue. Its use within a multidisciplinary assessment should be implemented by specific instruments able to establish the role of environmental factors. This new scale may represent an important instrument for the description of obesity-related disability and for planning and measuring the effectiveness of rehabilitation programs in obese subjects. A longitudinal study (Precilios et al. 2012) has shown its sensitivity to change at different time points. However, this and other instruments need to be further tested in multidisciplinary rehabilitation programs for monitoring health outcomes as well as for planning interventions on environmental factors that can reduce the impact of disability.

References

Al Snith S, Ottenbacher KJ, Markides KS, Kuo YF, Eschbach K, Goodwin JS (2007) The effect of obesity on disability versus mortality in older. Americans. Arch Intern Med 167:774–780

Alley DE, Chang VW (2007) The changing relationship of obesity and disability, 1988–2004. JAMA 298:2020–2027

American Thoracic Society (ATS) (2002) Statement: guidelines for the six-minute walk test. ATS committee on proficiency standards for clinical pulmonary function laboratories. Am J Respir Crit Care Med 166:111–117

Andreyeva T, Sturm R, Ringel JS (2004) Moderate and severe obesity have large differences in health care costs. Obes Res 12:1936–1943

Apovian CM, Frey CM, Wood GC, Rogers JZ, Still CD, Jensen GL (2002) Body mass index and physical function in older women. Obes Res 10:740–747

Backholer K, Wong E, Freak-Poli R, Walls HL, Peeters A (2012) Increasing body weight and risk of limitations in activities of daily living: a systematic review and meta-analysis. Obes Rev 13:456–468

Bergner M, Bobitt RA, Carter WB, Gilson BS (1981) The sickness impact profile: development and final revision of a health status measure. Med Care 19:787–805

Bogni M, Vigna L, Capodaglio P, Brunani A, Cimolin V, Donini LM, Riboldi L (2011) Impatto della disabilità correlata all'obesità nelle diverse categorie professionali. G Ital Med Lav Erg 33(Suppl):364–366

Borg G (1982) Psychophysical bases of perceived exertion. Med Sci Sports Exerc 14:377–381

Brunani A, Liuzzi A, Sirtori A, Raggi A, Berselli ME, Villa V, Ceriani F, Tacchini E, Vicari V, Parisio C, Vismara L, Zanini A, Vinci C, Contini F, Braga E, Ricappi A, Camerlengo M, Ristea M, Leonardi M (2010) Mapping an obesity clinical evaluation protocol to the international classification of functioning, disability and health. Disabil Rehabil 32:417–423

Bucks RS, Ashworth DL, Wilcock GK, Siegfried K (1996) Assessment of activities of daily living in dementia: development of the bristol activities of daily living scale. Age Ageing 25:113–120

Butler GS, Vallis TM, Perey B, Veldhuyzen van Zanten SJ, MacDonald AS, Konok G (1999) The obesity adjustment survey: development of a scale to assess psychological adjustment to morbid obesity. Int J Obes 23:505–511

Capodaglio P, Castelnuovo G, Brunani A, Vismara L, Villa V, Capodaglio EM (2010) Functional limitations and occupational issues in obesity: a review. Int J Ocuup Saf Ergon 16:507–523

Capodaglio P, De Souza S, Parisio C, Precilios H, Vismara L, Cimolin V, Brunani A (2012) Reference values for the six-minute walking test in obese subjects. Disabil Rehab. doi: 10.3109/09638288.2012.726313

Carpiniello B, Pinna F, Pillai G, Nonnoi V, Pisano E, Corrias S, Orrù MG, Orrù W, Velluzzi F, Loviselli A (2009) Obesity and psychopathology. a study of psychiatric comorbidity among patients attending a specialist obesity unit. Epidemiol Psichiatr Soc 8:119–127

Cash SW, Beresford SA, Henderson JA, McTiernan A, Xiao L, Wang CY, Patrick DL (2011) Dietary and physical activity behaviours related to obesity-specific quality of life and work productivity: baseline results from a worksite trial. Br J Nutr 6:1–9

Chen H, Guo X (2008) Obesity and functional disability in elderly Americans. J Am Geriatr Soc 56:689–694

Chen H, Bermudez OI, Tucker KL (2002) Waist circumference and weight change are associated with disability among elderly Hispanics. J Gerontol A Biol Sci Med Sci 57A:M19–M25

Chetta A, Zanini A, Pisi G, Aiello M, Tzani P, Neri M, Olivieri D (2006) References values for the 6-minute walk test in healthy subjects 20–50 years old. Respir Med 100:1573–1578

Chwastiak LA, Rosenheck RA, Kazis LE (2011) Association of psychiatric illness and obesity, physical inactivity, and smoking among a national sample of veterans. Psychosomatics 52:230–236

Claessen H, Arndt V, Drath C, Brenner H (2009) Overweight, obesity and risk of work disability: a cohort study of construction workers in Germany. Occup Environ Med 66:402–409

Claessen H, Brenner H, Drath C, Arndt V (2012) Repeated measures of body mass index and risk of health related outcomes. Eur J Epidemiol 27:215–224

Davies AR, Sherbourne CR, Peterson JR, Ware Jr JE (1988) Scoring manual: adult health status and patient satisfaction measures used in RAND's health insurance experiment. The RAND corporation (No. N-2190-HHS), Santa Monica

Donini LM, Cuzzolaro M, Spera G, Badiali M, Basso N, Bollea MR, Bosello O, Brunani A, Busetto L, Cairella G, Cannella C, Capodaglio P, Carbonelli MG, Castellaneta E, Castra R, Clini E, Contaldo F, Dalla Ragione L, Dalle Grave R, D'Andrea F, Del Balzo V, De Cristofaro P, Di Flaviano E, Fassino S, Ferro AM, Forestieri P, Franzoni E, Gentile MG, Giustini A, Jacoangeli F, Lubrano C, Lucchin L, Manara F, Marangi G, Marcelli M, Marchesini G, Marri G, Marrocco W, Melchionda N, Mezzani B, Migliaccio P, Muratori F, Nizzoli U, Ostuzzi R, Panzolato G, Pasanisi F, Persichetti P, Petroni ML, Pontieri V, Prosperi E, Renna C, Rovera G, Santini F, Saraceni V, Savina C, Scuderi N, Silecchia G, Strollo F, Todisco P, Tubili C, Ugolini G, Zamboni M (2010) Obesity and eating disorders. Indications for the different levels of care. An Italian expert consensus document. Eat Weight Disord 15(1–2 Suppl):1–31

Donini LM, Brunani A, Sirtori A, Savina C, Tempera S, Cuzzolaro M, Spera G, Cimolin V, Precilios H, Raggi A, Capodaglio P; SIO-ISDCA Task Force (2011) Assessing disability in morbidly obese individuals: the Italian society of obesity test for obesity-related disabilities. Disabil Rehabil 33:2509–2518

Dumortier M, Brandou F, Perez-Martin A, Fedou C, Mercier J, Brun JF (2003) Low intensity endurance exercise targeted for lipid oxidation improves body composition and insulin sensitivity in patients with the metabolic syndrome. Diabetes Metab 29:509–518

Dupuy HJ (1984) The psychological general well-being (PGWB) index. In: Wenger NK, Mattsson ME, Furberg CD, Elinson J (eds) Assessment of quality of life in clinical trials of cardiovascular therapies. Le Jacq Publishing Inc, pp 170–183

Elloumi M, Makni E, Ounis OB, Moalla W, Zbidi A, Zaoueli M, Lac G, Tabka Z (2011) Six-minute walking test and the assessment of cardiorespiratory responses during weight-loss programmes in obese children. Physiother Res Int 16:32–42

Enright PL (2003) The six-minute walk test. Respir Care 48:783–785

Ezzati M, Lopez AD, Rodgers A, Vander Hoorn S, Murray CJ (2002) Comparative risk assessment collaborating group: selected major risk factors and global and regional burden of disease. Lancet 360:1347–1360

Fletcher CM (1952) The clinical diagnosis of pulmonary emphysema-an experimental study. Proc R Soc Med 45:577–584

Ford ES, Li C, Zhao G, Pearson WS, Capewell S (2009) Trends in the prevalence of low risk factor burden for cardiovascular disease among United States adults. Circulation 120:1181–1188

Ford ES, Li C, Zhao G, Tsai J (2011) Trends in obesity and abdominal obesity among adults in United States from 1999–2008. Int J Obes 35:736–743

Gerlach Y, Williams MT, Coates AM (2012) Weighing up the evidence-a systematic review of measures used for the sensation of breathlessness in obesity. Int J Obes. doi:10.1038/ijo.2012.49

Gibbons WJ, Fruchter N, Sloan S, Levy RD (2001) Reference values for a multiple repetition 6-minute walk test in healthy adults older than 20 years. J Cardiopulm Rehabil 21:87–93

Gibson GJ (1998) Obesity, respiratory function and breathlessness. Chest 114:1372–1377

Guallar-Castillón P, Sagardui-Villamor J, Banegas JR, Graciani A, Fornés NS, López García E, Rodríguez-Artalejo F (2007) Waist circumference as a predictor of disability among older adults. Obesity 15:233–244

Hamilton BB, Granger CV, Sherwin FS, Zielezny M, Tashman JS. (1987) A uniform national data system for medical rehabilitation. In: Fuhrer MJ (ed) Rehabilitation outcomes: analysis and measurement. Baltimore

Han JH, Park HS, Shin CI, Chang HM, Yun KE, Cho SH, Choi EY, Lee SY, Kim JH, Sung HN, Kim JH, Choi SI, Yoon YS, Lee ES, Song HR, Bae SC (2009) Metabolic syndrome and quality of life (QOL) using generalised and obesity-specific QOL scales. Int J Clin Pract 63:735–741

Haslam DW, James WPT (2005) Obesity. Lancet 366:1197–1209

Herpertz S, Kielmann R, Wolf AM, Langkafel M, Senf W, Hebebrand J (2003) Does obesity surgery improve psychosocial functioning? A systematic review. Int J Obes Relat Metab Disord 27:1300–1314

Hirsch C, Anderson ML, Newman A, Kop W, Jackson S, Gottdiener J, Tracy R, Fried LP, Cardiovascular Health Study Research Group (2006) The association of race with frailty: the cardiovascular health study. Ann Epidemiol 16:545–553

Houston DK, Stevens J, Cai J, Morey MC (2005) Role of weight history on functional limitations and disability in late adulthood: the ARIC study. Obes Res 13:1793–1802

Hulens M, Vansant G, Claessens AL, Lysens R, Muls E (2003) Predictors of 6-minute walk test results in lean, obese and morbidly obese women. Scand J Med Sci Sports 13:98–105

Jain NB, Al-Adawi S, Dorvlo ASS, Burke DT (2008) Association between body mass index and functional independence measure in patients with deconditioning. Am J Phys Med Rehabil 87:21–25

Karlsson J, Taft C, Sjostrom L, Torgerson JS, Sullivan M (2003) Psychosocial functioning in the obese before and after weight reduction: construct validity and responsiveness of the obesity related problems scale. Int J Obes Relat Metab Disord 27:617–630

Karlsson J, Taft C, Ryden A, Sjostrom L, Sullivan M (2007) Ten-year trends in health-related quality of life after surgical and conventional treatment for severe obesity: the SOS intervention study. Int J Obes 31:1248–1261

Katz S, Ford AB, Moskowitz RW, Jackson BA, Jaffe MW (1963) Studies of illness in the aged: the index of ADL: a standardized measure of biological and psychosocial function. JAMA 185:914–919

Kolotkin RL, Crosby RD (2002) Psychometric evaluation of the impact of weight on quality of life-lite questionnaire (IWQOL-lite) in a community sample. Qual Life Res 11:157–171

Kolotkin RL, Head S, Hamilton M, Tse CK (1995) Assessing impact of weight on quality of life. Obes Res 3:49–56

Kolotkin RL, Crosby RD, Kosloski KD, Williams GR (2001) Development of a brief measure to assess quality of life in obesity. Obes Res 9:102–111

Kolotkin RL, Crosby RD, Williams GR (2002) Health-related quality of life varies among obese subgroups. Obes Res 10:748–756

Kolotkin RL, LaMonte MJ, Litwin S, Crosby RD, Gress RE, Yanowitz FG, Hunt SC, Adams TD (2011) Cardiorespiratory fitness and health-related quality of life in bariatric surgery patients. Obes Surg 21:457–464

Kotseva K, Wood D, De BG, De BD, Pyorala K, Keil U (2009) Cardiovascular prevention guidelines in daily practice: a comparison of EUROASPIRE I, II, and III surveys in eight European countries. Lancet 373:929–940

Larsson UE, Mattsson E (2001) Perceived disability and observed functional limitations in obese women. Int J Obes 25:1705–1712

Larsson UE, Mattsson E (2003) Influence of weight loss programmes on walking speed and relative oxygen cost (%VO2max) in obese women during walking. J Rehabil Med 35:91–97

Larsson EU, Reynisdottir S (2008) The six-minute walk test in outpatients with obesity: reproducibility and known group validity. Physiother Res Int 13:84–93

Lawton MP, Brod EM (1969) Assessment of older people: self-maintaining and instrumental activities of daily living. Gerontologist 9:179–186

Le Pen C, Levy E, Loos F, Banzet MN, Basdevant A (1998) 'Specific scale' compared with 'generic' scale: a double measurement of the quality of life in a French community sample of obese subjects. J Epidemiol Commun Health 52:445–450

Ling SM, Fried LP, Garrett ES, Fan MY, Rantanen T, Bathon JM (2003) Knee osteoarthritis compromises early mobility function: the Women's Health and Aging Study II. J Rheumatol 30:114–120

Ling C, Kelechi T, Mueller M, Brotherton S, Smith S (2012) Gait and function in class III obesity. J Obes. doi:10.1155/2012/257468

Locher JL, Roth DL, Ritchie CS, Cox K, Sawyer P, Bodner EV, Allman RM (2007) Body mass index, weight loss, and mortality in community-dwelling older adults. J Gerontol A Biol Sci Med Sci 62:1389–1392

Mahler DA, Weiberg DH, Wells CK, Feinstein AR (1984) The measurement of dyspnea. Contents, interobserver agreement, and physiologic correlates of two new clinical indexes. Chest 85:751–758

Mannucci E, Ricca V, Barciulli E, Di Bernardo M, Travaglini R, Cabras PL, Rotella CM (1999) Quality of life and overweight: the obesity related well-being (Orwell 97) questionnaire. Addict Behav 24:345–357

Marchesini G, Bellini M, Natale S, Belsito C, Isacco S, Nuccitelli C, Pasqui F, Baraldi L, Forlani G, Melchionda N (2003) Psychiatric distress and health-related quality of life in obesity. Diab Nutr Metab 16:145–154

Marks R (2007) Obesity profiles with knee osteoarthritis: correlation with pain, disability, disease progression. Obesity 15:1867–1874

McHorney CA, Ware JE Jr, Raczek AE (1993) The MOS 36-item short-form health survey (SF-36): II. Psychometric and clinical tests of validity in measuring physical and mental health constructs. Med Care 31:247–263

Mindi A, Wang J, Acharya SD, Yang K, Chasen ER, Choo J, Ye L, Burke LE (2011) Health-related quality of life among participants in the SMART weight loss trial. Appl Nurs Res. doi:10.1016/j.apnr.2011.08.001

Neovius K, Johansson K, Rössner S, Neovius M (2008) Disability pension, employment and obesity status: a systematic review. Obes Rev 9:572–581

Niero M, Martin M, Finger T, Lucas R, Mear I, Wild D, Glauda L, Patrick DL (2002) A new approach to multicultural item generation in the development of two obesity-specific measures: the obesity and weight loss quality of life (OWLQOL) questionnaire and the weight-related symptom measure (WRSM). Clin Ther 24:690–700

Oria HE, Moorehead MK (1998) Bariatric analysis and reporting outcome system (BAROS). Obes Surg 8:487–499

Peyrot N, Thivel D, Isacco L, Morin J, Duche P, Belli A (2009) Do mechanical gait parameters explain the higher metabolic cost of walking in obese adolescents? J Appl Physiol 106:1763–1770

Precilios H, Brunani A, Cimolin V, Tacchini E, Donini LM, Fabris De Souza S, Capodaglio P, Precilios H (2012) Measuring changes after multidisciplinary rehabilitation of obese individuals. J Endocrinol Invest 31. doi: 10.3275/8240

Raggi A, Brunani A, Sirtori A, Liuzzi A, Berselli ME, Villa V, Ceriani F, ICF-Obesity Group, Leonardi M (2010) Obesity-related disability: key factors identified by the international classification of functioning, disability and health (ICF). Disabil Rehabil 32:2028–2034

Raggi A, Sirtori A, Brunani A, Liuzzi A, Leonardi M (2009) Use of the ICF to describe functioning and disability in obese patients. Disabil Rehabil 31(S1):S153–S158
Reuser M, Bonneuxl LG, Willekens FJ (2009) Smoking kills, obesity disables: a multistate approach of the US health and retirement survey. Obesity 17:783–789
Reynolds SL, Saito Y, Crimmins EM (2005) The impact of obesity on active life expectancy in older American men and women. Gerontologist 45:438–444
Rieger E, Wilfley DE, Stein RI, Marino V, Crow SJ (2005) A comparison of quality of life in obese individuals with and without binge eating disorder. Int J Eat Disord 37:234–240
Ryden A, Karlsson L, Persson O, Sjostrom L, Sullivan M (1999) The obesity-related coping questionnaire (OCQ) from the Swedish obese subjects (SOS) study. Qual Life Res 8:580A
Ryden A, Karlsson J, Persson LO, Sjostrom L, Taft C, Sullivan M (2001) Obesity-related coping and distress and relationship to treatment preference. Br J Clin Psychol 40:177–188
Rydén A, Karlsson J, Persson L-O, Sjöström L, Taft C, Sullivan M (2001) Obesity-related coping and distress and relations to treatment preference. Br J Clin Psychol 40:177–188
Schultz Y, Kyle UU, Pichard C (2002) Fat-free mass index and fat mass index percentiles in Caucasians aged 18–98 y. Int J Obes 26:953–960
Scott K, Raj P, Anderson C, Philip J (2006) Population-based analysis of obesity and workforce participation. Obesity 14:920–927
Sin DD, Jones RL, Man SF (2002) Obesity is a risk factor for dyspnea but not for airflow obstruction. Arch Intern Med 162:1477–1481
Sirtori A, Brunani A, Villa V, Berselli ME, Croci M, Leonardi M, Raggi A (2012) Obesity is a marker of reduction in QoL and disability. Sci World J 2012:167520
Sjo¨berg L, Svensson E, Persson L (1979) The measurement of mood. Scand J Psychol 20:1–18
Stenholm S, Koster A, Alley DE, Houston DK, Kanaya A, Lee JS, Newman AB, Satterfield S, Simonsick EM, Visser M, Harris TB, Ferrucci L and for the Health, Aging, and Body Composition Study (2010) Joint association of obesity and metabolic syndrome with incident mobility limitation in older men and women—results from the health aging, and body composition study. J Gerontol A Biol Sci Med Sci 65A(1):84–92
Stenholm S, Alley D, Bandinelli S, Griswold ME, Koskinen S, Rantanen T, Guralnik JM, Ferrucci L (2009) The effect of obesity combined with low muscle strength on decline in mobility in older persons: results from the InCHIANTI study. Int J Obes 33:635–644
Stewart ST, Cutler DM, Rosen AB (2009) Forecasting the effects of obesity and smoking on US life expectancy. N Engl J Med 361:2252–2260
Stucki A, Daansen P, Fuessl M, Cieza A, Huber E, Atkinson R, Kostanjsek N, Stucki G, Ruof J (2004) ICF core sets for obesity. J Rehabil Med 44 (Suppl):107–113
Sullivan M, Karlsson J, Sjostrom L, Taft C (2001) Why quality-of-life measures should be used in the treatment of patients with obesity. In: Bjorntorp P (ed) International textbook in obesity. Chichester
Teixeira PJ, Going SB, Sardinha LB, Lohman TG (2005) A review of psychosocial pre-treatment predictors of weight control. Obes Rev 6:43–65
Tunceli K, Li K, Williams LK (2006) Long-term effects of obesity on employment and work limitations among U.S. Adults, 1986 to 1999. Obesity 14:1637–1646
Vincent HK, Vincent KR (2008) Obesity and inpatient rehabilitation outcomes following knee arthroplasty: a multicenter study. Obesity 16:130–136
Vincent KR, Lee LW, Weng J, Alfano AP, Vincent HK (2006) A preliminary examination of the CMS eligibility criteria in total-joint arthroplasty. Am J Phys Med Rehabil 85:872–881
Vincent HR, Vincent KR, Lamb KM (2010) Obesity and mobility disability in the older adult. Obes Rev 11:568–579
Wadden TA, Phelan S (2002) Assessment of quality of life in obese individuals. Obes Res 10(Suppl 1):50–57
Ware JE, Snow KK, Kosinski M, Gandek B (1993) SF36 health survey: manual and interpretation guide. The Health Institute, New England Medical Center
Weiner S, Sauerland S, Fein M, Blanco R, Pomhoff I, Weiner RA (2005) The Bariatric quality of life index: a measure of well-being in obesity surgery patients. Obes Surg 15:538–545

Wise RA, Brown CD (2005) Minimal clinically important differences in the six-minute walk test and the incremental shuttle walking test. J Chronic Obstr Pulm Dis 2:125–129
World Health Organization (2001) International classification of functioning, disability and health: ICF. World Health Organization, Geneva
World Health Organization (2006) Obesity and overweight (WHO Fact Sheet No. 311) WHO Media centre 2006. http://www.who.int/mediacentre/factsheets/fs311/en/print.html
World Health Organization (2011) Obesity and overweight. Report No. 311
Zamboni M, Mazzali G, Zoico E, Harris TB, Meigs JB, Di Francesco V, Fantin F, Bissoli L, Bosello O (2005) Health consequences of obesity in the elderly: a review of four unresolved questions. Int J Obes 29:1011–1029
Zamboni M, Mazzali G, Fantin F, Rossi A, DiFrancesco V (2007) Sarcopenic obesity: a new category of obesity in the elderly. Nutr Metab Cardiovasc Dis 18:388–395
Zigmond AS, Snaith RP (1983) The hospital anxiety and depression scale. Acta Psychiatr Scand 67:361–370

Chapter 5
Critical Aspects in Nursing

Daniela Imperiali, Roberta Cirillo, Amelia Brunani, Edda Maria Capodaglio and Paolo Capodaglio

Abstract The severely obese patient with disabling conditions is a "complex" patient, whose nursing imposes higher loads for healthcare providers than assisting normal-weight patients. Four main critical aspects related to bariatric nursing arise: (a) the need for adequate figures of healthcare providers, (b) the concern for both the patient's and the provider's safety, (c) the lack of bariatric equipment, (d) the patient-professional relationship, both at personal and professional level. Being consistently present in the Unit and responsible for the patient's physical and psychological needs, the nurse represents the reference professional. The different levels of nursing interventions in bariatric patients and their critical aspects are described in the chapter. Caring for those patients implies higher biomechanical risks for the healthcare provider. Patients handling algorithms, the specific class of equipment with expanded capacity and the specific environmental considerations for hospitals admitting bariatric patients are discussed. The healthcare operators should be consistently educated and updated about biomechanics of movements and postures, and the use of technological aids. Being organizatively and culturally unprepared for admitting those patients aggravates the level of risk of injuries both for the healthcare provider and the patient.

D. Imperiali (✉)
Nursing Office, Istituto Auxologico Italiano IRCCS, Piancavallo, Verbania, Italy
e-mail: imperiali@auxologico.it

R. Cirillo · A. Brunani
Rehabilitation Unit, Istituto Auxologico Italiano IRCCS, Piancavallo, Verbania, Italy
e-mail: r.cirillo@auxologico.it

A. Brunani
e-mail: brunani@auxologico.it

E. M. Capodaglio
Ergonomics Unit, Fondazione Maugeri IRCCS, Pavia, Italy
e-mail: edda.capodaglio@fsm.it

P. Capodaglio
Rehabilitation Unit and Research Lab in Biomechanics and Rehabilitation,
Istituto Auxologico Italiano IRCCS, Piancavallo, Verbania, Italy
e-mail: p.capodaglio@auxologico.it

P. Capodaglio et al. (eds.), *Disabling Obesity*, DOI: 10.1007/978-3-642-35972-9_5,

5.1 Introduction

There has been in the past years a noticeable increase in the number of admissions due to obesity to healthcare facilities. Nursing of such patients imposes higher loads for healthcare providers than assisting normal-weight patients (VISN8 2007; Hahler 2002). An exhaustive assessment of the complexity of nursing should take into account three fundamental elements of the severely obese patient: (1) clinical stability/instability, (2) capacity to perform tasks independently, (3) adequate solutions for the patient and the rehabilitation team. According to the Agency for Health Care Research and Quality (AHRQ) in the US, we can define as "complex" a patient affected by two or more chronic conditions which could mutually influence their outcome, by limiting life expectancy or the use of full therapeutic dose because of contraindications, etc. Given the frequency of existing comorbidities, we can define the severely obese patient with disabling conditions a "complex patient". To measure this complexity we should use multiple specific tools to grade the severity of obesity (BMI) and the functional consequences of the comorbidities (specific disability scales, see Chap. 4 of this book). From the organizational perspective, it should be borne in mind that nursing bed-ridden obese patients is more time-consuming than assisting lean patients, particularly in demanding tasks such as bodily hygiene, where nurses can spend up to twice the time. This represents a challenge to healthcare providers striving to provide dignified care that is effective and safe for both the patient and the provider. Besides that, even more independent obese patients require assistance also in numerous activities of daily living (Muir and Archer-Heese 2009). From a review of the specific literature, four main critical aspects related to bariatric nursing arise: (a) the need for adequate figures of healthcare providers, (b) the concern for both the patient's and the provider's safety, (c) the lack of bariatric equipment, (d) the patient-professional relationship, both at personal and professional level (Drake et al. 2005; Rose et al. 2006).

5.2 Role of the Nurse

Nurses collaborate within the rehabilitation team in the development of a rehabilitative individual project in order to identify the needs related to the patient's disability and to promote independence in activities of daily living. Information about the on-going rehabilitative project should be consistently provided by the nurse to the patient throughout the rehabilitation stay.

Obese persons are often stigmatized and have to cope with different forms of discrimination and prejudice: the Stigma Situations Questionnaire revealed that "comments from children", "negative assumptions by others", and "physical barriers" were among the negative perceptions experienced by extremely obese individuals referred for bariatric surgery (Sarwer et al. 2008). Even hospitals and medical settings are not immune to negative attitudes toward these patients (Puhl and Brownell 2001). Such attitudes could influence the quality of care and healthcare professionals tend to devote inadequate time to overweight patients, despite scientific guidelines underlining the importance of providing information to improve the patients' lifestyles. This can be a source of frustration (Mercer and Tessier 2001) with negative impact on quality of life (Kolotkin et al. 2001). Obese patients have a distorted perception of body image with consequent social handicap. This represents a problem which needs to be addressed and the nurse should acquire competences to help the patient strengthening his/her social skills and promoting social interaction and also to create mutual trust and a comfortable environment for the patient to make him/her adhere to the rehabilitation program (Golay et al. 2003; Latner 2001). Being consistently present in the Unit and responsible for his/her physical and psychological needs, the nurse represents the reference professional for the patient. Therefore, he/she should encourage and motivate the patient to health-wise virtuous behaviour promoting positive attitudes toward personal health and well-being. The nurse should have a holistic approach to the patient, helping him/her to maximize independence, quality of life and well-being. A trustworthy relationship with patients who lack of self-esteem is not an easy achievement. It is important that the nurse respects the following principles: relationship with patients based on trust, loyalty, and transparency; consistent positive attitude, availability, and coherence; counselling and education of patient and caregivers; stimulation of new abilities and capacities in the patient; empowerment of the patient.

5.3 Levels of Nursing Interventions

5.3.1 Self-Care

When the patient is partially dependent in self-care, the nurse has to define his/her capacity in the different areas of feeding, personal hygiene, dressing, toileting, to stimulate improvements in independence. Difficulties in personal care and hygiene are mainly due to movement limitations or secondary to architectural barriers or lack of perceived need for hygienic care. The interventions planned by the nurse range from ergonomic adaptations of the environment to the assistance in execution of hygienic care tasks. The nurse provides the minimum degree of supervision or direct assistance needed by the patient to safely and correctly perform daily tasks. Personal hygiene is often hindered by the excessive body mass and the deep skin folds, specially in part of the body which are not easily accessible to inspection: perineal, breast, abdominal, and lower limb folds are common sites for developing pressure sores and skin conditions. To provide adequate hygiene, an adequate number of healthcare providers and bariatric aids are needed (Holland et al. 2001).

5.3.2 Mobility Limitations

Walking, capacity of negotiating stairs, transferring from bed or moving around with the wheelchair, using the toilet and the shower are impaired in the severely obese patient. To counteract these mobility limitations, the nurse, together with the rehabilitation team, should evaluate architectural barriers, plan and perform postural changes in bed, educate to the correct use of aids and to injury and fall prevention. Reduced tolerance to physical activity in obese subjects can also be related to mechanical breathing difficulties and early fatigue. The nurse together with physiotherapists and adapted physical activity specialists should help promoting physical activity bouts on a daily basis, teaching "energy saving" and "joints unloading" techniques to increase activity levels and tolerance to effort. Reduced tolerance to effort can hinder the execution of basic daily activities, and simple gestures such as leaning forward to pick up objects, reaching upwards, lacing shoes may become very difficult.

5.3.3 Risk of Fall

Risk of fall is higher in obese patients as compared to their lean counterparts and mainly due to instability and excess of body mass (Menegoni et al. 2009). High scores on specific balance scales (i.e., a score >2 on the Conley scale) document the increased risk of fall in these patients. The nurse, together with the occupational therapist, should plan dedicated prevention and education interventions for patients and relatives/care givers. Teaching patients how to perform steadily and safely postural changes, how to use toilet/shower/tub are crucial aspects of these interventions. The nurse should also monitor and assist patients treated with diuretics or laxatives, specially at night time. Educating to the correct use of transferring and mobility aids is a very important duty of nurses and physiotherapists. As for risk of falls, patients have to be reminded to use non-slippery shoes and to avoid brisk movements and postural changes, architectural barriers should be removed and adequate night lighting provided.

5.4 Management of Comorbidities

5.4.1 Diabetes

Obese patients often present with diabetes. It is important that the nurse monitor glycemia in order to capture signs and symptoms of hypo/hyperglycemia, evaluate hydration, pharmacological therapy and nutrition. Specifically, the nurse educates the patient to inspect the skin and the eventual diabetic foot, to self-monitor glycemia, to correctly adhere to the hypoglycemic therapy, to follow a correct nutritional scheme, to accomplish consistent exercise bouts and to recognize early signs and symptoms of hypo/hyperglycemia.

5.4.2 Skin Conditions

Among the associated conditions, the risk of infection and skin conditions, such as dermatitis, ulcers, and delayed cicatrization play an important role. A literature review in obesity revealed the following data: the facility-acquired pressure ulcer rates for critical care units range between 7.14 and 14.5 %; a score of 10 to 12 on the Braden scale (Bergstrom et al. 1998) indicates that a patient is at high risk for developing a pressure ulcer; alteration in pressure and microclimate can cause skin breakdown. A possible "High Risk for Pressure Ulcer Protocol" suggested some strategies: use fewer barriers between the critically ill patients and bed to reduce friction; use a flat sheet instead of the fitted sheet. Use a moisture wicking underpad, instead of the heavy pink pads, to reduce friction and enable easier movement of the patients. Reposition patient by sliding flat sheet along a maximum inflated mattress (Racco and Phillips 2010).

5.4.3 Incontinence

Incontinence correlated to obesity also needs to be assessed. It is usually related to high intra-abdominal pressure, but can also be related to external factors, such as difficulty of sitting on an adequately sized commode. Bladder catheterism can also be more difficult than normal-weight patients.

5.4.4 Sleep Apnea

Medical management of severely obese patients include an assessment for sleep apnea (OSAS). Patients with OSAS require assisted ventilation at night, which may improve a range of aspects of quality of life, related cardiac arythmia, hypertension, daily drowsiness with the related risks at work and during driving. C-PAP or BI-PAP treatment implies that the nurse together with physiotherapists choose the most suited facial or nasal mask and help the patient understand the importance of ventilation in order to achieve a satisfactory quality of life. Compliance to ventilation and management of all the components of the device at home should also be part of the assessment. The nurse should check the patient's position during sleep, the correct adhesion of the mask to the face to avoid ineffective ventilation and prevent pressure sores. Effectiveness of ventilation has to be measured with oxygen saturation level at night and eventually oxygen therapy can be added to ventilation. Cardiac arythmias should be excluded with Holter ECG recordings.

5.5 The Nursing Assessment

At admission, the nurse is responsible of defining the health status and the problems of the bariatric patient in order to identify targeted nursing interventions. In general, the initial *assessment* aims at defining the patient's status and independence and the healthcare goals. The chart (Table 5.1) helps ranking the complexity of nursing tasks (Cavaliere and Snaidero 1999). The nurse collects the patient's medical and personal history, including habits, available social and family contacts, by means of standard templates for *planning* (Carpenito-Moyet 2010). The next phase is the *execution* of effective nursing interventions targeted to the achievement of the goals set. Parameters of effectiveness are chosen and monitored during the rehabilitation period. The main aspect remains the centrality of the patient: standardized programs have not shown to be equally effective as individual programs. The program must also adapt to the changes in patient's ability which may change daily in acute care settings. The *evaluation* phase then describes whether the outcomes are in line with what previously planned, thus allowing eventual changes in the healthcare plan.

5.5.1 Patient's Assessment

Vital parameters including oxygen saturation should be measured and the observation of skin, teeth, tongue, hair conditions carried out. Body weight and BMI are measured the day after admission and at least once a week. Patient's reaction to

Table 5.1 Indicator of complexity of nursing tasks

Needs	Medium–low complexity	Medium complexity	Medium–high complexity	High complexity
Feeding/hydration				
Urinary/bowel function				
Hygiene				
Transferring				
Breathing				
Cardiocirculatory function				
Therapeutic procedures				
Diagnostic procedures				
Social relationships				
Rest and sleeping				

Range complexity (21–77): Medium–low 21–29; Medium 30–36; Medium–high 37–45; High 46–77

measurement are monitored (anxiety and strategies adopted according to the value measured). The interview should also provide information about: type of eventual surgical procedure, post-surgical wound status, qualitative and quantitative food intake, presence of feeling rituals and phobias, binge eating.

Physical (weight, size, center of mass position, abnormal distribution of masses), psychological (collaboration, mood, motivation), and clinical characteristics of the patients affect the approach of the healthcare provider and safety of the tasks performed. Those characteristics orientate the choice of intervention and level of assistance (number of operators, time, number of transferring, etc.). The following points should be carefully evaluated both during the admitting assessment and on a regular basis:

- **BMI**: if >35 risk of injury increases both for operator and patient; if >45 refer to specific bariatric handling program
- **weight distribution** can severely affect safety and manoeuvres the patient is able to perform and is crucial in choosing appropriate aids
- **collaboration** during the manoeuvre
- **weight-bearing capability** on the lower limbs. Incapacity to bear weight on the legs increases the risk of fall for the patient and of injury for the operator. Mechanical aids can alleviate the problem and a targeted strengthening program should be initiated
- **upper** *limb strength*, which is needed in various manoeuvres (lean and push for performing sit-to-stand, grab handles, lean on aids for balance)
- **capacity** *to understand instructions* and simple commands, which are needed when collaborating to manoeuvres
- **pain** (i.e., presence of wounds, contractures, etc.) can affect the manoeuvre and limit movements causing an extra-effort for the operator, who in turn should know in advance where pain is localized in order to reach comfortable positions for the patient
- **orthostatic** *hypotension* or anxiety should be borne in mind, since its presence suggests cautious and slow movements
- **tolerance** *to effort* is reduced in obese patients, who may suddenly feel the need to sit down or find support, thus increasing the risk of fall
- **respiratory problems** can become more evident during manoeuvres with aids and in certain positions (Twedell 2003); this can increase anxiety in patients who may respond with brisk reaction movements
- **orthopedic conditions** (hip, knee replacements, etc.) require avoiding manoeuvres at risk of dislocating the joint, but also amputations, osteoporosis, spine instability require special attention
- **neurological conditions** (i.e., paralysis post-stroke, multiple sclerosis, etc.) require careful evaluation.

The overall evaluation of the patient at admission and during the hospital stay can only stem from the competent vision of the operator considering the above-mentioned points and it cannot merely result from the calculation of a numeric score.

5.5.2 Critical Measurements

5.5.2.1 Measuring Blood Pressure

Measuring blood pressure with standard-size cuffs (13 cm-wide, 32 cm-long) is one of the most common causes of error in the indirect measurement of blood pressure. The cuff should be wrapped round the intermediate part of the nondominant upper limb. Specific bariatric cuffs (15 cm × 30–35 cm) should be used.

5.5.2.2 ECG Recording

ECG recording can be difficult because the adipose tissue in the thoracic region hinders the correct positioning of the electrodes on the anatomical landmarks. Abnormal positioning can indeed lead to formulate wrong diagnoses. Copious sweating and large breasts can also hinder electrodes positioning. The supine position is often uncomfortable and dyspnoea may arise, also as a result of anxiety and mechanical difficulty in breathing, with secondary abnormalities or artifacts on the ECG tracks.

5.5.2.3 Measuring Body Weight, Height, Waist Circumference, and BMI

Measuring body weight, height, waist circumference, and BMI is crucial to define the nutritional status. Body composition (lean mass, fat mass, water), as assessed by bio-impedentiometry, should also be measured. Body weight can be measured with both electronic and mechanic weighing scales: floor-mounted standard scales with lateral supports for measuring weight in standing position, chair- or bed-scales, lifts with built-in scales. The choice of equipment depends on the patient's residual functional capacities and the frequency of measurements. The following points should be addressed when weighing patients:

- fully inform them
- weighing in the morning before breakfast and always at the same time and under the same conditions
- help them to get undressed and onto the scale
- avoid leaning on any surface other than the scale during the measurement
- include aids when weighing patients who regularly use canes, walkers, or braces.

5.5.3 Patient Handling Algorithms

The bariatric handling program aims at standardizing modalities of care among health professionals during the hospital stay, providing scientific background for the healthcare modalities. This facilitates the choice of intervention, equipment ,and the number of healthcare operators (Baptiste et al. 2008). The program must be related to work organization (i.e., adequate number of operators in the shift) and technology assessment (i.e., adequate number of effective aids). The policy of the hospital/institution should therefore be in this line and aimed at minimizing risk and optimizing quality of care (Cowley and Legget 2011). Essential specific components of a bariatric patient handling program include the following (Muir and Archer-Heese 2009): patient assessment tools, operational procedure and patient handling algorithms and guidelines, equipment needs, space and environment considerations, staff training and education including promotion of a safety culture. Safe patients handling programs are available online at Patient Care Ergonomics Resource Guide, Veterans Administration (2006) or www.visn8.med.va.gov/patientsafetycenter (see Ref. Safe bariatric patient handling toolkit.VISN 8 Patient Safety Center of Inquiry, Tampa, FL, 2007). Algorithms for the operators have been developed since 2001 (Patient Safety Center of Inquiry, Tampa FL) to help selecting the safest modality of handling based on the patient's characteristics. These algorithms address issues related to body mechanics and patient handling ergonomics and provide a sequential approach to asking questions related to patient abilities, the answers of which guide the healthcare provider in making the safest patient handling decision. There are various algorithms available in the literature for different types of patient's handling (Nelson 2006). As an example, two of them are reported in Figs. 5.1 and 5.2. The operator should not only know how to use the available equipment but also be able to assess their appropriateness according to the patient's characteristics. This will guarantee the patient adequate assistance according to his/her functional level. The key point is to avoid manual handling as much as possible and encourage patient's collaboration and participation, also by using adequate aids. By referring to algorithms, operators can minimize the risk of musculoskeletal injury related to handling of bariatric patients. They serve to standardize practice in an evidence-based modality, avoiding individual experience-based behaviors and guiding the operator in planning interventions. The algorithms are particularly recommended when assisting bariatric patients with BMI > 45, where risk is elevated and planning in advance is mandatory. They allow to check the requisites for a safe manoeuver (type of device, number of operators, number of phases, compatibility with spaces, etc.) and avoid errors due to negligence or improvisation. Specific algorithms refer to: vertical transferring (i.e., bed-chair, chair-WC, wheelchair-car seat, etc.), lateral transferring (bed-stretcher), repositioning in bed or in the wheelchair, moving body parts to access certain areas (abdomen, lower limbs, etc.), transport with stretcher, hygiene tasks.

5.6 Settings of Care

The different reasons why the obese patients have to be admitted to various settings are:

- any acute condition, in the emergency and accident unit, or admission for routine procedures or treatment (e.g., cardiac or abdominal surgery)
- managing patients who are obese in the intensive care unit is associated with a variety of problems; few data are available and most information are extrapolated from the bariatric surgery literature. Jbeili et al. (2007) described increased

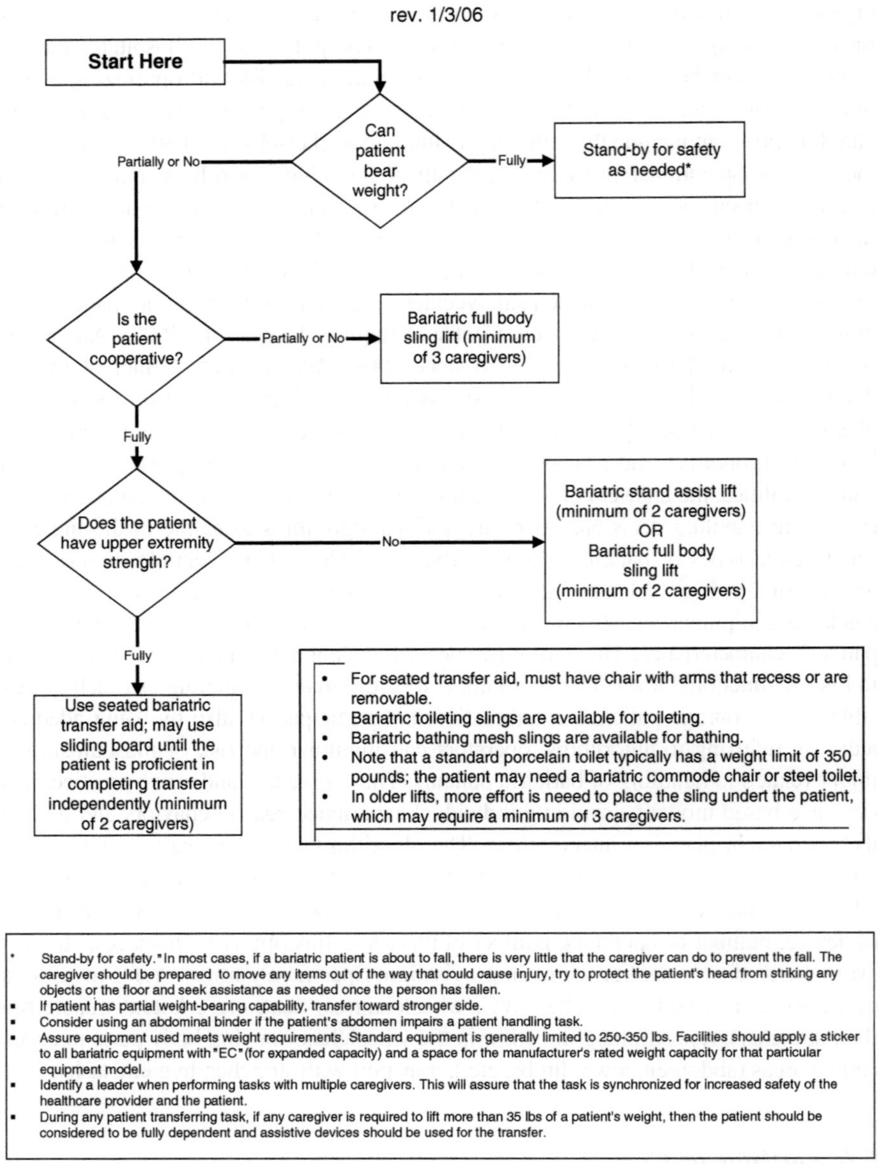

Fig. 5.1 Bariatric algorithm 1: bariatric transfer to and from: bed/chair, chair/toilet, or chair/chair. Rev. 1/3/06

difficulties in airway management, peripheral cannulation, and blood pressure recording in a pre-hospital study. After hospital admission, handling of obese patients need an additional rescuer for 33 % of patients transported to hospital and 47 % of bariatric patients had a reduced access to specialized diagnostic

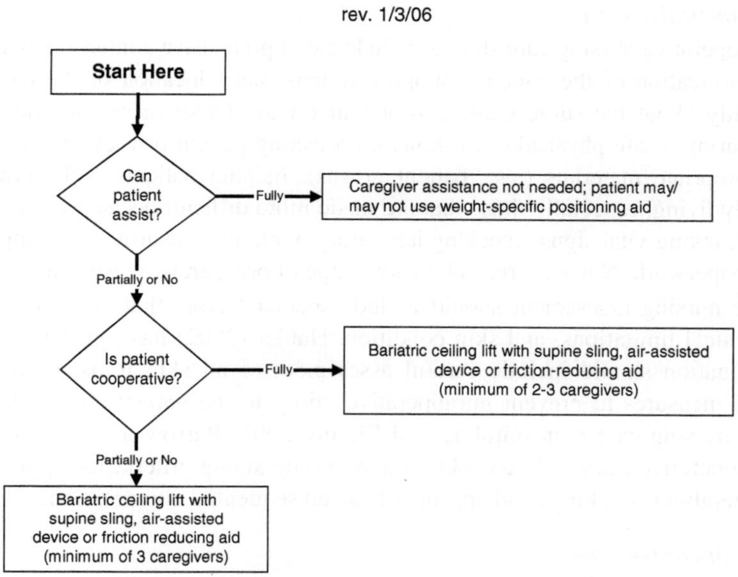

- When pulling a patient up in bed, place the bed flat or in a Trendelenburg position (if tolerated and not medically contraindicated) to aid in gravity; the side rail should be down.
- Avoid shearing force.
- Adjust the height of the bed to elbow height.
- Mobilize the patient as early as possible to avoid weakness resulting from bed rest. This will promote patient independence and reduce the number of high risk tasks caregivers will provide.
- Consider leaving a friction-reducing device covered with drawsheet, under patient at all times to minimize risk to staff during transfers as long as it doesn't negate the pressure relief qualities of the mattress/overlay.
- Use a sealed, high-density, foam wedge to firmly reposition patient on side. Skid-resistant texture materials vary and come in set shapes and cut-your-own rolls. Examples include:
  - Dycem (TM)
  - Scoot-Guard (TM): antimicrobial; clean with soap and water, air dry.
  - Posey-Grip (TM): Posey-Grip does not hold when wet. Washable, reusable, air dry.

- If patient has partial weight-bearing capability, transfer toward stronger side.
- Consider using an abdominal binder if the patient's abdomen impairs a patient handling task.
- Assure equipment used meets weight requirements. Standard equipment is generally limited to 250-350 lbs. Facilities should apply a sticker to all bariatric equipment with "EC" (for expanded capacity) and a space for the manufacturer's rated weight capacity for that particular equipment model.
- Identify a leader when performing tasks with multiple caregivers. This will assure that the task is synchronized for increased safety of the healthcare provider and the patient.
- During any patient transferring task, if any caregiver is required to lift more than 35 lbs of a patient's weight, then the patient should be considered to be fully dependent and assistive devices should be used for the transfer.

Fig. 5.2 Bariatric algorithm 3: bariatric reposition in bed: side-to-side, up in bed. Rev. 1/3/06

and therapeutic devices (Geary and Collins 2012). Obese trauma patients need oversized or dual backboards, cervical collars to maintain cervical spine stabilization. Sandbags, blanket rolls, and tape are alternate resources that can be used for neck stabilization (VanHoy and Laidlow 2009)

- surgical treatment of obesity (Mulligan et al. 2005):

Preoperative Care

Preoperative nursing care should include a comprehensive admission assessment, identification of the patient's support system, and education of the patient and family about the surgery and postoperative care. Other responsibilities include ensuring a safe physical environment; protecting patient privacy; providing size-appropriate materials (e.g., patient gowns); helping patients with activities of daily living, especially those that are made more difficult because of severe obesity; taking vital signs; checking laboratory work; and ensuring the completeness of paperwork. Nurses carry out a wide range of preoperative responsibilities.

The nursing assessment should include special factors that affect positioning, physical limitations, and skin condition. Hahler (2002) has suggested that skin evaluation should include careful assessment of all skin folds. Data indicate that measures to prevent intraoperative injury to the patient should be part of the nursing care plan (Graling and Elariny 2003; Barrow and Roux-en 2002). Preoperative nurses should obtain appropriate sizing criteria for application of antiembolic stockings and appropriate sized sequential compression.

Perioperative Care

The nurse's assessment should help to secure an appropriate bed designed to facilitate recovery and to ensure the safe transfer of the patient to the operating room. Careful consideration with respect to preoperative sedation should be taken if the patient must ambulate to the operating table or play a more active role in the transfer process. Graling and Elariny (2003) suggest that nurses should be prepared to review the planned procedure with the patient and provide the patient with ample opportunity to ask questions.

Intraoperative

Nurses involved in abdominal prep should give special attention to manipulating between the skin folds and under the panniculus. The risk for postoperative surgical infection is higher in weight loss surgery patients. Nurses must be prepared to administer all necessary medications, including: antibiotics, heparin or low-molecular weight heparin, prophylactic antiemetics, H2 blockers, sedatives, and intravenous fluids. In dealing with severely obese patients, proper drug absorption can be an issue if medications are administered using the subcutaneous or intramuscular route. The environment requires nurse managers to assess the need for extra staff: it may take as many as three staff members to retract the panniculus and thighs for urine catheter insertion. Extra assistance may also be necessary to position and place extra electrosurgical unit grounding pads to prevent burns.

Post-anesthesia Care Unit

Davidson et al. (2003) have noted that safe transfer of an unconscious patient may require the efforts of up to five trained staff members. Blackwood (2004) has reported the need to pay special attention to airway stability, hemodynamic stability, and pain management in severely obese patients. Davidson et al. (2003) have reported that the head of the patient's bed should be maintained at 35–40° angle to reduce abdominal pressure and increase tidal flow. Voelker (2004) has noted that severely obese patients are at heightened risk for quick oxygen desaturation and require close monitoring. Barrow and Roux-en (2002) has demonstrated the need to recognize lipophilic drugs that would place patients at risk for resedation. Sandlin (2003) has also described the need for increased awareness regarding over-sedation of severely obese weight loss surgery patients. The same author has reported the need for nurses to assess weight loss surgery patients for postoperative nausea and vomiting; in the case of gastric banding, prevention and/or treatment of postoperative nausea or vomiting could prevent band slippage. Sullivan (2004), Ferraro (2003) and Davidson et al. (2003) have documented various ways to transfer weight loss surgery patients, including use of a transfer mat or board or the use of several trained staff to transfer the patient laterally. These authors have reported that passive and active range of motion exercises are imperative for weight loss surgery patients who are bedridden on a long-term basis. The nurse is responsible for monitoring patients according to hospital standards of care.

Discharge and Follow-Up

Patients are typically discharged on postoperative day 3 or 4 but may require more monitoring due to risk from obesity-related comorbid conditions (Blackwood 2004; Gallagher 2004). Collaboration among providers is necessary in planning a course of care for these patients. The important area of assistance are: local medication for the risk of wound dehiscence; diets considerations particularly to drink and to eat dry meats and nutritional implementation for deficit nutrient absorption, drug administration with alternative ways (intravenously or intramuscular).

- managing obesity in primary care

  primary care is generally first level healthcare approach (usually provided by generalists rather than specialists) with an orientation to prevention and health promotion. The lifestyle-related interventions for obesity may be suitable for low technology prevention activities for nurses. In a previous review, it is reported that nurses's role mainly consists in providing one-to-one support to within surgery or clinic consultations and to a lesser extent group-based support sessions. The interventions could be characterized as general oral advice about nutrition and lifestyle. In some contexts, oral advice extended to calorie deficit diets and more developed referral options about physical activity. The evidence of positive outcome is mixed and around 10 % of patients entering a nurse led support program achieve a clinically significant weight loss (Brown and Psarou 2007).

5.7 Equipment

Despite increasing numbers of obese patients admitted to acute care facilities for surgery or treatment of nonsurgical conditions, there is little evidence of the problems nurses face in providing care to these patients. In a previous study, nurses reported concerns about the increased staffing needs for care obese for such normally simple tasks as catheterizing, turning, and ambulating and in particular the availability, placement in adequate room size, and use of specialized equipment. Finally, nurses perceived safety issues, both for themselves and their patients (Drake et al. 2005).

Bariatric is the specific class of equipment with expanded capacity, in contrast with standard equipment for normal-weight patients. It should accommodate body weight up to 500 kg as compared to a standard weight of 270 kg. Hospitals should either possess such equipment or they should be able to have them available within 8 h from the patient's admission, providing that there is enough space in the hospital room for their effective use. The aim of using this equipment is to reduce biomechanical load on the exposed operators and increase level of safety during patients' handling. They are mostly used when patients are dependent or the load imposed to the operator is excessive, but also when patients are partially collaborating, thus supporting their residual functional capacities and minimizing the load for the operator. The versatile equipment available in the market requires knowledge and competence from the operators, who should be instructed to proper use. Familiarity with the equipment is fundamental, but also operators should take part in the continuous research process toward the development of simple and effective aids. The indication of the maximum load should be present on each equipment to avoid improper and risky use. The equipment should be able to accommodate size and mass distribution (width and depth of the seat for both androgyn and gynoid obesity) under safe and comfortable conditions (Muir and Archer-Heese 2009). This equipment will include:

- electric adjustable bariatric bed,
- stretcher with extra width,
- lift system,
- full body sling, thoracic belt (to be used with sit-stand lift) and specialty sling, such as limb sling and pannus slings,
- low-air-loss pressure reducing mattress for comfort, easy of bed mobility, and prevention of pressure sores,
- lateral transfer device (i.e., inflatable air mattress with handles),
- trapeze bar system over the bed,
- wheelchair,
- walking aids,
- sit-to-stand device,
- commode up to 42 inches wide,
- shower chair or shower stretcher.

Other equipment considerations include *patient-care supplies*. A readily available bariatric cart supplied with the following items allows for a smoother admission and provides better and more dignified care: larger hospital gowns, pants, housecoats, slippers, blood pressure cuffs that will fit around a bariatric arm, longer needles, and catheters. As for compressive elastic socks, garments, and beddings should be adequately sized.

5.7.1 Bariatric Beds

Bariatric beds (Fig. 5.3a–b) are designed to multi-task and accommodate weight up to 500 kg. Electric adjustable bariatric beds have power-assisted drive for moving and a built-in weigh scale; an expandable deck allows for the deck of the bed to be pushed inwards to allow the bed to pass through doorways, and then increased in width once in a room, to provide adequate support for the patient. Bed dimensions are important not only to accommodate the patient's body but also to allow he/she to collaborate to the manoeuver (i.e., rotating on the side while changing from the lying to the sitting position) and minimize nurse reaching distance (Holland et al. 2001). Lateral sides of the bed favor patient's independence providing a stable support. As for manoeuvrability, the structure of the frame (lightness, compactness, dimensions), the wheels (materials, dimensions, number,

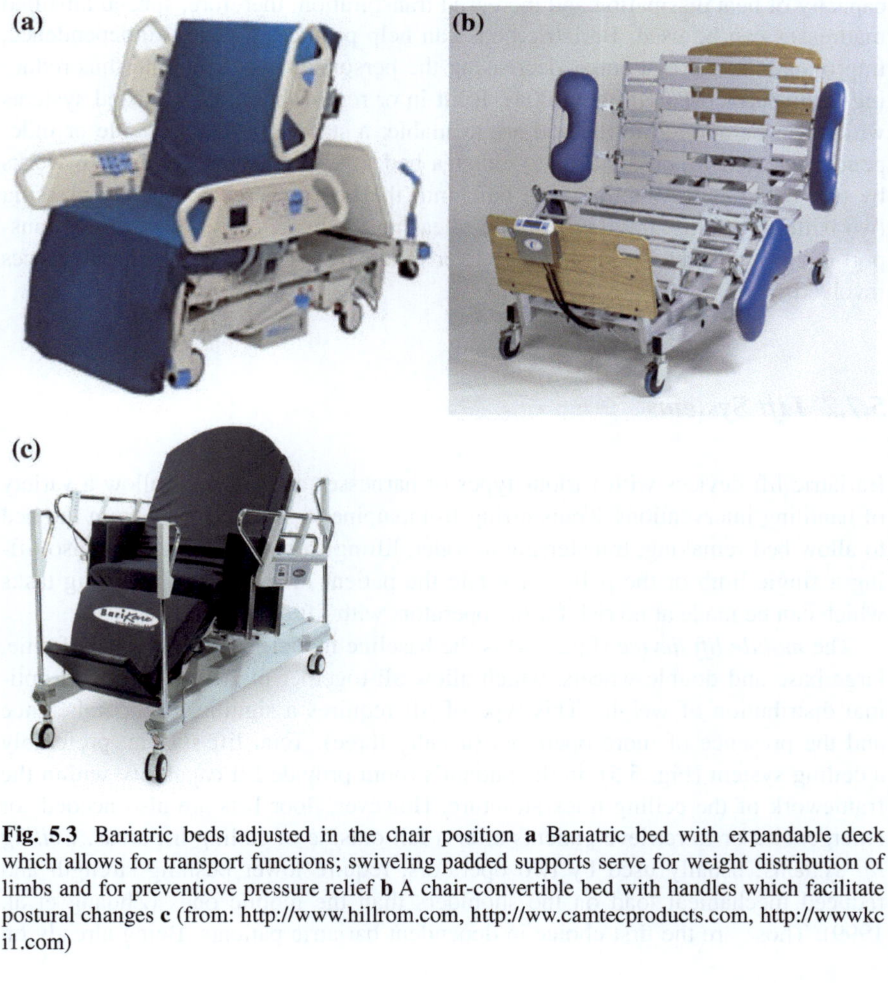

Fig. 5.3 Bariatric beds adjusted in the chair position **a** Bariatric bed with expandable deck which allows for transport functions; swiveling padded supports serve for weight distribution of limbs and for preventiove pressure relief **b** A chair-convertible bed with handles which facilitate postural changes **c** (from: http://www.hillrom.com, http://ww.camtecproducts.com, http://wwwkci1.com)

disposition), and the handles (conformation, dimensions, orientation, materials) are of particular importance. The presence of a fifth wheel facilitates manoeuvrability. Brakes, either activated by hand or pedals, should be easy to find. Pushing during transport can be easier if friction on the ground is reduced and application of force can be optimized if proper shoes, correct postures, and technical skills of the operator are correct. The electric regulation allows to easily change the patient's position, saving time and use of personnel. Complications secondary to immobility, pain, and intolerance to prolonged positions are therefore reduced, while comfort and well-being are increased, also due to special mattresses with differentiated pressure sections. The built-in weighing scale for instant body weight measurements allows to avoid additional transferring. Special features offered are turn assist up to 20° to help in patient positioning, percussion therapy, pulsating air suspension therapy, pressure relief therapy, and cardiac chair positioning. Some beds can be converted into chairs, allowing the postural change to sitting position without additional handling. Bariatric patients may show reduced capacity of heat dissipation and increased transpiration; therefore, special air-filled mattresses can be used. Bariatric beds can help promoting patient independence, improving clinical outcomes, decreasing the personnel workload, and thus reducing healthcare costs (Kramer 2004). Built-in or removable motor-assisted systems which can be applied to the bed are available: a stand alone, detachable or independent device which attaches to either a bed, linen cart or trolley, which works by two batteries, or another type built into the bed powered by simply pressing two buttons, unplugging the bed, and releasing the brakes. Use of powered transport devices reduces the risk of caregiver injury by reducing the push/pull forces involved.

5.7.2 Lift Systems

Bariatric lift devices with various types of harnesses and supports allow a variety of handling interventions. Transferring from supine to seated, lifting from the bed to allow bed remaking, transferring to toilet, lifting from the ground, and also lifting a single limb or the pelvis or rotate the patient in bed are all handling tasks which can be made at no risk for the operators with lift devices.

The *mobile lift device* (Fig. 5.4) is the baseline model, with sturdy steel frame, large base and double wheels, which allow all together higher stability and optimal distribution of weight. This type of lift requires a significant encomberence and the presence of more operators (usually three). Total lift system, preferably a ceiling system (Fig. 5.5), in the patient's room provide lift capability within the framework of the ceiling track structure. However, floor lifts are also needed for emergencies or to retrieve patients after a fall outside of ceiling lift areas. *Ceiling lift systems*, usually used by two operators, require lower pushing strength and reduced mechanical load on the shoulders than the mobile ones (Zhuang et al. 1999). Those are the first choice in dependent bariatric patients. Being already on

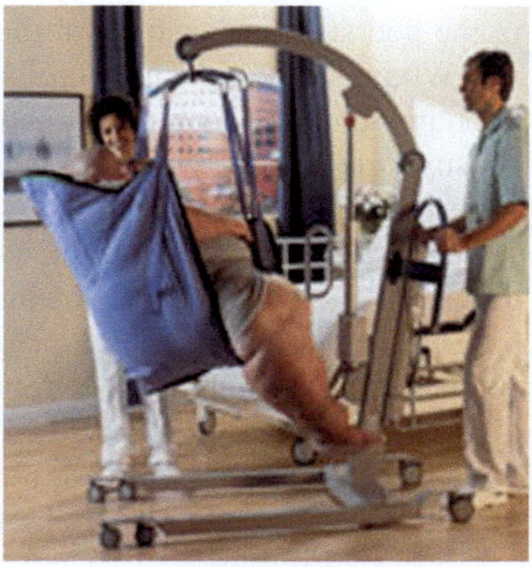

Fig. 5.4 Bariatric mobile lifting device (from: http://www.arjohuntleigh.com/)

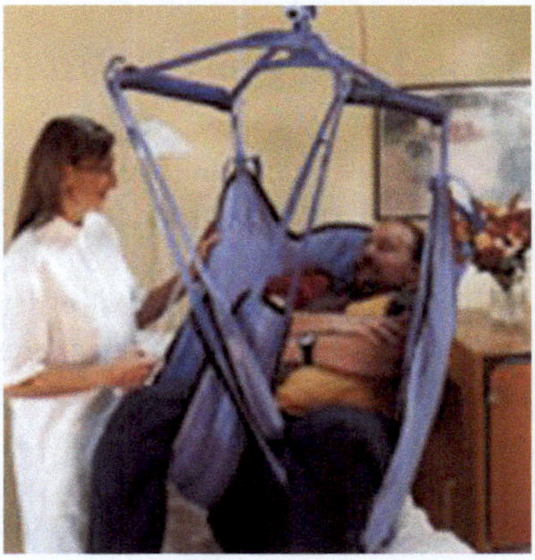

Fig. 5.5 Bariatric ceiling lifting device (from: http://www.arjohuntleigh.com)

site, time for preparation is minimized and effort for pushing and moving around the lift is annulled because horizontal and vertical manoeuvres are totally mechanized. Additional advantages of ceiling lift systems are: ready to use, no storage space required, reduced maintenance, automatic batteries charge when not used, multiple use possible (motor can be removed from the runner and used in different predisposed rooms), no need to move the lift around obstacles, better access to the patient, transferring a patient from a room to another (room-bathroom; room-corridor) can be fixed once and for all with single or transverse tracks.

A recurrent problem related to lift systems is the effort needed to apply/remove the sling from under the patient, which has to be repeated two times for each transferring. Applying the sling implies the following sequence: rotation to the side of the supine patient, insertion of the rolled up sling, return to supine position and rotation to the other side, complete spreading out of the sling under the patient, return to supine position, attachment of the sling to the lift.

Both insertion and extraction of the sling involve high postural loads and use of strength. Alternatively, the inclusion of sling, made of transpirant materials, in the bed structure or even in the patient's garments has been explored (Lloyd and Wikinson 1999). The wore sling should prevent localized pressure or friction and the related posture should be comfortable with regard to the area of support (head, forearms, legs) and transpiration (padding and covering materials).

The *active lift system* can be used for collaborative patients who need support during sit-to-stand, gait or standing. It can be used with lumbar corsets only in patients with a certain degree of strength, trunk control, and partial body weight support. Other simple nonelectronic aids (such as standing devices) can be used in assisted mode by the operator for sit-to-stand in patients who need support because of muscle weakness or after surgery; however, the patient should be able to grab, exert force with the limbs, and control movements.

5.7.3 Turning and Repositioning Slings

Two tasks which prove very challenging and place caregivers at a high risk for injury include insertion of patient care slings under bariatric patients, repositioning a patient up in bed or turning to the side. Currently, manufacturers are working on designing slings that can be left under patients who have to be moved frequently. The advantages to this would be less strain on the caregivers, less time taken to perform the task and most important, less risk and exposure to injury. Although there is no evidence or literature on leaving slings under patients and patient outcomes, this decision should be carefully weighed. Questions for consideration could be: "is the patient's skin compromised?", "how breathable is the sling?", "does the sling present rough uneven edges which can cause pressure points if left underneath?", "can the sling be left under the sheet and tucked into mattress when not in use?". Leaving a sling under a bariatric patient can be advantageous to caregiver safety but may be detrimental to the patient and clinical judgment should be used to determine the safest course of action.

5.7.4 Lateral Transferring

Minor aids for lateral transferring of the supine patient (i.e. bed to stretcher) include air systems which can be spread deflated under the dependent bariatric patient, even with pressure sores or skin conditions, then connected to an air compressor and inflated (Fig. 5.6). The low-friction material allow easy and effective lateral transferring with handles. Effectiveness of these systems is calculated around 85 %: to transfer a 200 kg patient the force required should be 30 kg, which would be a value compatible with the capacity of 90 % of the female population for tasks repeated during the work shift by 2 operators (Lloyd

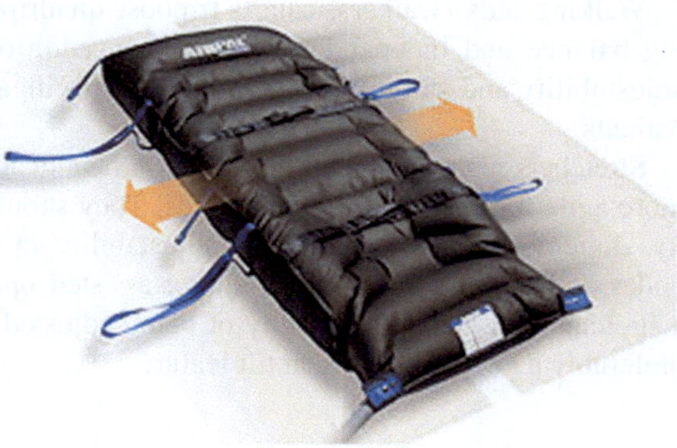

Fig. 5.6 Air-assisted lateral transfer system (from: http://www.kci1.com)

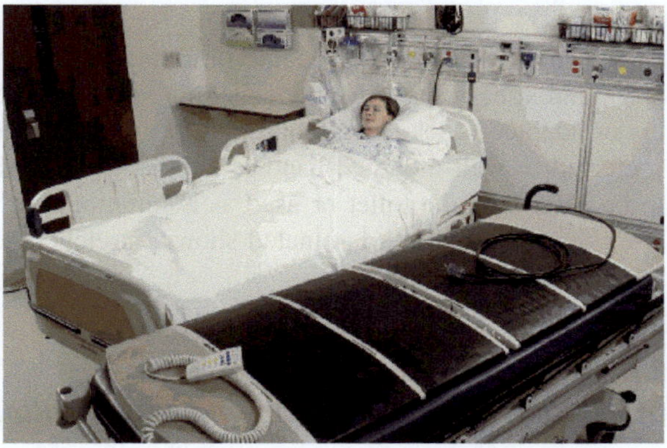

Fig. 5.7 A motor-assisted transfer technology (from: http://www.astirtechnologies.com/brochure/astir%20brochure%205.pdf)

and Baptiste 2006). For the same purpose, motor assisted transportable devices assisting lateral transferring of the patient by rolling a mat connected to a motor can be used (Fig. 5.7). They are more expensive than air systems and require two operators.

Pelvic belts with handles, simple low-friction sheets, and plastic/wooden transferring boards are not recommended for bariatric use in the hospital for their limited weight capacity. However, critical tasks, such as getting into a car do often benefit, lacking valid and affordable alternatives, from the use of those minor aids, which can bear up to 130–150 kg at the maximum.

5.7.5 Mobility Aids

Wheelchairs should have adequate dimensional requisites in width, depth, back support, arm rests, and height of the seat. Head and leg support may help unloading in part body weight.

Walking aids (walkers, canes, tripods, quadripods) provide support enhancing balance and independent mobility. In addition to the standard features of adjustability and stability, they should be specifically manufactured for bariatric patients.

Stretchers are fundamental in bariatric healthcare, both in emergency and as a more agile transport alternative to beds. They should have bariatric weight capacity, dimensions, manoeuvrability, adjustability in height and inclination, safety (sides and brakes), eventually a motor-assisted operating system. A stretchair is a stretcher that has the capability of being adjusted into a chair with armrests and preferably a knee gatch or seat tilt feature.

5.7.6 Equipment for Hygiene

In the bariatric bathroom hygienic furniture should be floor-mounted, preferably in steel, and at easy reach for patients on a wheelchair. Steel transferring aids (walkers or adapted toilet seats) facilitating postural changes in critical points are to be preferred to wall-mounted handles. Hygienic chair (Fig. 5.8a, b) is a steel frame to be placed on the toilet or used as a commode for whole body wash. Hygienic stretcher (Fig. 5.9) is a laminated shower stretcher, adjustable in height by means of a hydraulic system.

5.7.7 Space and Environment Considerations

The following requisites should be matched in hospital spaces in proportion to the number of bariatric patients admitted:

- accessibility (absence of barriers, wide walkways, clearly signed and equipped with wall-mounted support handles, wide doors and access points)
- benches and seats for resting
- lifts with adequate weight capacity and access
- reception and waiting rooms suited for obese visitors
- adequate number of rooms with expanded dimensions as compared to the 9.3 mq standard ones with 1.5 m free space around the bed (Wignall 2008)
- equipped adapted bathrooms (with steel bariatric furniture, floor-mounted supports, free space for wheelchair manoeuvering with two operators).

The estimated costs of an equipped bariatric room (bed, ceiling lift, wheelchair, commode) is around 32000€ (43000$). According to the "2004 Obese Patient Care Survey Market Research Report" (Novation LLC 2004), hospitals are now purchasing also devices for lateral transferring and positioning of bariatric patients. The bariatric room and bathroom spaces are expanded as

(a)

(b)

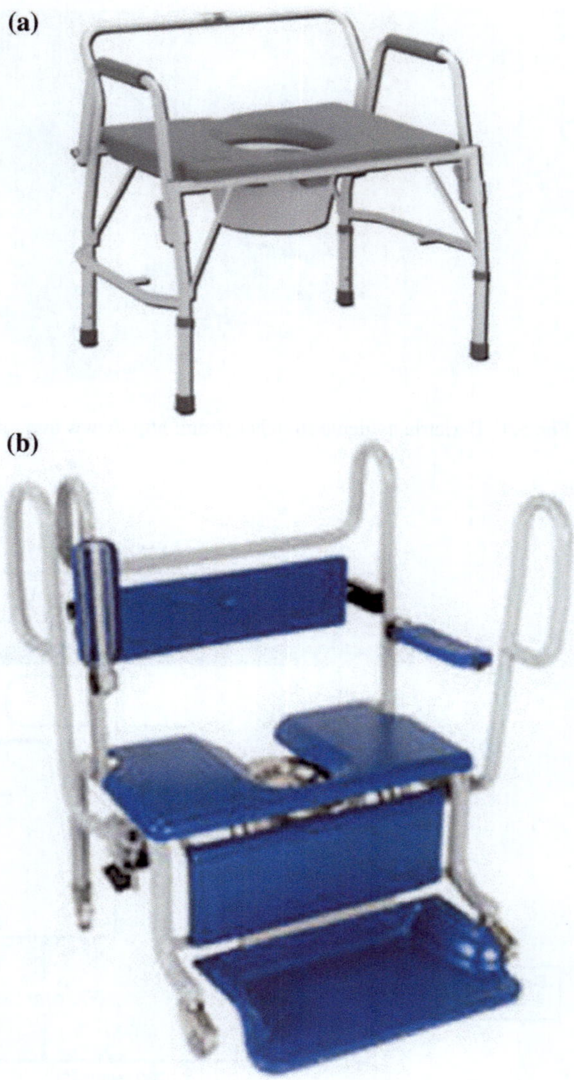

Fig. 5.8 Bariatric hygienic chairs (from: http://www.invacare.com, www.arjohuntleigh.com)

compared to the standard ones (Figs. 5.10 and 5.11) due to the greater encombrance of the specific equipment (Table 5.2) and the presence of more operators (Table 5.3). In the room, spaces must be large enough to accommodate the bed, the lift system, the wheelchair, the stretcher, the commode, the chair, and the presence of up to 6 operators performing the healthcare manoeuvres (3). Spaces and technologies are interconnected aspects: adequate spaces are essential requisites for the safe use of specific equipment (Table 5.4). Clearly, the bariatric environment should be *a priori* planned in the development

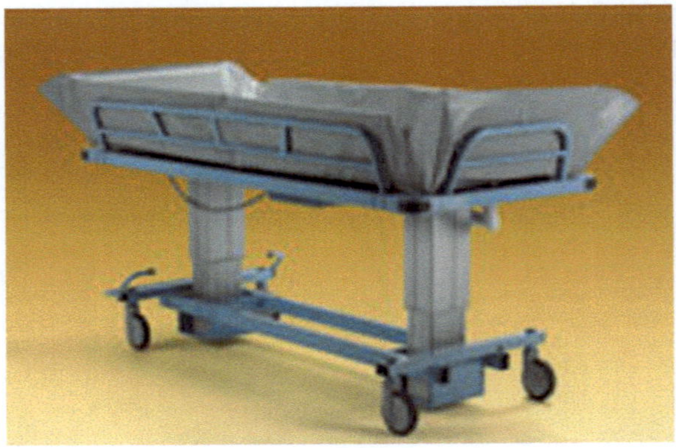

Fig. 5.9 Bariatric hygienic stretcher (from: http://www.trequipment.com)

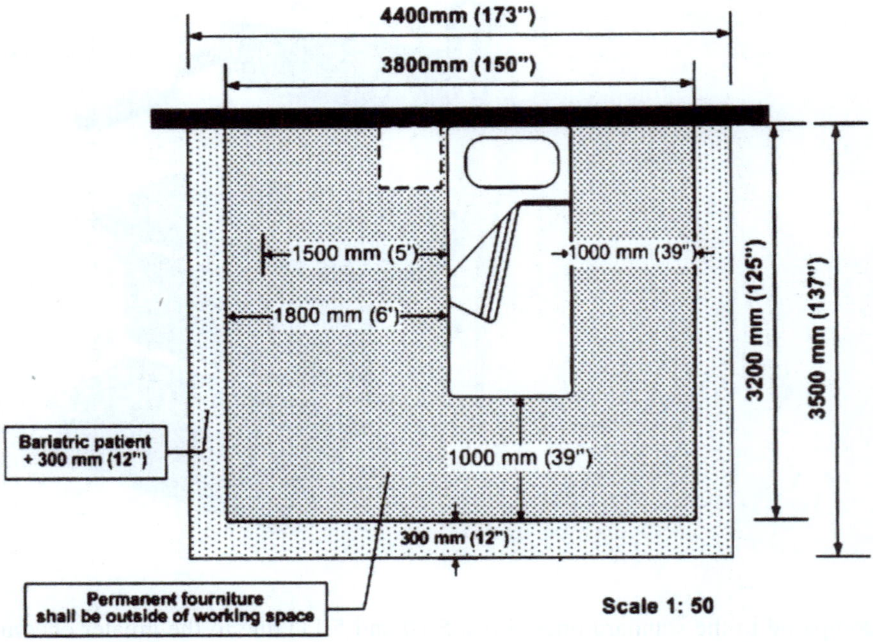

Fig. 5.10 Standard versus bariatric room's area

of clinical activities and, at admission, operators need to verify the immediate availability of the range of bariatric equipment for the patient's safety and quality of care.

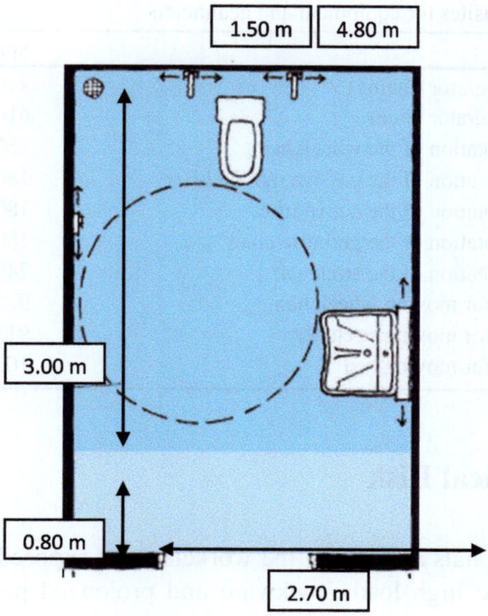

Fig. 5.11 Recommended dimensions for the bariatric bathroom

Table 5.2 Standard and bariatric dimensions

Equipment	Standard		Bariatric		Difference	
	Width (mm)	Length(mm)	Width (mm)	Length (mm)	Width (mm)	Length (mm)
Bed	1000	2260	1000–1370	2220–2444	380	229
Armchair	640	810	810	915	150	100
Stretcher	810	1980	960	2080	150	100
Passive mobile lifter	630–1120	1245	630–1980	1245–1625	864	380
Wheelchair	700	1220	850	1320	150	100
Electric wheelchair	820	1060	1000	1120	180	60
Commode/ shower chair	450	400	760	760	150	150

Table 5.3 Spatial requisites for equipment and healthcare

	Task	Space (mm)
1	One operator (facing)	810–1000
2	One operator (side)	610–760
3	180° rotation of the wheelchair	1500
4	180° rotation of the passive mobile lifter	1800 (2440 mm bariatric)
5	180° rotation of the active lifter	1800
6	180° rotation of the geriatric chair	1800
7	180° rotation of the stretcher	2400
8	Space for moving wheelchair	915–1000
9	Space for moving stretcher	915–1000
10	Space for moving bed	1060–1220

5.8 Biomechanical Risk

Healthcare professionals are among the workers most exposed to the risk of musculoskeletal injuries: high loads, awkward and prolonged postures, with special reference to spine flexion, rotation and extension, fatigue, monotonous repetitive tasks contribute to the onset of both acute and cumulative trauma. Risk related to manual handling, transferring, lifting, repositioning, and nursing is due either to peak (sudden high-intensity) or cumulative (medium-intensity but prolonged) loads. Caring for the bariatric patients implies even higher biomechanical risks for the healthcare provider (Collins 2010; Randall et al. 2009; Jang et al. 2007). The overall load on the spine while handling normal-weight patients in two operators corresponds to 470 and 650 kg when only one operator is involved. Such loads already exceed the safety limit (350 kg) and are close to the maximum limit (650 kg) recommended by NIOSH (Waters et al. 1993). Many nursing activities readily exceed safe working loads and can lead to musculoskeletal injuries. For instance, lifting a patient's leg, which is often necessary for completing a dressing change, represents approximately 16 % of total body weight (if the patient's weight is 175 kg, that of a leg would be 31 kg). This already exceeds a safe lifting load and therefore healthcare personnel should be thought to use a mechanical device with a limb sling. Moreover, several recent studies (Han et al. 2011, Bogossian et al. 2012; Malinauskiene et al. 2011; Pietroiusti et al. 2010) show a higher prevalence of obesity and metabolic syndrome in the nurse work force, with negative consequences on health status, increased rate of work-related musculoskeletal injuries and on work capacity. Notable characteristics of injured employees include advancing age, female gender, long working hours, increased BMI, history of prior back and upper extremity injures, no health and wellness activity attendance, and sick leave. Back and shoulder strain, fall-related and repetitive motion injures were the most severe and costly causes of sick leave (Brown and Thomas 2003; Lee and Chiou 1994). Work capacity of the nurse work force is also jeopardized by organizational and social issues (i.e., prevalence of female nurses, economical crisis, delayed retirement age, etc.).

Table 5.4 Check list for safe use of bariatric equipment and spaces (from "Safe bariatric patient handling toolkit", *Ref.* Baptiste et al. 2008)

Hospital bed	*Bathroom*
☐ Weight limit ______ lbs	☐ Doorframe width ____ inches
☐ Side rail support ______ lbs	☐ Shower door width ____ inches
☐ Bed Scale? Yes ____ if yes, weight limit ____ lbs. No ____	☐ Toilet weight bearing limit ____ lbs
☐ Width of bed ______ inches	☐ Wall mounted grab bars
☐ Bed adjustable for patient height? Yes____ No____	☐ weight limit ____ lbs
☐ Mattress type: Pressure relief____ Pressure reduction ____ Alternating____ Rotational ____	
☐ Other______________________________	
Wheelchair	*Patient care environment*
☐ Weight limit ____ lbs	☐ Patient care weight limit ____ lbs. (basic seating chair not Geri/Cardiac chair)
☐ Width ____ inches	☐ Patient chair width ____ inches
☐ Seat height ____ inches	☐ Geri/Cardiac chair weight limit ____ lbs
☐ Handle width ____ inches	☐ Geri/Cardiac chair width ____ inches
☐ Powered? Yes ____ No ____	☐ Geri/Cardiac seat height ____ inches
	☐ Step stool weight limit ____ lbs
Stretcher	*Transfer devices*
☐ Weight limit ____ lbs	☐ Lateral transfer devices weight limit ____ lbs
☐ Width ____ inches	☐ Lateral transfer devices width ___ inches
☐ Length ____ inches	☐ Powered? Yes ____ No ____
☐ Side rail support ____ lbs	☐ Full Body (sling) weight limit ____ lbs
☐ Powered? Yes ____ No ____	☐ Powered? Yes ____ No ____
Bedside commode/shower chair	☐ Full Body (sling) goes to the floor? Yes___ No ___
☐ Weight limit ____ lbs	☐ Sit to stand devices weight limit ___ lbs/
☐ Seat width ____ inches	☐ Sit to stand devices width ____ inches
☐ Adjustable height? Yes ____ No ____	☐ Powered? Yes. ____ No ____

(continued)

Table 5.4 (continued)

Hospital bed	*Bathroom*
Scales	*Ancillary departments*
☐ Weight limit ____ lbs	☐ Door widths ____ inches
☐ Width ____ inches	☐ X-ray table weight limit ____ lbs, width ____ inches
Walker	☐ CT Scan weight limit ____ lbs, width ____ inches
☐ Weight limit ____ lbs	☐ OR table limit ___ lbs, width ___ inches
☐ Width ____ inches	☐ ER equipment weight limit ____ lbs, width ____ inches
	☐ Exam room table weight limit ____ lbs, weight ____ inches

5.8.1 Hazardous Tasks

Identifying hazardous tasks represents the first step toward a policy of safety at work. The most bariatric hazardous healthcare tasks are related to daily (hygiene, bathing, ambulation, and dressing changes) or specific care (i.e., lifting one lower limb for hygiene) (Galinsky et al. 2010). Bariatric nursing can occur in various settings (hospital, physiotherapy, home), modalities (acute care, rehabilitation, long stay), and for different age ranges (pediatric, adult, geriatric). Some settings are particularly dangerous from a biomechanical point of view. In the *operating theatre*, for instance, hazardous tasks for the operator are prolonged standing, awkward postures, transferring the patient to the surgical bed, handling surgical devices (Wicker 2000; Garb and Dockery 1995). In the *acute ward*, the hazardous tasks may be patient transferring, hygiene on dependent patients, lifting patients from the ground, weighing patient on a scale, putting on elastic compressive socks, repositioning in bed, changing patient's undergarments/incontinence pads. In the *psychiatric unit*, restraining patients, specially if in a state of confusion or agitation, may represent a hazardous task due to the unpredictability, aggressiveness of some patients (brisk, sudden and powerful movements), and their lack of collaboration. Use of aids is often difficult (time needed for preparation, inadequate posture of the patient) and manual handling can be equally risky (close contact with an aggressive patient). In *rehabilitation units* and particularly in spinal units, where patients are dependent and often intensive daily healthcare and physiotherapy are provided, transferring, lifting, and repositioning of the patients are usually performed more often. Spaces in the rooms should allow two or more healthcare providers to operate with ease. Rehabilitation units should have adequate number of walking and mobility (walkers, tripods, quadripods and canes, lifters) and lifting aids. Minor aids for horizontal transferring (low-friction mats, air-filled mattresses) and major aids (ceiling-fixed lifters), adjustable bariatric stretchers and electric beds that can be converted into armchairs should be adequately proportioned to the number of patients. In *intensive care* units, electric adjustable beds are mandatory both to alternate postures in bed-ridden patients and perform daily care, therapy, and transferring. Spaces around the beds should have adequate dimensions to allow the presence of cumbersome equipment and more operators. In *emergency situations*, bariatric stretchers and chairs for transportation up/down stairs or in narrow spaces, bariatric examination couches should be available. Morbidly obese nonambulatory people in *long-term care facility* frequently are unable to perform self-care (Rotkoff 1999). Shower-chairs, special tubs, ceiling-fixed lifters (also for transferring to the bathroom) with weighing scale included, active lifters should be part of the equipment. As for *home care*, a wealth of situations often with inadequate spaces and equipment or lack of help/collaboration to perform the care tasks may arise (Gallagher 1998). Manual handling involving use of strength and awkward postures should be minimized by an extensive use of aids.

5.9 Staff Training and Education

Training and education of operators for bariatric assistance should be aimed to all the range of generic healthcare facilities that may admit bariatric patients to avoid improvisation and hazardous behavior when handling such patients. The operators should be consistently educated and updated about biomechanics of movements and postures, and the use of technological aids. Educational videos or written procedures may help refreshing the specific know-how, specially in facilities where admission of bariatric patients is infrequent. Conversely, when those admissions are frequent, upgrading, continuos practice, and education, the exchange of experiences with similar facilities are fundamental (see, National Association of Bariatric Nurses www.bariatricnurses.org/; The Patient Safety Center of Inquiry http://www.visn8.med.va.gov/patientsafetycenter/). Competence of the operators is essential to face complex situations and choose the most adequate assistance modality based on safety and clinical-rehabilitative criteria. It should be borne in mind that larger numbers of operators are required in the care of bariatric as compared to lean patients. Health managers are responsible for providing adequate personnel and maintaining efficient equipment to minimize work-related risks and promote quality of life and dignified care for the patients. The care of obese patients is a concern for all kind of hospitals and healthcare facilities. Being organizatively and culturally unprepared for admitting those patients aggravates the level of risk of injuries both for the healthcare provider and the patient.

References

Baptiste A, McCleerey M, Matz M, Evitt CP (2008) Proper sling selection and application while using patient lifts. Rehabil Nurs 33(1):22–32

Barrow CJ, Roux-en Y (2002) Gastric bypass for morbid obesity. AORN J 76:593–604

Bergstrom N, Braden B, Kemp M, Champagne M, Ruby E (1998) Predicting pressure ulcer risk: a multisite study of the predictive validity of the Braden scale. Nurs Res 47:261–269

Blackwood HS (2004) Obesity: a rapidly expanding challenge. Nurs Manage 35:27–35

Bogossian FE, Hepworth J, Leong GM, Flaws DF, Gibbons KS, Benefer CA, Turner CT (2012) A cross-sectional analysis of patterns of obesity in a cohort of working nurses and midwives in Australia, New Zealand, and the United Kingdom. Int J Nurs Stud 49(6):727–738

Brown ND, Thomas NI (2003) Exploring variables among medical center employees with injuries:developing interventions and strategies. AAOHN Journal 51(11):470–481

Brown I, Psarou A (2007) Literature review of nursing practice in managing obesity in primary care: developments in the UK. J Clin Nurs 17:17–28

Cavaliere B, Snaidero D (1999) Metodologia per la rilevazione della complessità assistenziale infermieristica: calcolo dell'indice di complessità assistenziale. Management Infermieristico 1:32–36

Carpenito-Moyet LJ (2010). Diagnosi infermieristiche. Applicazione alla pratica infermieristica. 5°edizione, CEA

Collins JW (2010) Safe patient handling lifting standards for a safer American Workforce. CDC Congressional Testimony

Cowley SP, Leggett S (2011) Manual handling risks associated with the care, treatment and transportation of bariatric (severely obese) clients in Australia. Work 39(4):477–483

Davidson JE, Kruse MW, Cox DH, Duncan R (2003) Critical care of the morbidly obese. Crit Care Nurs Q 26:105–116
Drake DJ, Dutton K, McAuliffe M, Engelke M, Rose MA (2005) Challenges that nurses face in caring for morbidly obese patients in the acute setting. Surg Rela Dis 1:462–466
Ferraro DR (2003) Laparoscopic adjustable gastric banding for morbid obesity. AORN J 77:923–940
Galinsky T, Hudock S, Streit J (2010) Addressing the need for research on bariatric patient. Rehabil Nurs 35(6):242–247
Gallagher SM (1998) Needs of the homebound morbidly obese patient: a descriptive survey of home health nurses. Ostomy Wound Manage 44(4):32–34
Gallagher S (2004) Taking the weight off with bariatric surgery. Nursing 34:58–63
Garb JR, Dockery CA (1995) Reducing employee back injuries in the perioperative setting. AORN J 61(6):1046–1052
Geary B, Collins N (2012) Are we prepared for a growing population? Morbid obesity and its implications in Irish emergency departments. Eur J Emergency Med 19:117–120
Golay A, Fossati M, Deletraz M, de Luzy F, Howles MN, Ybarra J (2003) Multydisciplinary approach to obesity treatment. Diabetes Obes Metab 5:274–279
Graling P, Elariny H (2003) Perioperative care of the patients with morbid obesity. AORN J 77:802–805
Hahler B (2002) Morbid obesity: a nursing care challenge. Dermatol Nurs 14(4):249–256
Han K, Trinkoff AM, Storr CL, Geiger-Brown J (2011) Job stress and work schedules in relation to nurse obesity. J nurs adm 41(11):488–495
Holland DE, Krulish HKR, Roche JD (2001) How to creatively meet care needs of the morbidly obese patient with morbidly deficits challenging the entire health care team. Nurs Manage 32:39–41
Jbeili C, Penet C, Jabre P (2007) Out of hospital management characteristics of severely obese patients. Ann Francaises d'Anesthesie et de Reanimation 26:921–926
Latner JD (2001) Self-help in the long-term treatment of obesity. Obesity review 2:87–97
Lee YH, Chiou WK (1994) Risk factors for low back pain, and patient handling capacity of nursing personnel. J Saf Res 25(3):135–145
Malinauskiene V, Leisyte P, Romualdas M, Kirtiklyte K (2011) Associations between self-rated health and psychosocial conditions, lifestyle factors and health resources among hospital nurses in Lithuania. J Adv Nurs 67(11):2383–2393
Menegoni F, Galli M, Tacchini E, Vismara L, Cavigioli M, Capodaglio P (2009) Gender-specific effect of obesity on balance. Obesity 17(10):1951–1956
Mercer S, Tessier S (2001) A qualitative study of general practitioners' and practice nurses' attitudes to obesity management in primary care. Health Bull 59:248–253
Muir M, Archer-Heese G (2009) Essentials of a bariatric patient handling program. Online J issues Nurs 14:Manuscript 5
Mulligan A, Young LS, Randall S, Raiano C, Velardo P, Breen C, Bushee L (2005) Best practices for perioperative nursing care for weight loss surgery patients. Obes Res 13:267–273
Jang R, Karwowski R, Quesada PM, Rodrick D, Sherehiy P, Cronin SN, Layer JK (2007) Biomechanical evaluation of nursing tasks in a hospital setting. Ergonomics 50(11):1835–1855
Kolotkin R, Meter K, Williams G (2001) Quality of life and obesity. Obes Rev 2:219–229
Kramer KL (2004) WOC nurses as advocates for patients who are morbidly obese: a case study promoting the use of bariatric beds. J Wound Ostomy Continence Nurs 31(6):379–384
Lloyd JD, Baptiste A (2006) Biomechanical evaluation of friction reducing devices for lateral patient transfer. AAOHN J 54(3):113–119
Lloyd J, Wikinson S (1999) Integrated sling for patient lift manipulation. Dept of Veteran's Affairs Patents and Licensing Office
Nelson A (ed) (2006) Safe patient handling and movement: a practical guide for health care professionals. Springer Publishing Company, New York
Novation LLC (2004) Obese patient care survey market research report. Available from www.novationco.com

Pietroiusti A, Neri A, Somma G, Coppeta L, Iavicoli I, Bergamaschi A, Magrini A (2010) Incidence of metabolic syndrome among night-shift healthcare workers. Occup Environ Med 67(1):54–57

Puhl R, Brownell K (2001) Bias, discrimination and obesity. Obes Res 9:788–805

Racco M, Phillips B (2010) Developing a protocol for intensive care patients at high risk for pressure ulcers. Crit Care Nurse 30:77–80

Randall SB, Pories WJ, Pearson A, Drake DJ (2009) Expanded Occupational Safety and Health Administration 300 log as metric for bariatric patient-handling staff injuries. Surg Obes Relat Dis 5(4):463–468

Rose MA, Baker G, Drake D, Engelke M, McAuliffe M et al (2006) Nursing requirements for care of morbidly obese patients in the acute care setting. Nurs Surg Patient Care 2006 1(2):115–121

Rotkoff N (1999) Care of the morbidly obese patient in a long-term care facility. Geriatr Nurs. 20(6):309–313

Sandlin D (2003) Perianesthesia nursing considerations for laparoscopic adjustable gastric banding minimally invasive surgical patients. J Perianesth Nurs 18:272–276

Sarwer DB, Fabricatore AN, Eisenberg MH, Sywulak LA, Wadden TA (2008) Self-reported stigmatization among candidates for bariatric surgery. Obesity (Silver Spring). 16(Suppl 2):S75–S79

Sullivan EE (2004) The safe transfer of care. J Perianesth Nurs 19:108–110

Twedell D (2003) Physical assessment of the bariatric person. J Continuining Educ Nurs 34(4):147–148

VanHoy SN, Laidlow VT (2009) Trauma in obese patients: implications for nursing practice. Crit Care Nurs Clin N Am 21:377–389

VISN8 (2007) Patient safety center of inquiry. Safe bariatric patient handling toolkit. Bariatric Nurs Surg Patient Care 2007 2(1):7–45

Voelker M (2004) Assessing quality of life in gastric bypass clients. J Perianest Nurs 19:89–101

Waters TR, Putz-Anderson V, Garg A, Fine LJ (1993) Revised NIOSH equation for the design and evaluation of manual lifting tasks. Ergonomics 36(7):749–776

Wicker P (2000) Manual handling in the perioperative environment. British J Perioperative Nurs 1(5):255–259

Wignall D (2008) Design as a Critical Tool in Bariatric Patient Care. J Diabetes Sci Technol 2(2):263–267

Zhuang Z, Stobbe TJ, Hsiao H, Collins JW, Hobbs GR (1999) Biomechanical evaluation of assistive devices for transferring residents. Appl Ergon. 30(4):285–294

Chapter 6
Effectiveness of Multidisciplinary Inpatient and Outpatient Rehabilitation on Functional Outcomes in Obese Patients with Orthopedic Comorbidities

Paolo Capodaglio, Heather K. Vincent and Jason Ludlow

Abstract Outpatient and inpatient rehabilitation programs are designed to minimize or abolish disability associated with obesity, which may be predominantly due to cardiorespiratory and musculoskeletal issues, and general deconditioning. This chapter discusses the efficacy of outpatient multidisciplinary rehabilitation in obese patients, and presents emerging evidence of the effects of inpatient rehabilitation on functional outcomes in obese patients with disabling orthopedic conditions. In inpatient setting, obesity and its related comorbidities may prolong or complicate hospital stays and reduce efficiency of functional gain. Because of the relative scarcity of data regarding inpatient rehabilitation programs, we performed a large prospective investigation on obese patients with orthopedic conditions ($N = 464$). Intensive rehabilitation administered to obese patients with orthopedic conditions and motor difficulties consisted of a 4-week multidisciplinary program covering nutritional, motor, and psychological aspects. Our data provide initial evidence that multidisciplinary intensive rehabilitation is effective in minimizing the obesity-related disability and enhancing functional capacities in obese patients with motor disability. Outpatient rehabilitation programs in obese patients focus on chronic musculoskeletal conditions resulting from obesity, specifically joint misalignment, balance and postural issues, and osteoarthritis. Common goals include decreasing body weight, reducing metabolic disease risk and musculoskeletal pain severity, and improving functional strength and mobility.

P. Capodaglio (✉)
Rehabilitation Unit and Research Laboratory in Biomechanics and Rehabilitation,
Istituto Auxologico Italiano IRCCS, Piancavallo, Verbania, Italy
e-mail: p.capodaglio@auxologico.it

H. K. Vincent
Human Performance Laboratory, Department of Orthopaedics and Rehabilitation,
Divisions of Research and Physical Medicine and Rehabilitation, University of Florida,
Gainesville, FL, USA
e-mail: vincehk@ortho.ufl.edu

J. Ludlow
College of Medicine, University of Florida, Gainesville, FL, USA
e-mail: vincehk@ortho.ufl.edu

P. Capodaglio et al. (eds.), *Disabling Obesity*, DOI: 10.1007/978-3-642-35972-9_6,

The chapter discusses the randomized controlled trials which have previously examined the efficacy of a variety of outpatient exercise programs using variations of resistance and/or aerobic exercise modalities at achieving these goals.

6.1 Introduction

As obesity rates rise, it is expected that the global burden of obesity and related degenerative musculoskeletal diseases will become unsustainable. Obesity adversely influences functional capacity across the lifespan, by inducing physical discomfort, limiting exercise capacity, and eliciting painful disability over time. Obesity is characterized by comorbidities ranked among the most burdensome diseases for high-income countries (World Health Organization 2008). This somatic fragility affects independence and difficulties in daily activities including personal hygiene and dressing, indoor mobility, household chores, lifting and carrying bags, walking and standing in a queue have been reported (Menegoni et al. 2009; Xu et al. 2008; Capodaglio et al. 2010). Disability associated with obesity may be predominantly due to cardiorespiratory and musculoskeletal issues, and general deconditioning. In inpatient setting, obesity and its related comorbidities may prolong or complicate hospital stays and reduce efficiency of functional gain.

Rehabilitation strategies that reduce pain symptoms, increase strength, and optimize function and mobility are urgently needed to offset an impending epidemic of obesity-induced disability. Outpatient and inpatient rehabilitation programs are designed to minimize or abolish disability. Obese acute care patients may be referred to receive care in the *inpatient rehabilitation* setting in which health care teams commonly use a multidisciplinary approach to accomplish these goals. Teams generally consist of physiatrists, endocrinologists, cardiologists, pneumologists, physiologists, physical and occupational therapists, adapted physical activity experts, nurses, clinical psychologists, and psychiatrists. Common goals include achieving maximal independent functional ability before discharge by increasing strength, coordination, endurance, balance, cognition, and reducing body weight and cardiovascular risk factors. *Outpatient programs* typically focus on exercise protocols, manual therapies, and other adjuvant components that enhance joint range of motion and muscle strength. Common goals for obese patients with disability include: reduce painful movement, correct muscle strength imbalance, and deficits and increase functional mobility. For the obese patient, multidisciplinary rehabilitation programs should also include components that reduce cardio-metabolic risk factors, decrease fat mass, improve the lean-to-fat-mass ratio, and increase physical conditioning and increase residual functional capacity. Education and psychological counseling together with diet and physical activity play a key role in counteracting this chronic condition, since a life style change is needed to maintain benefits and minimize health risks in the long term.

This chapter discusses the efficacy of outpatient multidisciplinary rehabilitation in obese patients, and presents emerging evidence of the effects of inpatient rehabilitation on functional outcomes in obese patients with disabling orthopedic conditions. Relatively, few papers (Farrell et al. 2008; Vincent et al. 2006a, 2009) have shown that multidisciplinary rehabilitation in severely obese patients with metabolic disease risk factors improves physical performance, lowers body weight, and reduces cardiovascular risk factors. There is, however, a relative lack of studies of in-patient rehabilitation outcomes for patients with severe obesity and related disabling musculoskeletal conditions. Attestation of this significant deficit was recently addressed by the Italian Ministry of Health (Ministero della Salute 2011), when it initiated processes to organize rehabilitation pathways across different settings, acknowledging the need for metabolic rehabilitation. This has direct relevance to obese patients who can derived functional and metabolic benefits from rehabilitation programs.

6.2 Considerations of Efficacy Assessment of Rehabilitation Effects in Obese Patients

One of the major challenges in research available on obesity-related disability is the lack of obesity-specific scales reflecting relevant physical difficulties and limitations in everyday life. Standardized functional assessments are often non-specifically validated in the obese population and are not designed to measure clinical appropriateness, which contributes to underestimation of the functional difficulties. In 2002, the United States task force on developing obesity outcomes and learning standards (TOOLS) (Ropka 2002) recommended a summary of outcome measures for the obese patient, which included the use of the Medical Outcome Survey 36-Item Short-Form Health Survey (SF-36) or its 12-items version (SF-12) (Ware et al. 1996) and the use of an existing symptom-specific scale in addition to the more general measure when the goal is to provide an in-depth evaluation of one specific obesity-related symptom. An exhaustive review of the kinds of assessments that can be used to measure obesity-related disability has been provided in Chap. 4. Measuring disability with instruments reflecting the full range of functional status appears, therefore, crucial in order to initiate rehabilitation programs for obese subjects and to assess their effectiveness. In 2009, it was with this goal in mind that the Italian society of obesity (SIO) developed a new scale for assessing disability correlated to obesity in adult populations, the short-form questionnaire the Obesity-related Disability test (*Test SIO Disabilità Obesità Correlata*; *TSD.OC*) (Donini et al. 2010). This instrument is intended to target the most important obesity-specific functional status dimensions such as physical, emotional functioning, and social functioning. Minimizing disability and educating individuals to cope with everyday life tasks in the presence of an excessive body mass are the major goals of rehabilitation. Subjective and objective tests can be used to determine efficacy of functional and metabolic improvement achieved during rehabilitation. Table 6.1 provides a list of these tests relating to physical independence, activities of daily living (ADL), self care and transfer capacity, and functional mobility.

Table 6.1 Subjective and objective tests of physical function and metabolic capacity to assess rehabilitation efficacy

Test	Description
Functional independence measure (FIM) rating	7: *Complete independence*. The patient performs all the tasks described as making up the activity within a reasonable amount of time, and does so without modification, assistive devices, or aids 6: *Modified independence*. The activity requires an assistive device aid, the activity takes more than reasonable amount of time, or the activity involves safety (risk) considerations; requires another person for either supervision or physical assistance to perform the activity 5: *Supervision or setup*. The patient requires no more help than standby, cueing, or coaxing, without physical contact; alternatively, the helper setup needed items or applies orthoses or assistive/adaptive devices 4: *Minimal contact assistance*. The patient requires no more help than touching, and expends 75 % or more of the effort 3: *Moderate assistance*. The patient requires more help than touching, or expends between 50 and 74 % of the effort 2: *Maximal assistance*. The patient expends less than half of the effort. Maximal or total assistance is required 1: *Total assistance*. The patient expends less than 25 % of the effort; Maximal or total assistance is required. Total score (possible score range 18–126 points) has FIM Motor and Cognitive subscores
Functional visual analog score (VAS)	Using a scale from 0–100 mm, patients rate the level of disability
Obesity-related disability test (TSD.OC)	Questionnaire is composed of 36 items divided into seven sections (pain: 5 items; stiffness: two items; ADL and indoor mobility: seven items; housework: seven items; outdoor activities: five items; occupational activities: four items; social life: six items), which reflect the domains in which individuals experience the most common problems. A disability score has been defined as (raw total score/max total score) × 100
Symptom limited exercise testing	Using a treadmill or stationary cycle, the time to exhaustion can be used to indicate aerobic capacity, cardiopulmonary function, and responses
Timed up and go test (TUG)	This test measures the time to complete a chair rise, walk 3 m, turn around walk back and sit down. This assesses balance and transfer capacity, and strongly related to functional mobility, and static and dynamic balance
Six-min walk test	This test measures the walking distance covered in 6 min; disease-specific symptoms are monitored (pain, dyspnea, and heart rate); this test reflects aerobic ability and endurance
Gait parameters	Cadence, stride and step lengths, step widths, base of support, and toe out angles can be used to show changes in dynamic balance and ability to perform mobility tasks

(continued)

Table 6.1 (continued)

Test	Description
Gait speed at comfortable or fastest speeds	The time to complete a walking distance of 4–10 m is measured and is used to reflect power in obese persons and short distance walk speed is related to lifespan; Capture of the fastest walking speed has prognostic ability to predict an increased incidence of persistent physical limitations
Chair rise time	The time to rise from a chair without using arms can be used to reflect lower extremity strength; modifications include performing five in a row, or as many as can be completed in 30 s
Short physical performance battery test (SPPB)	This series of tests measures lower extremity functioning in older persons; timed measures of standing balance, walking speed, and ability to rise from a chair. Each of the three performance measures was assigned a score ranging from 0 to 4, with four indicating the highest level of performance and 0 the inability to complete the tests

6.3 Inpatient Rehabilitation Programs

There are a few studies that report functional outcomes in cohorts with musculoskeletal impairments and obesity. Most inpatient rehabilitation studies have performed patient stratification of outcomes based on body mass index (BMI) values after elective orthopedic surgeries (Vincent et al. 2006b, c, 2008, 2009; Stickles et al. 2001) or musculoskeletal trauma (Vincent et al. 2012a). Other studies have stratified outcomes by BMI in metabolically complex conditions such as stroke (Padwal et al. 2012) or burns (Farrell et al. 2008). In general, physical function and independence improve with aggressive inpatient rehabilitation. Functional independence measure (FIM) rating changes captured during inpatient rehabilitation reveal that a high BMI does not prevent FIM gains in persons with total knee arthroplasty (Vincent et al. 2006a; Vincent and Vincent 2008; Stickles et al. 2001) or total hip arthroplasty (Vincent et al. 2006b, 2009; Stickles et al. 2001). Vincent et al. recently showed that even with severe orthopedic trauma (no central nervous system impairment or spinal cord injury), obese persons can achieve significant improvement in FIM ratings from admission to discharge, with similar length of stay; FIM efficiency, however, was slightly less in the obese patients compared with non-obese patients (3.1 vs. 2.6 points gained/day) with similar improvements in walking distance and stair climbing (Vincent et al. 2012a). Obese patients are able to fully participate in physical therapy, occupational therapy, and speech-language therapy sessions like non-obese patients with orthopedic conditions (Vincent et al. 2012b; Padwal 2012). However, the existing evidence consistently shows that functional gains are achieved less efficiently and at higher costs than in leaner patients (Vincent et al. 2012a; Padwal et al. 2012; Stickles et al. 2001).

In populations with burn injury, the entire patient cohort improved with inpatient rehabilitation, but patients with BMI of >25.15 kg/m^2 achieved a lower FIM locomotor score and were less likely to be discharged home (Farrell et al. 2008). No associations between BMI and FIM ratings in post-acute stroke rehabilitation patients were found (Padwal et al. 2012).

6.3.1 Inpatient Rehabilitation with Obesity Considerations

Because of the relative scarcity of data regarding inpatient rehabilitation programs in obese patients with orthopedic conditions, we attempted to fill this scientific void. We performed a large prospective investigation on obese patients ($N = 464$) admitted to our Rehabilitation Unit with orthopedic conditions for multidisciplinary rehabilitation (Table 6.2). Patients were categorized into groups based on disability rankings: "low" = disability score range 0–19.4 points, "moderate" = score range 19.4–47.2 points, "moderately severe" = 47.3–68.4 points and "severe" = 68.4–100 points as represented by TSD.OC scores. The comorbidities present in our sample were: osteoarticular pathologies (100 %), cardiovascular pathologies (23.4 %), diabetes (6.8 %), cardiovascular, and diabetes (7.6 %). Exclusion criteria were acute cardiovascular, respiratory, neurological, and psychological conditions contraindicating intensive rehabilitation program.

6.3.2 Rehabilitation Intervention

Intensive rehabilitation administered to obese patients with orthopedic conditions and motor difficulties consisted of a 4-week multidisciplinary program covering nutritional, motor, and psychological aspects. Patients underwent rehabilitation exercises for ~3 h per day to optimize muscle strength and the lean to fat mass ratio, increase joint range of motion and enhance cardio-respiratory conditioning. Supervised aerobic exercise (walking, recline cycle, and arm ergometer) was

Table 6.2 General characteristics of the inpatient rehabilitation cohort ($n = 464$) at admission

Variable	Whole sample number (%)
Sex	
Female	372 (80.2)
Male	92 (19.8)
Age (years)	
Mean (SD)	63.58 (10.40)
Range	21–80
BMI (kg/m^2)	
Mean (SD)	41.61 (6.06)
Range	30–59.7

performed 30–60 min daily, 5 days/week, on the basis of the individual capacities as assessed by a 6-min walking test or, in selected cases, an ergometric test, selecting a work load of approximately 40 % of the individual maximal oxygen uptake. Resistance exercises were performed 30–60 min daily in individual sessions. Intensities of isotonic strengthening exercise were set at 40 % of the individual 10-repetition maximum during the first week, 50 % during the second week and 60 % during the remaining rehabilitation period. The muscle groups to be trained were identified during the initial physiatrist evaluation. Strengthening exercises with arms and legs under supervision included dynamic standing and floor calisthenics using body weight as movement resistance.

All subjects underwent a caloric restriction of the caloric intake (1200–1800 kcal/day) containing 21 % proteins, 53 % carbohydrates, and 26 % lipids. Dieticians tailored the personal balanced nutritional program (75 % of the resting metabolic rate) in order to achieve a 5 % reduction in body weight at discharge and to reinforce a healthy life style in the long term. Patients also underwent nutritional education consisting of daily lectures, demonstrations, and group discussions aimed enhancing compliance to diet. Psychological counseling, consisting of two sessions/week of cognitive-behavioral strategies, was aimed at educating individuals to control their eating habits, stress and anxiety, improve problem solving, self-perception, involvement in care, and compliance to therapies.

6.3.2.1 Outcome Measures

At admission and discharge (at 4 weeks), the following outcome measures were collected: body weight, BMI, FIM motor, and cognitive ratings (Hamilton et al. 1994) obesity-related disability test (TSD.OC) (Donini et al. 2010) the functional visual analog score (VAS), and the timed-up-go (TUG) (Mathias et al. 1986). The disability level was determined using the TSD.OC score (low level score range 0–19.4, moderate level score range 19.4–47.2, moderately severe 47.2–68.4 and severe 68.4–100 points).The functional VAS (0–10 mm) was meant to measure subjective perception of functioning.

6.4 Results

Patient characteristics are shown in Table 6.2. We have stratified results according to the four categories of disability (low to severe; Table 6.3). All of the main outcome parameters, with the exclusion of FIM Cognitive, improved significantly at discharge. The percentage change in FIM (physical and total ratings) and the change in TUG scores were statistically different among the four disability strata. The change in functional VAS was different between moderately severe *versus* severe level and body weight change of the low disability was higher if compared to the severe disability level ($p < 0.05$). Correlations existed

Table 6.3 Anthropometric and functional parameters (mean, standard deviation and % change) at admission and discharge (after a 4-week rehabilitation program) in the four levels of disability severity

	BW (Kg)	FIM (physical)	FIM (cognitive)	FIM (Total)	TSD.OC (%)	VAS	TUG (s)
Low disability (#30)							
Admission	111.98 (13.25)	87.20 (4.57)	34.93 (0.25)	122.47 (4.45)	12.05 (4.59)	64.50 (26.59)	12.58 (16.27)
Discharge	106.41 (12.89)*	89.57 (3.00)*	34.93 (0.25)	124.50 (3.13)*	7.99 (8.06)*	87.00 (14.94)*	11.59 (15.24)*
% change	−4.97	2.86	0	1.72	−35.28	−22.5	−7.71
Moderate disability (#103)							
Admission	105.46 (17.26)	84.53 (5.01)	35.42 (5.31)	119.39 (5.18)	35.43 (7.67)	57.72 (20.55)	13.94 (8.06)
Discharge	100.96 (16.52)*	87.37 (6.35)*	34.91 (0.49)	122.82 (3.64)*	24.88 (17.33)*	80.48 (13.72)*	10.79 (4.57)*
% change	−4.24	3.52	−0.54	2.97	−29.58	−22.76	−17.78
Moderately severe disability (#140)							
Admission	104.69 (19.12)	81.82 (5.71)	34.82 (0.76)	116.73 (5.84)	58.49 (5.97)	50.48 (20.69)	15.64 (5.12)
Discharge	100.25 (17.79)*	86.47 (3.96)*	34.77 (1.22)	121.21 (4.49)*	37.66 (17.52)*	75.28 (14.95)*	11.90 (3.78)*
% change	−4.13	5.97	−0.09	3.96	−35.29	−24.79	−21.59
Severe disability (#191)							
Admission	103.56 (15.48)	75.15 (9.08)	34.55 (1.18)	109.52 (9.29)	80.67 (7.82)	37.25 (24.17)	22.11 (12.68)
Discharge	99.29 (14.52)*	81.57 (7.32)*	34.57 (1.15)	116.14 (7.75)*	55.18 (20.19)*	66.79 (20.19)*	14.84 (6.53)*
% change	−4.04	9.31	0.05	6.34	−31.89	−29.74	−27.61

* = $p < 0.05$, admission versus discharge

BW body weight; *FIM* functional independence measure; *VAS* visual analog scale; *TUG* timed up and go

between anthropometric and functional measures; age was mildly correlated with FIM (both physical and total value, respectively $r = -0.36$ and -0.37, $p < 0.05$), TSD.OC scores ($r = 0.23$, $p < 0.05$) and TUG scores ($r = 0.42$, $p < 0.05$) at admission and with the FIM (both physical and total value, respectively $r = 0.19$ and 0.20, $p < 0.05$) and the TSD.OC change ($r = 0.18$, $p < 0.05$). A mild correlation existed between body weight and FIM ratings (both physical and total value, respectively $r = -0.12$ and -0.13, $p < 0.05$) and TSD.OC ($r = 0.18$, $p < 0.05$) at admission was evident.

The comparison among the four disability levels demonstrated that the percent changes in FIM (physical and total score) and TUG changes were statistical different among groups. The change in VAS was different between moderately severe *versus* severe level and body weight change of the low disability was higher if compared to the severe disability level ($p < 0.05$). No other statistical differences were observed. The research for correlations between anthropometric and functional measure showed that the patients' age had a mild correlation with FIM (both physical and total value, respectively $r = -0.36$ and -0.37, $p < 0.05$), TSD.OC ($r = 0.23$, $p < 0.05$) and TUG ($r = 0.42$, $p < 0.05$) at admission and with the FIM (both physical and total value, respectively $r = 0.19$ and 0.20, $p < 0.05$) and the TSD.OC change ($r = 0.18$, $p < 0.05$). As for their body weight, a mild correlation with FIM (both physical and total value, respectively $r = -0.12$ and -0.13, $p < 0.05$) and TSD.OC ($r = 0.18$, $p < 0.05$) at admission was evident. No other statistical correlations were found.

6.5 Benefits of Inpatient Rehabilitation

The aim of rehabilitation is to gain health and functional independence. To achieve true empowerment by the end of the program, rehabilitation should assess individuals according to their wishes and context and then implement all of the required therapeutic interventions to ensure the achievement of the highest possible level of functioning and participation. Disability threshold values could operate as appropriateness criteria for admitting obese individuals to different rehabilitation settings. We have previously shown that the TSD-OC, validated against the SF-36 (Donini et al. 2010), is sensitive to change, capturing differences at post-rehabilitation in a longitudinal design (Precilios et al. 2012). Given the lack of specific measures able to capture the disability associated with morbid obesity, the TSD-OC questionnaire appears an interesting tool for measuring outcomes in this area, since it considers all the domains of disability which are reported in the literature to be obesity related. The obesity-specific disability scale appears sensitive to changes among groups with different grades of disability and the percentage of change does not differ among groups. Based on our data, multidisciplinary rehabilitation interventions appear effective in reducing both mild and severe disabilities related to obesity with comorbidities. The variation in body weight does not appear to be related to the changes in disability score, suggesting that functional improvements maybe independent from weight loss and related to other factors. The TUG test seems to improve more in the most severely disabled obese. It is known that frail lean elderly people because of their lower initial functional level yield a potential to larger improvements in functional capacities as compared to non-frail elderly (Buchner and de Lateur 1991; Capodaglio et al. 2006). In contrast with this, younger obese individuals seem to benefit more from rehabilitation interventions than the older ones, where the deterioration at different functional levels (i.e., reduced muscle strength secondary to sarcopenia, fat infiltration of the muscles), combined with negative factors such as an increased inertia from excessive body mass, and aggravated by the aging process itself, may have possibly determined a downward shift of the strength-function relationship. Thus, the physiological gains not reach a threshold sufficient to elicit a dramatic functional change. Also, we know that sarcopenic obesity in old age is more strongly associated with disability in daily activities than either sarcopenia or obesity *per se* (Jensen and Hsiao 2010). Longer exposure to obesity, functional deterioration at various levels, and greater peripheral tissue damage in old age may therefore hinder significant responses to training and rehabilitation. Among the young patients, those with higher BMI showed the greatest functional improvements, probably due to greater weight reduction than their counterparts with lower BMI. Our results from a relatively large obese population sample with a representative range of BMI and age corroborate the use of TSD-OC as a global assessment of the obesity-related disability and also defined some initial reference values to stage the degree in disability severity. The disability grading proposed could help the decision making of allocating patients to appropriate rehabilitation settings tailored to their rehabilitative needs.

Further, longitudinal studies are needed to assess whether short-term functional gains can be maintained in the medium and long term. Sensitivity to change for each section of the TSD-OC (pain; stiffness; ADL and indoor mobility; housework; outdoor activities; occupational activities; and social life) will need further investigations. Future studies should address different intervention modalities (i.e. moderate vs. higher intensity resistance training etc.) and compare the effectiveness of different rehabilitation interventions (e.g., moderate versus higher intensity resistance training integrated with dietary changes on physical function, body and muscle composition).

Our data provide initial evidence that multidisciplinary intensive rehabilitation is effective in minimizing the obesity-related disability and enhancing functional capacities in obese patients with motor disability. The disability grading proposed needs to be confirmed in larger studies but could help the decision making of allocating patients to appropriate rehabilitation settings tailored to their rehabilitative needs. Our data on the effectiveness of multidisciplinary rehabilitation of severe obesity with comorbidities provide some initial support to the indications of the Italian Ministry of Health for such rehabilitation pathway. Further, cost-effectiveness data are needed to help allocating patients to the most appropriate rehabilitative setting.

6.6 Outpatient Rehabilitation Programs

Outpatient rehabilitation programs in obese patients typically focus on chronic musculoskeletal conditions resulting from obesity, specifically joint misalignment, balance and postural issues, and osteoarthritis. Irrespective of age, obese patients demonstrate need for rehabilitative strategies to combat orthopedic issues.

6.6.1 Younger Obese Individuals

In younger persons, obesity and fat distribution is associated with joint pain. As the discomfort of movement increases, voluntary physical activity is diminished and obesity is exacerbated. Children with high adiposity are up to four times as likely to have joint pain than non-obese counterparts (Podeszwa et al. 2006). In children up to 18 years of age, common sites for joint pain include the low back and the knee (Sjolie 2004; Taylor et al. 2006). Evidence shows that up to 75 % of obese children with pain symptoms avoid physical activities including recreational play, sports, and activities of daily living (Wilson et al. 2010). Weight accumulation is related to increased systemic inflammation, joint compression, joint misalignment, and onset of degenerative joint disease. Even among children, functional test battery performance (push/pulling activity, long jump, vertical jump, and ball throwing) and walking endurance performance is inversely associated with fat mass and body mass index (Riddiford-Harland et al. 2006). Studies have also shown that obese children have compromised lower body transfers, jumps, and chair rise times (Macfarlane et al. 2011). Investigators have, therefore, encouraged early interventions for prevention of joint disease and related disability.

6.6.2 Transitional-Aged Obese Adults

Prospective evidence demonstrates that as obese young individuals transition to middle age, their risk for developing knee pain nearly doubles (Heo et al. 2010). The Behavioral Risk Factor Surveillance Survey outcomes revealed a J-shaped relationship between BMI values and the presence of functional impairment and the use of assistive devices (wheelchairs, canes or special beds/phones) in adults 18–64 years of age (Peytremann-Bridevaux and Santos-Eggimann 2008). The odds risk of experiencing physical disability with stair climbing ($\geq$1 flights), activities of daily living, and walking 100 meters increased is higher in persons 50 years or older compared with those less than 50 years. Also, greater disability values occur in women compared to men (Duvigneaud et al. 2008). Additionally, obese adults demonstrate patterns of less physical activity and this may perpetuate muscle atrophy and declines in muscle strength (Stenholm et al. 2008). Muscle atrophy, in turn, attenuates resting and exercise metabolic rates which can work in concert with sedentary behavior to promote weight gain (Thomas 2007).

6.6.3 Older Obese Adults

As the obese population ages, musculoskeletal degeneration, pain, and related complications will increase in prevalence. Body composition has unfavorably shifted to that of less muscle mass relative to accumulation of fat mass, and strength and exercise capacity diminish (Visser 1998a); functional disability follows suit (Visser et al. 1998b; Ramsay et al. 2006; Lebrun et al. 2006; Sternfeld et al. 2002; Vincent et al. 2012a). Because aging and obesity both independently contribute to muscle inadequacy and deterioration of mobility, the combination of the two likely exacerbates negative physiological changes in the musculoskeletal system and shapes an environment ideal for sarcopenia and functional disability.

6.7 Outpatient Rehabilitation Programs

Similar to inpatient populations, outpatient programs have different goals depending on the individual or cohort. For obese patients, common goals include decreasing body weight, reducing metabolic disease risk and musculoskeletal pain severity, and improving functional strength and mobility. Numerous randomized controlled trials have previously examined the efficacy of a variety of outpatient exercise programs at achieving these goals (Vincent et al. 2006a). These trials have used variations of resistance exercise (RX) and/or aerobic exercise (AX) modalities. Programs that have used RX for strengthening ranged from 2.5 to 6 months, with training generally occurring at a frequency of 3 days a week using a variety of weight machines or body weight exercises depending on the musculoskeletal condition. Most often, large muscle group activity was included and incorporated knee extensors/flexors, elbow extensors/flexors, pushing and pulling exercises (seated row and chest press), and trunk strengthening exercises (abdominal curls and lumbar extensions). Programs incorporating AX typically used AX training frequencies of 3–5 days per week for 20–45 min at moderate intensity (60–85 % max HR or VO2 max). AX regimens provide a stimulus that involves sustained large muscle activity such as stair climbing, walking, cycling ergometry, or aquatic aerobic exercise (Avila et al. 2010; Bouchard et al. 2009, 2010; Davison et al. 2002; Lim et al. 2010). Some trials have combined RX and AX into a multimodal approach where participants perform aerobic, resistive, and flexibility activities within a single exercise session. Multimodal activity programs have been implemented for durations lasting 3 months–1 year (Grant et al. 2004; Manini et al. 2010; Schlenk et al. 2011). Importantly, many of the multimodal activity programs often incorporated dietary modification as part of the rehabilitation strategy. Dietary interventions ranged from nutritional and behavioral modification counseling, to structured meal plans with caloric restriction (typically 750 kcal deficit) and partial meal replacements (bars and shakes) (Frimel et al. 2008; Miller et al. 2006; Villareal et al. 2006; Anton et al. 2011).

These studies show that comprehensive rehabilitation programs lasting 3–18 months in duration that incorporate AX and RX (2–3 days per week) produce clinically meaningful changes in body composition, strength, and functional mobility in obese individuals. Vincent et al. (2006b) examined the effects of 6 months of RX training in obese older adults and found that after the rehabilitation period, the RX group lost 2 % fat mass and demonstrated a 15 % increase in walk time to exhaustion and an 18 % increase in overall body strength compared to no significant changes in the control group. Metabolic improvements included reduction of cardiovascular disease biomarkers in the RX group. Similarly, Davidson et al. (2009) compared the effects of 6 months of RX, AX, and combined AX + RX to a control group and found that the number of chair stands completed in 30 s increased in all intervention groups, with greater changes in the RX and AX + RX than the AX group. Additionally, the greatest loss in fat mass occurred in the AX + RX group, and gain in skeletal muscle in the RX and AX + RX groups, highlighting the enhancement of functional improvements with the addition of RX to exercise programs (Davidson et al. 2009).

Of all the programs studied, multimodal combined exercise programs coupled with dietary intervention produced the greatest improvements in functional performance (Villareal et al. 2006, 2011). In a 12-month study comparing exercise (AX + RX), diet, and exercise + diet against a control group in older obese patients, Villareal et al. (2011) showed that exercise + diet induced the greatest improvements in PPT (Modified Physical Performance test consisting of timed trials of walking and other measures of physical function) scores and peak fitness values compared with the other three groups. Interestingly, scores on the Functional Status Questionnaire, in which higher scores indicate a higher ability to perform activities of daily living, increased more in the diet-exercise group than in the diet group. Given the potential for diet-induced weight loss therapy to exacerbate sarcopenia in older obese patients, Frimel et al. tested whether adding multimodal exercises to a diet program could attenuate declines in muscle mass and physical function and found that the reduction in lean mass in both the upper and lower extremities were significantly less in the exercise + diet group than in the diet group (Frimel et al. 2008).

6.8 Additional Benefits of Rehabilitation for Obese Patients

In addition to directly improving body composition and physical function, regular exercise can favorably alter many of the physiological pathways that contribute to degenerative disease and sarcopenia in the obese, older adult (Fig. 6.1). Rehabilitation programs have the capacity to offset diabetes pathways by increasing skeletal muscle insulin sensitivity, controlling blood glucose, and reducing glycated hemoglobin levels (Balducci et al. 2010). Favorable shifts in blood pressure and lipoprotein levels also occur with regular exercise. Several independent studies have shown that combinations of diet and exercise can reduce both systemic

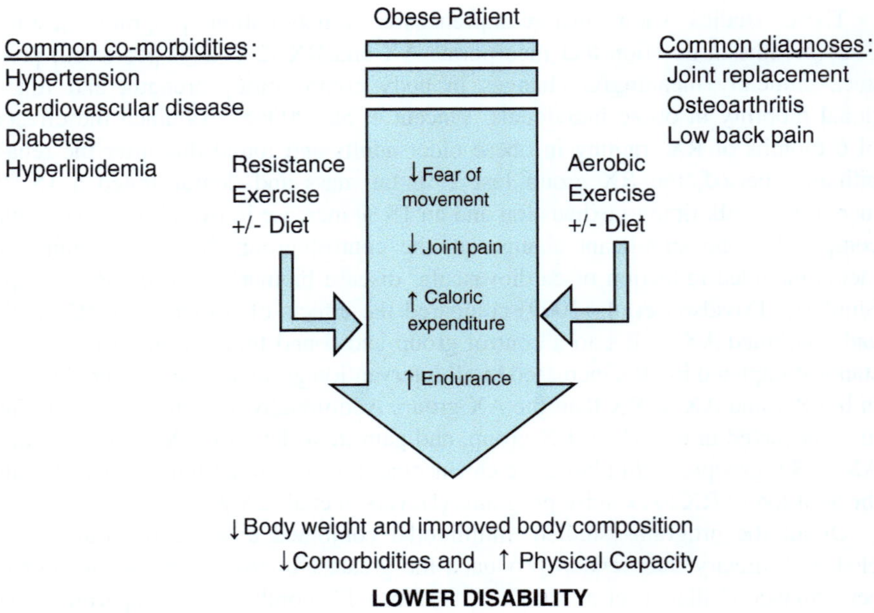

Fig. 6.1 Comprehensive benefits of outpatient rehabilitation in obese patients

inflammation and oxidative stress, both of which have been implicated in obesity-related metabolic diseases such as diabetes and cardiovascular disease. One study combining 12 weeks of low glycemic index diet with AX showed a significant decrease in cytokine production by monocytes (Kelly et al. 2011).

Perhaps of greater clinical importance is the ability of exercise to treat degenerative joint disease itself while reducing disability in the obese. The concurrent use of AX (for caloric expenditure and joint health) and RX (for strengthening the musculature supporting the joints and promoting cartilage health) is a balanced program that may combat joint diseases. Exercise can be applied to this population at any disease stage to help strengthen muscles that surround the arthritic joint, and help control or reduce body weight, the latter being a main modifiable factor underlying OA (Suri et al. 2012). Importantly, regular exercise training can reduce joint pain by as much as 30 % (Vincent et al. 2006c): in a recent study on older obese adults with knee osteoarthritis, pain severity, and functional outcomes such as walking performance and stair climb demonstrate the greatest improvements after exercise + diet interventions compared to exercise or diet alone. Additionally, in a similar study comparing exercise (AX + RX), diet, and exercise + diet against a Control group in patients with knee osteoarthritis, Focht et al. showed that exercise + diet induced the greatest improvements in pain, stair climb time, and 6-min walk distance compared with the other three groups (Messier et al. 2004). Exercise interventions can be a cost effective adjuvant to traditional knee OA therapies in obese adults.

6.9 Obesity-Related Considerations in Outpatient Rehabilitation

In the outpatient setting, there are unique challenges faced by the obese patient. Weight stigma related to body weight and embarrassment of being unable to get into exercise equipment can also deter the obese from participating in outpatient rehabilitation programs. Additionally, obese individuals may, over time, develop motor patterns secondary to chronic pain. For example, alterations to gait (e.g., center of mass excursions, toe out angle, and step length or width) could add to musculoskeletal complications and prolong the rehabilitation process (Cimolin et al. 2011). Our laboratory has also shown that fear of movement due to pain (kinesiophobia) is higher in morbidly obese patients who are seeking physical therapy programs for knee (Vincent et al. 2010a) or back pain (Vincent et al. 2010b). Kinesiophobia ratings correspond with higher perceived disability of tasks such as running, jumping, stair climb, and chair rise in persons with BMI values >40 kg/m^2. Interestingly, kinesiophobia ratings were elevated in persons even when no functional impairments in range of motion/strength existed, suggesting that overcoming fear is important in helping obese individuals achieve their potential in rehabilitation programs. To help overcome fear of movement due to pain, exercise may need to be supervised initially and periodically thereafter to help ensure that activity is performed at the appropriate training stimulus and not compromised because of fear.

A final consideration for outpatient rehabilitation programs may hinge on the palatability of continuing with rehabilitation exercise over the long term. In our experience, we find that if exercise programs are focused on primary symptoms (e.g., for knee or low back pain) and time efficient, the likelihood that patients will keep exercising when formal rehabilitation is complete is high. Interventions that can reduce joint pain and permit patients to increase daily physical activity may be good incentive to continue the rehabilitation exercises. Over time, daily activity becomes enjoyable and caloric expenditure increases; these contribute to modification of cardiovascular disease factors (blood pressure, weight, and blood lipids) and improve mental well-being. Hence, if as a field we can help obese patients to overcome these limitations and engage in diet and multimodal exercise programs, we can break the vicious cycle of pain, fear of movement/activity avoidance, weight gain, and functional disability.

References

Anton SD, Manini TM, Milsom VA et al (2011) Effects of a weight loss plus exercise program on physical function in overweight, older women: a randomized controlled trial. Clin Interv Aging 6:141–149

Avila JJ, Gutierres JA, Sheehy ME, Lofgren IE, Delmonico MJ (2010) Effect of moderate intensity resistance training during weight loss on body composition and physical performance in overweight older adults. Eur J Appl Physiol 109(3):517–525

Balducci S, Zanuso S, Nicolucci A et al (2010) Effect of an intensive exercise intervention strategy on modifiable cardiovascular risk factors in subjects with type 2 diabetes mellitus: a randomized controlled trial: the Italian diabetes and exercise study (IDES). Arch Intern Med 170(20):1794–1803

Bouchard DR, Soucy L, Sénéchal M, Dionne IJ, Brochu M (2009) Impact of resistance training with or without caloric restriction on physical capacity in obese older women. Menopause 16(1):66–72

Bouchard DR, Soucy L, Sénéchal M, Dionne IJ, Brochu M (2010) Changes in objective and self-reported measures of physical capacity after an intervention in obese older women. J Women Aging 22(1):34–36

Buchner DM, de Lateur BJ (1991) The importance of skeletal muscle strength to physical function in older adults. Ann Behav Med 13:91–98

Capodaglio P, Capodaglio EM, Facioli M, Saibene F (2006) Long-term strength training for community-dwelling people over 75: impact on muscle function, functional ability and life style. Eur J Appl Physiol. doi:10.1007/s00421-006-0195-8

Capodaglio P, Castelnuovo G, Brunani A, Vismara L, Villa V, Capodaglio EM (2010) Functional limitations and occupational issues in obesity. Int J Occup Saf Ergon 16:507–523

Cimolin V, Vismara L, Galli M, Zaina F, Negrini S, Capodaglio P (2011) Effects of obesity and chronic low back pain on gait. J Neuroeng Rehabil 26(8):55

Davidson LE, Hudson R, Kilpatrick K et al (2009) Effects of exercise modality on insulin resistance and functional limitation in older adults: a randomized controlled trial. Arch Intern Med 169(2):122–131

Davison KK, Ford ES, Cogswell ME, Dietz WH (2002) Percentage of body fat and body mass index are associated with mobility limitations in people aged 70 and older from NHANES III. J Am Geriatr Soc 50(11):1802–1809

Donini LM, Brunani A, Sirtori A et al (2010) Assessing disability in morbidly obese individuals: the Italian society of obesity test for obesity-related disabilities. Disabil Rehabil 1–10 (Early Online). doi:10.3109/09638288.2011.575529

Duvigneaud N, Matton L, Wijndaele K et al (2008) Relationship of obesity with physical activity, aerobic fitness and muscle strength in flemish adults. J Sports Med Phys Fitness 48(2):201–210

Farrell RT, Gameli RL, Aleem RF, Sinacore DR (2008) The relationship of body mass index and functional outcomes in patients with acute burns. J Burn Care Res 29:102–108

Frimel TN, Sinacore DR, Villareal DT (2008) Exercise attenuates the weight-loss-induced reduction in muscle mass in frail obese older adults. Med Sci Sports Exerc 40(7):1213–1219

Grant S, Todd K, Aitchison TC, Kelly P, Stoddart D (2004) The effects of a 12-week group exercise programme on physiological and psychological variables and function in overweight women. Public Health 118(1):31–42

Hamilton BB, Laughlin JA, Fiedler RC, Granger CV (1994) Interrater reliability of the 7-level functional independence measure (FIM). Scand J Rehabil Med 26(3):115–119

Heo M, Pietrobelli A, Wang D, Heymsfield SB, Faith MS (2010) Obesity and functional impairment: influence of comorbidity, joint pain, and mental health. Obesity (Silver Spring) 18(10):2030–2038

Jensen GL, Hsiao PY (2010) Obesity in older adults: relationship to functional limitation. Curr Opin Clin Nutr Metab Care 13:46–51

Kelly KR, Haus JM, Solomon TP et al (2011) A low-glycemic index diet and exercise intervention reduces TNF(alpha) in isolated mononuclear cells of older, obese adults. J Nutr 141(6):1089–1094

Lebrun CE, van der Schouw YT, de Jong FH, Grobbee DE, Lamberts SW (2006) Fat mass rather than muscle strength is the major determinant of physical function and disability in postmenopausal women younger than 75 years of age. Menopause 13(3):474–481

Lim JY, Tchai E, Jang SN (2010) Effectiveness of aquatic exercise for obese patients with knee osteoarthritis: a randomized controlled trial. PMR 2(8):723–731

Macfarlane GJ, de Silva V, Jones GT (2011) The relationship between body mass index across the life course and knee pain in adulthood: results from the 1958 birth cohort study. Rheumatology (Oxford) 50(12):2251–2256

Manini TM, Newman AB, Fielding R et al (2010) Effects of exercise on mobility in obese and nonobese older adults. Obesity (Silver Spring) 18(6):1168–1175

Mathias S, Nayak USL, Isaacs B (1986) Balance in elderly patients: the 'Get up and go' test. Arch Phys Med Rehabil 67:387–389

Menegoni F, Galli M, Tacchini E, Vismara L, Cavigioli M, Capodaglio P (2009) Gender-specific effect of obesity on balance. Obesity (Silver Spring) 16(2):232–239

Messier SP, Loeser RF, Miller GD et al (2004) Exercise and dietary weight loss in overweight and obese older adults with knee osteoarthritis: the arthritis, diet, and activity promotion trial. Arthr Rheumatol 50(5):1501–1510

Miller GD, Nicklas BJ, Davis C, Loeser RF, Lenchik L, Messier SP (2006) Intensive weight loss program improves physical function in older obese adults with knee osteoarthritis. Obesity (Silver Spring) 14(7):1219–1230

Ministero della Salute. Quaderni, del, Ministero, della. Marzo-Aprile 2011 Roma: Istituto Poligrafico e Zecca dello Stato;vol 8

Padwal RS, Wang X, Sharma AM, Dyer D (2012) The impact of severe obesity on post-acute rehabilitation efficiency, length of stay, and hospital costs. J Obes 2012:972365(Epub 2012 Feb 22)

Peytremann-Bridevaux I, Santos-Eggimann B (2008) Health correlates of overweight and obesity in adults aged 50 years and over: results from the survey of health, ageing and retirement in europe (SHARE). Obesity and health in europeans aged > or = 50 years. Swiss Med Wkly 138(17-18):261–266

Podeszwa DA, Stanko KJ, Mooney JFr, Cramer KE, Mendelow MJ (2006) An analysis of the functional health of obese children and adolescents utilizing the PODC instrument. J Pedatric Orthop 26(1):140–143

Precilios H, Brunani A, Cimolin V, Tacchini E, Donini LM, Fabris DeSouza S, Capodaglio P (2012) Measuring changes after multidisciplinary rehabilitation of obese individuals. J Endocrinol Invest (Jan 31 2012). [Epub ahead of print]

Ramsay SE, Whincup PH, Shaper AG, Wannamethee SG (2006) The relations of body composition and adiposity measures to ill health and physical disability in elderly men. Am J Epidemiol 164(5):459–469

Riddiford-Harland DL, Steele JR, Baur LA (2006) Upper and lower limb functionality: are these compromised in obese children? Int J Pediatr Obes 1(1):42–49

Ropka ME (2002) Symptom status and functional status outcomes: humanistic outcomes in obesity disease management. Obes Res 10(suppl 1):42S–49S

Schlenk EA, Lias JL, Sereika SM, Dunbar-Jacob J, Kwoh CK (2011) Improving physical activity and function in overweight and obese older adults with osteoarthritis of the knee: a feasibility study. Rehabil Nurs 36(1):32–42

Sjolie AN (2004) Low-back pain in adolescents is associated with poor hip mobility and high body mass index. ScandJ Med Sci Sports 14(3):168–175

Stenholm S, Harris TB, Rantanen T, Visser M, Kritchevsky SB, Ferrucci L (2008) Sarcopenic obesity: definition, cause and consequences. Curr Opin Clin Nutr Metab Care 11(6):693–700

Sternfeld B, Ngo L, Satariano WA, Tager IB (2002) Associations of body composition with physical performance and self-reported functional limitation in elderly men and women. Am J Epidemiol 156(2):110–121

Stickles B, Phillips L, Brox WT, Owens B, Lanzer WL (2001) Defining the relationship between obesity and total joint arthroplasty. Obes R` 9(3):219–223

Suri P, Morgenroth DC, Hunter DJ (2012) Epidemiology of osteoarthritis and associated comorbidities. PMR 4(5 Suppl):S10–S19

Taylor ED, Theim KR, Mirch MC et al (2006) Orthopedic complications of overweight in children and adolescents. Pediatrics 117(6):2167–2174

Thomas DR (2007) Loss of skeletal muscle mass in aging: examining the relationship of starvation, sarcopenia and cachexia. Clin Nutr 26(4):389–399
Villareal DT, Banks M, Sinacore DR, Siener C, Klein S (2006) Effect of weight loss and exercise on frailty in obese older adults. Arch Intern Med 17(2):172–173
Villareal DT, Chode S, Parimi N et al (2011) Weight loss, exercise, or both and physical function in obese older adults. N Engl J Med 364(13):1218–1229
Vincent HK, Vincent KR (2008) Obesity and inpatient rehabilitation outcomes following knee arthroplasty: a multicenter study. Obesity (Silver Spring) 16(1):130–136
Vincent KR, Vincent HK, Lee LW, Alfano AP (2006a) Inpatient rehabilitation outcomes in primary and revision total knee arthroplasty patients. Clin Orthop Rehabil 446:201–207
Vincent HK, Alfano AP, Lee L, Vincent KR (2006b) Sex and age effects on outcomes of total hip arthroplasty after inpatient rehabilitation. Arch Phys Med Rehabil 84(4):461–467
Vincent HK, Bourguignon C, Vincent KR (2006c) Resistance training lowers exercise-induced oxidative stress and homocysteine levels in overweight and obese older adults. Obesity (Silver Spring) 14(11):1921–1930
Vincent HK, Dejong G, Mascarenhas D, Vincent KR (2009) The effect of body mass index and hip abductor brace use on inpatient rehabilitation outcomes after total hip arthroplasty. Am J Phys Med Rehabil 88:201–209
Vincent HK, Lamb KM, Day TI, Tillman SM, Vincent KR, George SZ (2010a) Morbid obesity is associated with fear of movement and lower quality of life in patients with knee pain-related diagnoses. PMR 2(8):713–722
Vincent HK, Omli MR, Day TI, Hodges M, Vincent KR, George S (2010b) Fear of movement, quality of life and self-reported disability in obese patients with chronic lumbar pain. Pain Med 12(1):154–164
Vincent HK, Seay AN, Vincent KR, Atchison JW, Sadasivan K (2012a) Effects of obesity on rehabilitation outcomes after orthopedic trauma. Am J Phys Med Rehabil (Jun 3 29, Epub ahead of print)
Vincent HK, Raiser SN, Vincent KR (2012b) The aging musculoskeletal system and obesity-related considerations with exercise. Ageing Res Rev (in press)
Visser M, Langlois J, Guralnik JM et al (1998a) High body fatness, but not low fat-free mass, predicts disability in older men and women: the Cardiovascular Health Study. Am J Clin Nutr 68(3):584–590
Visser M, Harris TB, Langlois J et al (1998b) Body fat and skeletal muscle mass in relation to physical disability in very old men and women of the Framingham Heart Study. J Gerontol A Biol Sci Med Sci 53(3):M214–M221
Ware JJ, Kosinski M, Keller SD (1996) A 12-Item short-form health survey: construction of scales and preliminary tests of reliability and validity. Med Care 34(3):220–233
Wilson AC, Samuelson B, Palermo TM (2010) Obesity in children and adolescents with chronic pain: associations with pain and activity limitations. Clin J Pain 26(8):705–711
World health organization (2008) The global burden of disease: 2004 update. World Health Organization, Geneva
Xu X, Mirka GA, Hsiang SM (2008) The effects of obesity on lifting performance. Appl Ergon 39:93–98

Chapter 7
Cardiac Rehabilitation

Luca A. Gondoni

Abstract The combination of both sedentary lifestyle and unrestricted food availability increases the prevalence of obesity and type 2 diabetes and therefore the risk of cardiovascular morbidity and mortality due to coronary artery disease. A cardiomyopathy of obesity where the myocardial abnormalities cannot be ascribed to comorbidities is well recognized and its presence is supported both by experimental models and by epidemiological and clinical studies: left ventricular hypertrophy, reduced diastolic and systolic function, chronotropic incompetence are the typical features of the obesity cardiomyopathy. Moreover obese subjects have a high prevalence of diabetes, hypertension, atrial fibrillation, and sleep apnea. Cardiac rehabilitation programs include diet, physical exercise, and behavioral counseling as a strategy to fight obesity and its consequences and can possibly reverse the anomalies in cardiovascular function. The results are encouraging and obesity cardiomyopathy can partially be reversed by means of rehabilitation programs. Unfortunately, the obese often regain weight even after an early success. The adoption of an active lifestyle can have a favorable impact on prognosis in obese patients independently of weight loss. In this chapter, the modifications that occur in cardiac structure and function of obese subjects, the approach to cardiac rehabilitation, and the main results of weight loss and exercise are reviewed.

7.1 Introduction

Obesity is strongly related to heart disease and is part of a metabolic imbalance associated with an unfavorable cardiovascular risk profile. The prevalence of diabetes, arterial hypertension, and dyslipidemia is high in obese subjects who are therefore at high risk for coronary artery disease (Bogers et al. 2007). Moreover,

L. A. Gondoni (✉)
Istituto Auxologico Italiano, IRCCS-Ospedale San Giuseppe Cardiac Rehabilitation,
via Cadorna, 90, 28824 Piancavallo di Oggebbio, Verbania, Italy
e-mail: gondoni@auxologico.it

P. Capodaglio et al. (eds.), *Disabling Obesity*, DOI: 10.1007/978-3-642-35972-9_7,

a specific obesity cardiomyopathy has been described, which is a condition determined by a complex hemodynamic derangement: increased circulating blood volume causes ventricular overload and therefore ventricular remodeling. Endocrine abnormalities such as insulin resistance, increased activity of the renin–angiotensin–aldosterone system which in turn activates the sympathetic nervous system all give a relevant contribution to the development of the obesity cardiomyopathy; last a direct pathogenic effect of adiposity exists which acts through endothelial dysfunction, vasculopathy, and the proinflammatory status that derives from adiposopathy (i.e., adipose tissue dysfunction) that characterizes obesity (Bays 2011). The eventual picture is characterized by the presence of left ventricular hypertrophy, diastolic and systolic dysfunction, altered autonomic balance, altered heart rate response to exercise, and a high prevalence of atrial fibrillation.

Cardiac rehabilitation is an active multifactorial process aimed at favoring clinical stability, reducing disability, maintaining an active role in society, reducing cardiovascular events, improving quality of life, and survival. In recent years the concept of rehabilitation has evolved from a disease-centered process to a person-centered one. The patient is no longer regarded as a person who is sick but as a person who has rights. Even if the evidence supporting the reversibility of the obesity cardiomyopathy when a sustained weight loss is obtained is sufficiently clear, conventional cardiac rehabilitation programs are only marginally interested to this issue and do not describe specific interventions. Our program tries to tackle the problems of patients who add up obesity to their heart problems.

7.2 Obesity and Heart Disease

The first description of "adipositas cordis" has been published in the early nineteenth century (Cheyne 1818). Since then, much has been discovered on this topic and I will very briefly review the main features of this multifaceted relationship.

The activation of the renin-angiotensin-aldosterone and sympathetic nervous system contributes, together with sodium and volume retention, to the rise in blood pressure and exerts direct trophic effects on cardiac myocytes resulting in a progressive impairment of left ventricular filling and, subsequently, diastolic and systolic dysfunction that can eventually progress to heart failure. Insulin resistance has been linked to the development of early myocardial abnormalities in obese individuals, it promotes myocardial fatty acid uptake and can increase the levels of circulating angiotensinogen. Of note, angiotensin II is a potent growth factor for cardiac myocytes and favors hypertrophy and fibrosis. Myocardial dysfunction further activates the renin-angiotensin-aldosterone system and therefore sympathetic nervous system introducing a vicious cycle. Obese have increased levels of adipokines, free fatty acids, and inflammatory mediators which cause adipocyte dysfunction and induce adverse effects in the heart and vascular bed (de Ferranti and Mozarrafian 2008). Leptin resistance also characterizes the obese state and contributes to the increased sympathetic nervous system activity: it is associated with the development of hypertension and mediates the effects of growth hormone, inducing cardiac hypertrophy (Sweeney 2010). Lipotoxicity, that is typical to the obesity condition, is the process by which the excess of fatty acids and triglycerides in cardiac myocytes is metabolized in non-oxidative pathways that disrupt normal cellular signaling leading to organ dysfunction and in some cases apoptosis and eventual myocardial dysfunction (Wende and Abel 2010). Obesity is also one of the conditions associated with increased reactive oxygen species, i.e., chemically reactive molecules containing oxygen which are a natural by-product of the normal metabolism of oxygen but can increase markedly during environmental stress, causing significant damage to cell structures. They cause a decrease in the bioavailability of nitric oxide which is probably the common pathogenic mechanism of endothelial dysfunction (Capellini et al. 2010). Obese individuals finally have impaired microcirculatory responses to stress which might be related to the increased propensity of obese individuals to develop hypertension (Agapitov et al. 2002) and be an important pathophysiologic link to heart failure (Bagi et al. 2012).

Ventricular remodeling and myocardial abnormalities associated to altered hemodynamics and loading conditions were the first factors identified in the pathogenesis of obesity cardiomyopathy. The high metabolic activity of adipose tissue and the increased lean mass of obese subjects lead to a circulatory state characterized by increases both in preload and afterload, which predispose individuals to left ventricular remodeling and dilatation, increase wall stress and therefore promote compensative ventricular hypertrophy. The augmented wall stress causes a concomitant increase in myocardial oxygen consumption, with eventual left ventricular dysfunction. Myocyte contractile dysfunction activates cardiac adrenergic drive, which initially compensates for, but ultimately contributes to remodeling and progressive contractile dysfunction. This peculiar hemodynamic situation has been described by Alpert who proposed a model for the pathophysiology of the obesity cardiomyopathy (Alpert 2001) (Fig. 7.1).

Heart failure with preserved ejection fraction is frequently associated with hypertension and obesity: it is currently considered as a combination of diastolic dysfunction and subclinical systolic dysfunction and has a poor prognosis which is only slightly better than that of the classical systolic heart failure syndrome. The physio-pathological picture may be regarded as a good model to describe the obesity cardiomyopathy: its main features include abnormal diastolic filling, which is easily documented by Doppler techniques and tissue Doppler imaging, left ventricular hypertrophy or concentric remodeling, myocardial fibrosis, increased left ventricular stiffness which is the result of the combined effects of the passive elastic properties of the ventricle, the energy dependent active relaxation properties, and the atrial contribution to ventricular filling. This derangement increases left ventricular filling pressure and subjects patients to a marked dependence on preload and afterload: indeed even small ventricular volume increments cause a significant increase in filling pressure while an increase in blood pressure causes an afterload mismatch with a severe diastolic dysfunction that might even determine acute pulmonary edema (Zarich et al. 1991; Peterson et al. 2004; Wong et al. 2004; Wong and Marwick 2007; Lavie et al. 2007).

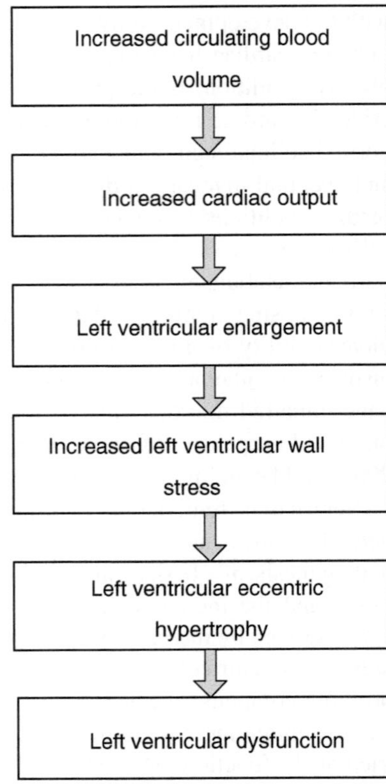

Fig. 7.1 Model for the pathophysiology of the obesity cardiomyopathy (modified from Ref. Alpert 2001)

The whole picture of obesity cardiomyopathy includes also alterations of the right ventricular structure and function (Wong et al. 2006). This is partly due to the functional dependence of the ventricles but also to the high prevalence of obstructive sleep apnea that is a condition characterized by repetitive interruption of ventilation during sleep, caused by the collapse of the pharyngeal airway. Obstructive apneas are very common in the obese and induce intermittent hypoxemia and hypercapnia disrupting the autonomic and hemodynamic responses to sleep, causing a heightened sympathetic drive, and acting also as an important mechanism for triggering systemic inflammation and endothelial dysfunction. During apneas the patient makes a forced inspiration against a closed upper airway thus generating a negative pressure in the chest cavity, which impairs atrial and ventricular filling. Almost every type of arrhythmia may be present in patients with sleep apnea: in particular ventricular ectopic beats, ventricular tachycardia, sinus arrests are prevalent (Selim et al. 2010). The same is true for atrial fibrillation which is another condition with a high prevalence in the obese population (Smith et al. 2010) and can be detrimental in the hemodynamics of patients who already have diastolic dysfunction. The excess risk for atrial fibrillation associated with obesity, besides the presence of apneas, is related to left atrial dilatation which is very frequent in the obese (Wang et al. 2004).

A peculiar response to exercise in obese patients has been described which is mediated by the altered autonomic balance: parasympathetic function is decreased while sympathetic nervous tone is augmented in obesity (Esler et al. 2001). Obese have however a reduced response to sympathetic stimuli, regardless of basal sympathetic tone. We studied heart rate behavior in a group of obese patients and we found that heart rate response to exercise is markedly abnormal in the obese (Gondoni et al. 2009): resting heart rate is similar in normal weight, overweight and obese individuals, but heart rate at peak exercise, chronotropic index (i.e. the ability to use chronotropic reserve) and heart rate recovery (i.e., the decrease in heart rate immediately after effort cessation) are all lower in obese subjects. Once again a parallel can be found with the subjects with heart failure and preserved ejection fraction who have a marked chronotropic incompetence which contributes to the reduced exercise capacity which characterizes this condition (Borlaug et al. 2006). Obese patients also have a blunted response during exercise in cardiac index and in systemic vascular resistance index: blood pressure always increases during physical effort, but has a more marked surge in obese than in lean subjects (Fornitano and Godoy 2010). In a group of healthy subjects, the increase in both systolic and diastolic blood pressure was higher in the obese group as opposed to controls (Burger et al 2009). Finally cardiac output, albeit being higher at rest, has a lower increase in obese during exercise as opposed to lean counterparts (Alauddin et al. 1990).

The reduced exercise tolerance of obese can therefore be explained by a set of cardiovascular variables, that include a greater cardiac workload at any level of physical activity (which is determined by the excess of body weight), a blunted heart rate response to exercise, a reduced vasodilatation, an excessive increase in blood pressure, a subclinical diastolic and systolic dysfunction with an elevated filling pressure that further increases during exercise, and a lower cardiac output that doesn't adequately increase during effort.

7.3 Cardiac Rehabilitation in the Obese

Cardiac rehabilitation traditionally has paid little attention to disability. Coronary artery disease is the first cause of death in the developed world but the issue of the degree of disability related to this condition has been addressed only by a few studies: nevertheless it is estimated that coronary artery disease is the fourth cause of disability accounting for 4.1 % of disability adjusted life years (WHO 2008). In patients with coronary artery disease, aged 63 $\pm$ 11 years, significant limitations in household and work activities are documented using the WHO-DAS II scores and the same is true in several domains of the ICF model (Racca et al. 2010).

If we look at the indications for an intensive residential rehabilitation program listed in the Italian guidelines (Istituto Superiore di Sanità 2008) we find a series of cardiac conditions such as recent cardiac surgery, recent myocardial infarction, recent coronary angioplasty, heart failure (NYHA class III-IV), and, what is particularly relevant to this chapter, patients with high or intermediate risk, even without an index event when a rehabilitation program can prevent or slow the progression of the disease: no mention at all of disability. As a consequence, quite often rehabilitation outcomes include mortality, need for revascularization interventions, dyspnea, and other clinical relevant parameters and they overlook the disability that the disease causes.

Even more importantly, when we are dealing with obese patients we have to consider that obesity per se is a relevant cause of disability (Backholer et al. 2012; DeCaria et al. 2012): the probability that an elderly cohort of patients has to develop disability has a U-shaped relationship with BMI, being lower in the overweight subset of patients and higher in the obese group and in the lowest BMI group (<18.5 kg/m^2). What is interesting is that weight cycling and weight variability (particularly weight loss) increase both mortality and the risk of limitations in activities of daily living and mobility (Arnold et al. 2010).

In a classical cardiac rehabilitation program the main components include management of lipid levels, management of hypertension, cessation of smoking, weight reduction, management of diabetes, psychosocial management, physical-activity counseling, and exercise training (Ades 2001).

In obese subjects, current guidelines indicate that weight loss is the main goal of rehabilitation programs and it has to be pursued by behavioral modification, diet, and exercise (Ades et al. 2010): the results of such behavioral interventions are often disappointing, but we have to consider that the available drugs are poorly effective and bariatric surgery, although potentially useful, is expensive and, by definition, very much invasive and is therefore reserved to the more difficult situations.

If we are to review the present scenario in cardiac rehabilitation of the obese we can identify some problems which affect the process: the percentage of patients who accesses the programs is low even after an acute event and a treatment gap exists between recommended therapies and the care that patients actually receive. Moreover, one of the critical issues that we have to consider is a methodological one that hinders rehabilitation research: we might have to strive for a paradigm shift from randomized controlled trials to observational studies. There is no doubt that randomized trials have been one of the key instruments in medical research for several decades, but, when we are dealing with behavioral interventions, the design of randomized controlled trial poses some problems: do we really have the possibility to identify a placebo group in such settings? For instance, if my aim is to measure the efficacy of physical training on cardiovascular risk factors in obese subjects how can I randomize a group to nonintervention? When I visit a patient and tell him/her that he/she will be recruited in a clinical trial that studies the effects of physical activity (and I have to do that, since the patient has to consent to data collection and use) how can I tell him/her not to exercise since he/she has been allocated to the nonintervention group? Definitely, if my intervention is a supervised one I will have a much higher compliance in the intervention group, but I am not comparing a treatment to a placebo. I can compare different methods of exercise, but, even if this could be an interesting point, the true problem is to have a high adherence to behavioral prescriptions and not to compare different diet and/ or exercise models. Such difficulties can at least partially explain the bad results of a recent trial (West et al. 2012). The need to make a careful selection of patients in randomized studies poses critical problems when we try to extend the results from the study population to the real world: for instance, the Multiple Risk Factor Intervention Trial (MRFIT) actually enrolled 12,679 patients, while the authors have screened over 300,000 subjects (MRFIT 1976). Are we sure that this subset is truly representative of the whole population? One of the first randomized trials on behavioral interventions for obesity demonstrated a significant improvement in weight and cardiovascular risk in obese subjects who underwent a one-year intensive lifestyle intervention consisting of diet and physical activity (Goodpaster et al. 2010). Nearly 600 subjects were considered eligible, but only 130 (21.8 %) were actually randomized to the study protocol: the reasons that stand behind the exclusion are various, the main issue being represented by 178 candidates (a number which is bigger than the actual study population) who were excluded because they "did not attend" or were "not interested": I am strongly convinced that these are the patient that we are presently unable to treat efficaciously. We have also to consider that obesity is a cause of social disparities and an unfavorable social environment is a potential contributor to the development of obesity (Clarke et al. 2010; Dugravot et al. 2010). It is probable that the subjects who are more predisposed to become obese are the same who are less inclined to attend a trial.

A non-randomized method gives the opportunity of studying variables in the setting of clinical intervention and has a more widespread and significant enforceability in clinical practice (McKee et al. 1999): therefore a possible way to overcome the difficulties that randomized trials created is to use observational data and to identify the patients that respond poorly to the existing programs, including subjects who are lost on their way: this will help us in designing more tailored plans or even in deciding that peculiar categories of patients are not suitable for a rehabilitation program. We have done such a study which describes the results of our program in a population of obese subjects with chronic ischemic heart disease (Gondoni et al. 2008): we found that, after a 23-day intervention, mean BMI decreased from 38.0 ± 4.9 to 36.7 ± 4.8 kg/m^2 and attained METs at a treadmill test increased from 6.2 ± 2.5 METs to 7.3 ± 2.7 ($p < 0.001$ for both), but, after a median interval of 358 days, BMI have increased by 1.0 ± 2.4 kg/m^2 and only 21 % of patients continued to lose weight. I think that the more interesting feature of our study has been the identification of the weakest subsets of population who are represented by the elderly, the diabetics, and the women: they do need a different protocol from what is currently used. Moreover, and this might be very relevant to the definition of better rehabilitation programs, bad results, both in weight loss and in exercise tolerance, were obtained by the less educated people, even if the program was strictly supervised: we do not exactly know why this category of patients did worse than the more educated ones, but this raises profound issues in both social and clinical aspects.

A word on drug prescription and treatment optimization is worth. This is a relevant point that can be addressed during a rehabilitation program: the issue is not negligible since quite often guidelines are not fully respected and patients are treated suboptimally. For instance, in 1,172 consecutive patients affected with obesity and coronary artery disease who underwent a cardiac rehabilitation program at our institution, the percentage of subjects taking a statin at admission was only 75 %. Moreover, the choice of treatment can be made, whenever possible, according to the effect on weight that some drugs have: just as an example, treatment of type 2 diabetes is preferably achieved with metformin that favors weight loss rather than with sulfonylureas that tend to cause weight increase. This topic has been recently reviewed (Bray and Ryan 2012): the authors also addressed the issue of the pharmacological treatment of non-complicated obesity.

I would like to concentrate on two other relevant aspects of cardiac rehabilitation in the obese: patient-centered care and physical training.

7.3.1 Patient-Centered Care

Patient-centered care is one of the goals of medicine in the third millennium. It applies to all specialties, but rehabilitation is a setting in which this concept can be more vigorously implemented because times are longer than the ones of acute medicine and therefore more suitable for the physician to cope with patients' necessities. Recently a statement has been published by the American College of Cardiology Foundation (American College of Cardiology Foundation 2012) which reviews the topic in detail and I will only briefly summarize the key points.

Clinician–patient communication, a must in every field of medicine, should be patient-centered to include the patient perspective and the psychosocial context along with shared understanding and responsibility. On the contrary, quite often clinicians do not allow patients to express their concerns during clinical interviews and redirect them toward the issue that the clinician himself deems interesting. Beyond time constraints and doctors' resistance, this idea brings forward a revolution in the patient/clinician relation shifting the pivotal point from the doctor toward the patient.

Health literacy improvement is one of the goals of rehabilitation programs that generally offer the patients the opportunity to participate in meetings where clinically meaningful problems are explained and discussed together with nurses or clinicians. Even highly educated adults may be challenged by processing, understanding, and using health-related information to make appropriate decisions about their health. Low health literacy can result in decreased adherence to medical recommendations, failure to engage in healthy behaviors, and inferior outcomes. Indeed a high percentage of the patients who participate in our program has a low literacy level: only 6 % of patients have a university degree, 19 % have a high school diploma and over 35 % of subjects has no education or has obtained only a 5 year elementary school qualification. All these patients can hopefully improve awareness of their health problems.

Ideally, within the patient-centric model of care, the life experiences and health attitudes of patients must be incorporated into communication and decision making: when clinicians are oblivious to individual patient needs, the chance that quality care or outcome improvements ensue is low, while the well informed patient is more likely to be actively engaged in his or her health management, including behavioral modifications that remain the cornerstone of the treatment of obesity.

Within this perspective, quality of life should be one of the standard measure which has to be performed in a rehabilitation program. Cardiovascular diseases manifest themselves through subjective symptoms that must be assessed explicitly. Traditional techniques include physicians' interpretation of health status with the Canadian Cardiovascular Society classification for angina and the New York Heart Association classification for heart failure which have been demonstrated to have poor inter-rater reliability. Disease-specific, patient-reported outcome measures have been developed that enable patients to report their health status in valid, reproducible, and sensitive metrics shifting the perspective once again from the doctor's to the patient's one (Spertus 2008).

As a consequence, the decision making process moves toward a shared plane thus empowering patients who no longer delegate physicians, but, ideally, become protagonist in their health management: they should be encouraged to accept responsibility for managing their health condition and work collaboratively with their healthcare team. Anyway a problem still exists not only for doctors, but also among patients who seem to have a long way to go: "doctors should get down from their pedestals, but patients must get up from their knees" R. Johnstone of the International Association of Patients Organization says.

7.3.2 Physical Training

Already in ancient Greece regular physical activity was considered an essential part of a healthy behavior. After a long time span during which physical fitness sank into oblivion, in the last 50 years a reappraisal took place and a large amount of evidence on the importance of physical activity has accumulated: the effects of regular aerobic training on cardiovascular risk and general wellbeing are clearly demonstrated and the same is true for resistance training that is recommended as a part of a comprehensive exercise program since it reduces inflammatory biomarkers in obese subjects and prevents sarcopenia (Ahmadi et al. 2011). In a study conducted in patients with heart failure and preserved ejection fraction and mean BMI around 30 kg/m^2, after a 4-month training period the study subjects improved their performance secondary to an improvement in peak heart rate, chronotropic reserve, and particularly in arterial-venous oxygen difference (Haykowsky et al. 2012). The beneficial effects of regular exercise comprise a reduction of blood pressure, a better glucose uptake by muscle cells, an increase in HDL cholesterol, and weight loss. A good review of the topic has been published by Kokkinos and Myers (2010). Exercise is safe and the risk of a cardiovascular event is low after both high-intensity exercise and moderate-intensity exercise in a cardiovascular rehabilitation setting (Rognmo et al. 2012).

Exercise prescription is one of the challenging issue when dealing with obese patients who are affected with a variety of comorbidities that make it difficult to design a tailored program. Intensity, duration, and type of exercise are the main characteristics on which we have to decide. As a preliminary remark, both intensity and duration are positively related to a reduction in incident coronary heart disease in healthy subjects (Tanasescu et al. 2002). Current guidelines recommend intensity range between 3 and 6 METs on most days of the week. The caloric expenditure should be at least 1000 kcal/week and this goal can be achieved by walking 30 min every day or walking 60 min every second day with an almost identical effect on survival and quality of life. It should be noted that paradoxically the obese are forced to exercise more than lean subjects: work is indeed defined as the product of distance and mass (the mass of an obese subject is obviously greater than the one of a lean person) and obese do a surplus of work only because their weight is bigger. Also cycling is a good exercise for obese subjects: it has several advantages over walking, i.e., the load is much lower since the patients is sitting and, if done on a cycle-ergometer, it does not force the patient to go out and that may increase compliance. Nevertheless walking has zero expenses and favors social interaction since the patients must necessarily get out to walk. Attending a gym program can be useful for improving muscular mass, articular mobility, body's posture, and motor schemata.

For a more precise definition of intensity, which may be a valuable point in heart patients, we can use two standard approaches: the subjective one, which is generally based upon fatigue perception and a more objective one which relies on heart rate during exercise. In our experience the Borg scale, which is widely used to grade the rate of perceived exertion has low reliability, at least in obese, being influenced by the education of the patient even more than by the actual attained level of effort (Gondoni et al. 2010). Ideally the program should be targeted to a direct measure on oxygen consumption, however this is not always feasible and a variety of methods have been proponed which use heart rate as a measure for exercise intensity. We think that target heart rate has to be based on two parameters, i.e., baseline heart rate and peak heart rate during a symptom limited treadmill stress test. This is the consequence of the abnormal behavior of heart rate in obese patients which has been discussed earlier and drug treatment that very often in coronary artery disease and/or heart failure includes beta-blockers (or other drugs) that reduce both resting and peak effort heart rate. We have utilized the formula Target HR $=$ Resting HR $+$ (Peak HR $-$ Resting HR) $\times$ 0.7 i.e., resting heart rate $+$ 70 % of the chronotropic reserve: that level of effort is generally associated with a "comfortable fatigue" sensation.

As a general rule, as we were noting before, exercise should include both aerobic and resistance training. However the debate on the type of exercise that carries the best results in obesity rehabilitation is still open: possibly an evolutionary approach can enlighten the picture. Human beings of the third millennium are genetically almost identical to their very far ancestors. The homo genus appeared on earth some 2,400,000 years ago. Only in the last 10,000 years human beings began to change the environment in which they lived instead to merely adapt them to it, beginning in the Middle East where the first transition from the hunter/gatherer model to the agriculture and breeding model took place.

Only very recently the industrial and computer revolutions changed dramatically the environment in which we live. 10,000 years are way too short a period to modify significantly our genetics. Therefore, the exercise we are programmed for is the same that hunters/gatherers did for hundreds of centuries (O'Keefe et al. 2011). In Paraguay, where several years ago the anthropologist Kim Hill had the opportunity to study the Ache people who still live as hunter-gatherers, the typical activity consisted of hunting almost every day (excepting when it rained, which generally happened twice a week): that means walking for 10 km on an uneven ground, with bursts of high intensity activity (i.e., running and chasing the quarry for 1–2 km). After a very hard day, they tried to have a quiet day, but the word "quiet" needs to be more precisely defined: the quiet day entailed house-keeping, gathering of vegetables and wood, butchery of captured animals, social activities (that included dancing) etc. Calories expenditure was therefore very high and food intake not always took place on a regular basis: indeed obesity was virtually absent in the past. With almost the same genetic configuration, we live in an environment where physical activity is no longer a necessity and the opportunity for exercise has to be actively sought; on the contrary food availability is superabundant and the consequence of this ominous combination is the dramatic prevalence of obesity.

A good activity program should therefore entail both aerobic training, small bursts of anaerobic high intensity activity, and strength exercise. In our institution patients, after attending a meeting during which the type of exercise are described and the beneficial effects on health are depicted, undergo an adapted physical activity program. Individualized aerobic activity consists of cycle ergometer, recline ergometer, or arm ergometer (the choice depending on orthopedic problems): on the first day the patient exercises until fatigue or dyspnea ensue. From the second day on, exercise duration is prolonged gradually until the patient reaches 30 min without interruption. The same increasing curve is adopted for intensity that increases progressively until the target heart rate is reached. The aim of such a duration and intensity of exercise is to optimize fat oxidation (Achten and Jeukendrup 2004). For patients who do not have a treadmill test and therefore do not have a precise indication for heart rate the choice is to exercise at a heart rate around 60–70 % of the maximum age-predicted value. Heart rate is monitored with pulse oximeter or ECG telemetry depending on patient's characteristics. When the patient is on beta-blocker therapy and therefore heart rate is no longer a useful tool to define exercise intensity, if a treadmill test is not available, we check the respiratory rate and do the talk test, which consists simply in asking the patient if he/she can talk comfortably while exercising. There are evidences that asking whether comfortable speech is possible results in exercise intensities that are within well-accepted guidelines for exercise prescription even without the necessity for preliminary exercise testing. Moreover the results are very similar in different modes of exercise (Persinger et al. 2004). For outdoor exercise the time increases progressively from 30 to 60 min: the intensity is generally measured trough the talk test since we do not have the possibility to use telemetry for ECG monitoring.

The purpose of aerobic exercise is multifaceted: to improve exercise tolerance through metabolic, cardiovascular, respiratory, and muscular effects, to improve energy expenditure, particularly through fat oxidation, to improve mood, socialization, and self-esteem.

Patients also undergo a strength exercise program that is individualized considering patient's characteristics. They work for 45–60 min on a daily basis at low intensity and the main purposes of this activity are to increase muscle strength, to balance muscle tone and mass, to reach a good equilibrium between flexor and extensor muscles, to improve joint stability and mobility, to improve motor coordination, to improve body and motor schemata, and to relax both physically and psychologically.

7.3.3 Results of Rehabilitation Programs

Overall cardiac rehabilitation has clinically relevant results in the obese, even if the improvement is generally less than that of the lean counterpart. A problem can possibly emerge since it is difficult to separate the effects of training and dieting because both interventions are almost always both included in programs. There is one study that compares the different effects of dieting and exercising (Villareal et al. 2011) in which the patients were assigned to one of four groups: a control group, a diet only group, a group that received only exercise training, and a group that received both weight management instruction and physical training. As it was easily foreseeable, the authors concluded that weight loss obtained with a combination of low calorie diet and exercise provides greater improvement in physical function than either intervention alone.

Considering this premise, that was widely accepted as a postulate even before its demonstration in a clinical trial, a wide amount of data supports the benefit of a combination of physical exercise and a low calorie diet.

Weight loss by any means (dieting or bariatric surgery) results in favorable hemodynamic changes and reverses heart remodeling left ventricular mass and chamber size decrease, ejection fraction slightly improves after a 12-week program (Schrauwen-Hinderling et al. 2010). In subjects with uncomplicated obesity an improvement of a tissue Doppler index of systolic function has been demonstrated (Gondoni et al. 2007; Brunani et al. 2008). Vagal function, and therefore heart rate variability, recovers after weight loss. Also heart rate behavior, baroreflex sensitivity, and vagal tone improve after sustained weight loss (Brinkworth et al. 2006; Alvarez et al. 2005; Poirier et al. 2003). Heart rate recovery is slow in obese subjects, but it improves after exercise (Kim et al. 2009). Indeed virtually every index of cardiac function and marker of cardiovascular risk improves after diet and physical training.

We know that both diet and physical activity are important but, at least for patients, weight loss is definitely the first goal of their efforts and sometimes they have an overoptimistic anticipation of the results.

7.3.4 The Obesity Paradox

Contrary to common sense, it should be kept in mind that already a 5–10 % decrease is associated with clinically meaningful improvements and, if we were able to obtain a sustained weight loss, we could reduce or even correct all comorbidities such as impaired glucose tolerance and dyslipidemia and improve the overall cardiovascular risk (Mourot et al. 2010).

Nevertheless a word of caution on weight loss might be safe. Obesity has a complex relationship with all-cause mortality: the ideal BMI is around 23 kg/m^2 (Berrington de Gonzalez et al. 2010; Whitlock et al. 2009) and we can say with certainty that obesity is associated with increased death rates for many cause-specific outcomes. However we know from many good quality studies that, in patients with congestive heart failure, low BMI is a risk factor for mortality and the presence of obesity is associated with lower mortality when compared to normally weighting patients (Fonarow et al. 2007). These findings are also applicable to patients with other cardiovascular diseases, in particular in patients with coronary artery disease. Furthermore, this obesity paradox has been reported in patients with noncardiac diseases such as chronic lung or kidney disease.

What make those data disturbing is the need to reconcile them with what is known about young and healthy people, in whom obesity is a risk factor for adverse events. Even if the data are strong, there are several observation that I would like to put forward. Obesity is possibly only a marker and not an independent factor of the protective effect: indeed obese patients with heart failure are younger, have more skeletal muscle mass, and have lower levels of natriuretic peptides. Moreover the possibility exists that some of these patients are erroneously labeled as heart failure patients and their symptoms are actually secondary to obesity that can cause dyspnea. Obese individuals have more fat free mass (i.e., more muscle mass) than patients who are not obese, and that in itself is beneficial.

What about weight loss in this category of patients? We must say there is no evidence that weight loss in heart failure results in an improvement in prognosis. Indeed probably no physician would advise weight loss for an obese patient (BMI 30–35 kg/m^2) with AIDS or a malignant cancer with metastases. So why are we suggesting obese patients with heart failure that they must lose weight? On the contrary there are evidences showing that weight loss in heart failure is associated with a poor outcome (Pocock et al. 2008). However, those considerations possibly do not apply to the morbid obese (i.e., BMI > 40 kg/m^2) subset in whom data are scarce. Finally, one should also consider that our data concern mortality and not severity of symptoms and consequent quality of life. In a patient with obesity, if the focus of the treatment is on mitigating symptoms, weight loss may be of great value.

The attention that patients, and unfortunately doctors as well, have put on weight loss could be misleading: weight loss can even worsen the prognosis because it focuses on food restriction, thus favoring weight regain, eating disorders and obesity stigmatization.

7.3.5 A New Goal

Cardiorespiratory fitness is a strong predictor of cardiovascular and all-cause mortality (Kodama et al. 2009) and is the result of the interaction of several risk factors. Age and sex are the main determinants, but BMI (particularly percent body fat) and leisure time physical activity are key factors in explaining exercise capacity (Gondoni et al. 2006; Lakoski et al. 2011). A tentative cutoff at 7.9 METs seems to separate patients with high or low rates of events. Obese subjects have a poor event-free life expectancy: in a consecutive sample of 3,728 obese patients who participated in our program, 732 were not able to perform a treadmill exercise test; of the remaining 2,996 the vast majority reached a peak exercise level of less than 7.9 METs and only 620 patients (21 %) exceeded that threshold.

A significant improvement in quality of life, as well as in life expectancy, might be achieved independently of weight change only because of regular exercise (Byberg et al. 2009). Cardiac rehabilitation causes a significant improvement in the cardiovascular risk profile at all levels of BMI, independently of weight loss. In a population-based study, interesting although not specifically dealing with obese patients, high intensity physical activity produced a 32 % reduction in mortality and even a moderate level of physical activity was enough to cause a 22 % reduction in mortality: those are big numbers. This issue whether it is weight or fitness that makes the difference has been addressed in an interesting study by Lee and coworkers (Lee et al. 1999): they studied over 14,000 men and found that, for each 1-MET improvement, a 19 % decrease in cardiovascular mortality was present, regardless of BMI or percent body fat change. Obese subjects who exercise regularly have a lower risk for metabolic abnormalities and their consequences than sedentary obese and represent the so called 'fat-but-fit' phenotype. In an analysis of the NHANES population (Wildman et al. 2008) more than 30 % of the obese were free of risk factor clustering: they were the more active fraction of the population. Apparently the category of metabolically healthy obese is constituted by the insulin sensitive subjects who are identified by the absence of fatty liver disease (Stefan et al. 2008) which is another condition that can be favorably influenced by physical activity.

Considering that obesity is a chronic disease and therefore a complete recovery is not foreseeable, we can try to adopt the same concepts that we use when we treat other chronic conditions such as, for example, hypertension or diabetes, and set the 'fat-but-fit' phenotype as a new possible target for obese subjects. Within this perspective exercise plays the most relevant role and could be considered as the main treatment of obesity, allowing a large number of obese patients to improve their health status even without the achievement of ideal body weight or of a significant weight loss. We can say that an obese subject who is fit and exercises regularly has a lower cardiovascular mortality than a normally weighing individual who is unfit and sedentary (Sui et al. 2007). Therefore our target, even if most of the features that characterize the obesity cardiomyopathy can be reversed by means of weight loss, should mainly focus on the adoption of an active life style which has a great impact on the prognosis of obese patients.

References

Achten J, Jeukendrup AE (2004) Optimizing fat oxidation through exercise and diet. Nutrition 20:716–727

Ades PA (2001) Cardiac rehabilitation and secondary prevention of coronary heart disease. New Engl J Med 345:892–902

Ades PA, Savage PD, Harvey-Berino J (2010) The treatment of obesity in cardiac rehabilitation. J Cardiopulm Rehabil Prev 30:289–298

Agapitov AV, Correia ML, Sinkey CA, Dopp JM, Haynes WG (2002) Impaired skeletal muscle and skin microcirculatory function in human obesity. J Hypertens 20:1401–1405

Ahmadi N, Eshaghian S, Huizenga R, Sosnin K, Ebrahimi R, Siegel R (2011) Effects of intense exercise and moderate caloric restriction on cardiovascular risk factors and inflammation. Am J Med 124:978–982

Alauddin A, Meterissian S, Lisbona R, MacLean LD, Forse RA (1990) Assessment of cardiac function in patients who were morbidly obese. Surgery 108:809–818

Alpert MA (2001) Obesity cardiomyopathy: pathophysiology and evolution of the clinical syndrome. Am J Med Sci 321:225–236

Alvarez GE, Davy BM, Ballard TP, Beske SD, Davy KP (2005) Weight loss increases cardiovagal baroreflex function in obese young and older men. Am J Physiol 289:E665–E669

American College of Cardiology Foundation (2012) Health policy statement on patient-centered care in cardiovascular medicine: a report of the American College of Cardiology Foundation Clinical Quality Committee. J Am Coll Cardiol 59:2125–2143

Arnold AM, Newman AB, Cushman M, Ding J, Kritchevsky S (2010) Body weight dynamics and their association with physical function and mortality in older adults: the cardiovascular health study. J Gerontol A Biol Sci Med Sci 65A:63–70

Backholer K, Wong E, Freak-Poli R, Walls HL, Peeters A (2012) Increasing body weight and risk of limitations in activities of daily living: a systematic review and meta-analysis. Obes Rev 13:456–468

Bagi Z, Feher A, Cassuto J (2012) Microvascular responsiveness in obesity—implications for therapeutic intervention. Br J Pharmacol 165:544–560

Bays HE (2011) Adiposopathy is "sick fat" a cardiovascular disease? J Am Coll Cardiol 57:2461–2473

Berrington de Gonzalez A, Hartge P, Cerhan JR, Flint, Hannan L, MacInnis RJ, Moore SC, Tobias GS, Anton-Culver H, Freeman LB, Beeson WL, Clipp CL, English ER, Folsom AR, Freedman DM, Giles G, Hakansson N, Henderson KD, Hoffman-Bolton J, Hoppin JA, Koenig KL, Lee IM, Linet MS, Park Y, Pocobelli G, Schatzkin A, Sesso HS, Weiderpass E, Willcox BJ, Wolk A, Zeleniuch-Jacquotte A, Willett WC, Thun MJ (2010) Body-mass index and mortality among 1.46 million white adults. N Engl J Med 363: 2211–2219

Bogers RP, Bemelmans WJ, Hoogenveen RT, Boshuizen HC, Woodward M, Knekt P, van Dam RM, Hu FB, Visscher TL, Menotti A, Thorpe RJ Jr, Jamrozik K, Calling S, Strand BH, Shipley MJ for the BMI-CHD Collaboration Investigators (2007) Association of overweight with increased risk of coronary heart disease partly independent of blood pressure and cholesterol levels: a meta-analysis of 21 cohort studies including more than 300,000 persons. Arch Intern Med 167:1720–1728

Borlaug BA, Melenovsky V, Russel SD, Kessler K, Pacak K, Becker LC, Kass DA (2006) Impaired chronotropic and vasodilator reserves limit exercise capacity in patients with heart failure and preserved ejection fraction. Circulation 114:2138–2147

Bray GA, Ryan DH (2012) Medical therapy for the patient with obesity. Circulation 125:1695–1703

Brinkworth GD, Noakes M, Buckley JD, Clifton PM (2006) Weight loss improves heart rate recovery in overweight and obese men with features of the metabolic syndrome. Am Heart J 152(693):e1–e6

Brunani A, Liuzzi A, Titon A, Graci S, Castagna G, Viberti GC, Gondoni LA (2008) Evaluation of rosiglitazone administration on cardiovascular function in severe obesity. Clin Cardiol 31:602–606

Burger JPW, Serne EH, Nolte F, Sumlders EH (2009) Blood pressure response to moderate physical activity in increased in obesity. Neth J Med 67:342–346

Byberg L, Mealhus H, Gedeborg R, Sundström J, Ahlbom A, Zethelius B, Berglund LG, Wolk A, Michaëlsson K (2009) Total mortality after changes in leisure time physical activity in 50 years old men: 35 year follow-up population based cohort. Br J Sports Med 43:482

Capellini VK, Celotto AC, Baldo CF, Olivon VC, Viaro F, Rodrigues AJ, Evora PR (2010) Diabetes and vascular disease: basic concepts of nitric oxide physiology, endothelial dysfunction, oxidative stress and therapeutic possibilities. Curr Vasc Pharmacol 8:526–544

Cheyne J (1818) A case of apoplexy in which the fleshy part of the heart was converted into fat. Dublin Hosp Rep 2:216–223

Clarke PJ, O'Malley PM, Schulenberg JE, Johnston LD (2010) Midlife health and socioeconomic consequences of persistent overweight across early adulthood: findings from a national survey of American adults (1986–2008). Am J Epidemiol 172:540–548

DeCaria JE, Sharp C, Petrella JP (2012) Scoping review report: obesity in older adults. Int J Obes 36:1141–1150

de Ferranti S, Mozaffarian D (2008) The perfect storm: obesity, adipocyte dysfunction, and metabolic consequences. Clin Chem 54:945–955

Dugravot A, Sabia S, Stringhini S, Kivimaki M, Westerlund H, Vahtera J, Guéguen A, Zins M, Goldberg M, Nabi H, Singh-Manoux A (2010) Do socioeconomic factors shape weight and obesity trajectories over the transition from midlife to old age? Results from the French GAZEL cohort study. Am J Clin Nutr 92:16–23

Esler M, Rumantir M, Wiesner G, Kaye D, Hastings J, Lambert G (2001) Sympathetic nervous system and insulin resistance: from obesity to diabetes. Am J Hypertens 14:S304–S309

Fonarow GC, Srikanthan P, Costanzo MR, Cintron GB, Lopatin M, ADHERE Scientific Advisory Committee and Investigators (2007) An obesity paradox in acute heart failure: analysis of body mass index and inhospital mortality for 108,927 patients in the acute decompensated heart failure national registry. Am Heart J 153:74–81

Fornitano LD, Godoy MF (2010) Exercise testing in individuals with morbid obesity. Obes Surg 20:583–588

Gondoni LA, Liuzzi A, Titon AM, Taronna O, Nibbio F, Ferrari P, Leonetti G (2006) A simple tool to predict exercise capacity of obese patients with ischaemic heart disease. Heart 92:899–904

Gondoni LA, Titon AM, Silvestri G, Nibbio F, Taronna O, Ferrari P, Leonetti G (2007) Short term effects of physical exercise and low calorie diet on left ventricular function in obese subjects: a tissue doppler study. Nutr Metab Cardiovasc Dis 17:358–364

Gondoni LA, Titon AM, Nibbio F, Caetani G, Augello G, Mian O, Tuzzi C, Averna E, Parisio C, Liuzzi A (2008) Short term effects of a hypocaloric diet and a physical activity program on weight loss and exercise capacity in obese subjects with chronic ischemic heart disease: an everyday practice study. Acta Cardiol 63:153–159

Gondoni LA, Titon AM, Nibbio F, Augello G, Caetani G, Liuzzi A (2009) Heart rate behavior during exercise stress test in obese patients. Nutr Metab Cardiovasc Dis 19:170–176

Gondoni LA, Nibbio F, Caetani G, Augello G, Titon AM (2010) What are we measuring? Considerations on subjective ratings of perceived exertion in obese patients for exercise prescription in cardiac rehabilitation programs. Int J Cardiol 140:236–257

Goodpaster BH, DeLany JP, Otto AD, Kuller L, Vockley J, South-Paul JE, Thomas SB, Brown J, McTigue K, Hames KC, Lang W, Jakicic JM (2010) Effects of diet and physical activity interventions on weight loss and cardiometabolic risk factors in severely obese adults: a randomized trial. JAMA 304:1795–1802

Haykowsky MJ, Brubaker PH, Stewart KP, Morgan TM, Eggebeen J, Kitzman DW (2012) Effect of endurance training on the determinants of peak exercise oxygen consumption in elderly

patients with stable compensated heart failure and preserved ejection fraction. J Am Coll Cardiol 60:120–128
Istituto Superiore di Sanità (ISS) (2008) Cardiologia riabilitativa e prevenzione secondaria delle malattie cardiovascolari. Available at http://www.snlg-iss.it/lgn_cardiologia_riabilitativa
Kim MK, Tanaka K, Kim MJ, Matsuo T, Ajisaka R (2009) Exercise training-induced changes in heart rate recovery in obese men with metabolic syndrome. Metab Syndr Relat Disord 7:469–476
Kodama S, Saito K, Tanaka S, Maki M, Yachi Y, Asumi M, Sugawara A, Totsuka K, Shimano H, Ohashi Y, Yamada N, Sone H (2009) Cardiorespiratory fitness as a quantitative predictor of all-cause mortality and cardiovascular events in healthy men and women. J Am Med Assoc 301:2024–2035
Kokkinos P, Myers J (2010) Exercise and physical activity: clinical outcomes and applications. Circulation 122:1637–1648
Lakoski SG, Barolw CE, Farrell SW, Berry JD, Morrow JR, Haskell WL (2011) Impact of body mass index, physical activity, and other clinical factors in cardiorespiratory fitness (from the Cooper Center longitudinal study). Am J Cardiol 108:34–39
Lavie CJ, Milani RV, Ventura HO, Cardenas GA, Mehra MR, Messerli FH (2007) Disparate effects of left ventricular geometry and obesity on mortality in patients with preserved left ventricular ejection fraction. Am J Cardiol 100:1460–1464
Lee SL, Blair SN, Jackson AS (1999) Cardiorespiratory fitness, body composition, and cardiovascular disease mortality in men. Am J Clin Nutr 69:373–380
McKee M, Britton A, Black N, McPherson K, Sanderson C, Bain C (1999) Methods in health services research. Interpreting the evidence: choosing between randomised and non-randomised studies. Br Med J 319:312–315
Mourot L, Boussuges A, Maunier S, Chopra S, Rivière F, Debussche X, Blanc P (2010) Cardiovascular rehabilitation in patients with diabetes. J Cardiopulm Rehabil Prev 30:157–164
O'Keefe JH, Vogel R, Lavie CJ, Cordain K (2011) Exercise like a hunter-gatherer: a prescription for organic physical fitness. Progr Cardiovasc Dis 53:471–479
Persinger R, Foster C, Gibson M, Fater DCW, Porcari JP (2004) Consistency of the talk test for exercise prescription. Med Sci Sports Exerc 36:1632–1636
Peterson LR, Waggoner AD, Schechtman KB, Meyer T, Gropler RJ, Barzilai B, Dávila-Román VG (2004) Alterations in left ventricular structure and function in young healthy obese women: assessment by echocardiography and tissue Doppler imaging. J Am Coll Cardiol 43:1399–1404
Pocock SJ, McMurray JJ, Dobson J, Yusuf S, Granger CB, Michelson EL, Ostergren J, Pfeffer MA, Solomon SD, Anker SD, Swedberg KB (2008) Weight loss and mortality risk in patients with chronic heart failure in the candesartan in heart failure: assessment of reduction in mortality and morbidity (CHARM) programme. Eur Heart J 29:2641–2650
Poirier P, Hernandez TL, Weil KM, Shepard TJ, Eckel RH (2003) Impact of diet-induced weight loss on the cardiac autonomic nervous system in severe obesity. Obes Res 11:1040–1047
Racca V, Spezzaferri R, Modica M, Mazzini P, Jonsdottir J, De Maria R, Ferratini M (2010) Functioning and disability in ischaemic heart disease. Disabil Rehabil 32:S42–S49
Rognmo O, Moholdt T, Bakken H, Hole T, Mølstad P, Myhr NE, Grimsmo J, Wisløff U (2012) Cardiovascular risk of high- versus moderate-intensity aerobic exercise in coronary heart disease patients. Circulation 126:1436–1440
Schrauwen-Hinderling VB, Hesselink MK, Meex R, van der Made S, Schär M, Lamb H, Wildberger JE, Glatz J, Snoep G, Kooi ME, Schrauwen P (2010) Improved ejection fraction after exercise training in obesity is accompanied by reduced cardiac lipid content. J Clin Endocrinol Metab 95:1932–1938
Selim B, Won C, Yaggi HK (2010) Cardiovascular consequences of sleep apnea. Clin Chest Med 31:203–220

Smith JG, Platonov PG, Hedblad B, Engström G, Melander O (2010) Atrial fibrillation in the Malmö diet and cancer study: a study of occurrence, risk factors and diagnostic validity. Eur J Epidemiol 25:95–102

Spertus JA (2008) Evolving applications for patient-centered health status measures. Circulation 118:2103–2110

Stefan N, Kntaritzis K, Machann J, Schick F, Thamer C, Rittig K, Balletshofer B, Machicao F, Fritsche A, Häring H (2008) Identification and characterization of metabolically benign obesity in humans. Arch Intern Med 168:1609–1616

Sui X, LaMonte MJ, Laditka JN, Hardin JW, Chase N, Hooker SP, Blair SN (2007) Cardiorespiratory fitness and adiposity as mortality predictors in older adults. J Am Med Assoc 298:2507–2516

Sweeney G (2010) Cardiovascular effects of leptin. Nat Rev Cardiol 7:22–29

Tanasescu M, Leitzmann MF, Rimm EB, Willet WC, Stampfer MJ, Hu FC (2002) Exercise and intensity in relation to coronary heart disease in men. JAMA 288:1994–2000

The Multiple Risk Factor Intervention Trial (MRFIT) (1976) A national study of primary prevention of coronary heart disease. J Am Med Assoc 235:825–827

Villareal DT, Chode S, Parimi N, Sinacore DR, Hilton T, Armamento-Villareal R, Napoli N, Qualls C, Shah K (2011) Weight loss, exercise, or both and physical function in obese older adults. N Engl J Med 364:1218–1229

Wang TJ, Parise H, Levy D, D'Agostino RB Sr, Wolf PA, Vasan RS, Benjamin EJ (2004) Obesity and the risk of new-onset atrial fibrillation. J Am Med Assoc 292:2471–2477

Wende AR, Abel ED (2010) Lipotoxicity in the heart. Biochim Biophys Acta 1801:311–319

West RR, Jones DA, Henderson AH (2012) Rehabilitation after myocardial infarction trial (RAMIT): multi-centre randomised controlled trial of comprehensive cardiac rehabilitation in patients following acute myocardial infarction. Heart 98:637–644

Whitlock G, Lewington S, Sherliker P, Clarke R, Emberson J, Halsey J, Qizilbash N, Collins R, Peto R (2009) Prospective studies collaboration. Body-mass index and cause-specific mortality in 900,000 adults: collaborative analyses of 57 prospective studies. Lancet 373:1083–1096

WHO (2008). The global burden of disease: 2004 update. WHO, Geneva

Wildman RP, Muntner P, Reynolds K, McGinn AP, Rajpathak S, Wylie-Rosett J, Sowers MR (2008) The obese without cardiometabolic risk factor clustering and the normal weight with cardiometabolic risk factor clustering. Arch Intern Med 168:1617–1624

Wong C, O'Moore-Sullivan T, Leano R, Byrne N, Beller E, Marwick TH (2004) Alterations of left ventricular myocardial characteristics associated with obesity. Circulation 110:3081–3087

Wong C, O'Moore-Sullivan T, Leano R, Hukins C, Jenkins C, Marwick TH (2006) Association of subclinical right ventricular dysfunction with obesity. J Am Coll Cardiol 47:611–616

Wong C, Marwick TH (2007) Obesity cardiomyopathy: diagnosis and therapeutic implications. Nat Clin Pract Cardiovasc Med 4:480–490

Zarich SW, Kowalchuk GJ, McGuire MP, Benotti PN, Mascioli EA, Nesto RW (1991) Left ventricular filling abnormalities in asymptomatic morbid obesity. Am J Cardiol 68:377–381

Chapter 8
Respiratory Rehabilitation, Work Capacity, and Sleep-Related Disorders in Obesity

Alberto Salvadori, Paolo Fanari and Alberto Braghiroli

Abstract Obesity has increased in the recent decades. It is the product of a multifactorial etiology and behavioral and social/cultural factors. Obesity implies a reduction in work capacity and a lower nocturnal sustainable level of ventilation. These conditions generally determine functional disabilities. Diet and physical activity are the classical therapeutic references of obesity, together with positive airway pressure in case of sleep-related disorders. This chapter describes the advisable indications and opportunities in recommending work programs at different power outputs considering the analysis in obese subjects of their work capacity and cardiopulmonary adaptation. Moreover, it describes biochemical modifications on some accepted drivers of cardiac and respiratory functions in normal subjects, and their likely effectiveness in conditioning work capacity in obesity. Finally, it talks about the most severe disorders related to sleep like obesity hypoventilation syndrome and obstructive sleep apnea syndrome and their standard treatment.

8.1 Introduction

Obesity is a condition characterized by an increase of body mass which requires a greater metabolic energy exchange at rest and especially during physical exercise. Obese subjects are generally less active than normal subjects. They must overcome a functional impairment caused by a decrease in compliance of the respiratory system, with a reduced functional residual capacity (FRC) and expiratory reserve volume (ERV) (Naimark and Cherniack 1960). Moreover, well-established

A. Salvadori (✉) · P. Fanari
Respiratory Rehabilitation, S Giuseppe Hospital, Istituto Auxologico Italiano IRCCS, Piancavallo, Verbania, Italy
e-mail: salvadori@auxologico.it

A. Braghiroli
Respiratory Rehabilitation, Maugeri Foundation IRCCS, Veruno, NO, Italy
e-mail: alberto.braghiroli@fsm.it

P. Capodaglio et al. (eds.), *Disabling Obesity*, DOI: 10.1007/978-3-642-35972-9_8,

indices of obesity, such as the body mass index (BMI), correlate with cardiac left ventricular mass and wall thickness (Lauer et al. 1991) and abnormalities of left ventricular diastolic filling have been observed in asymptomatic pathologically obese subjects at rest (Zarich et al. 1991).

Sleep is a privileged window for the assessment of respiratory function in obese patients and the treatment of sleep-related respiratory disorders is a cornerstone of respiratory rehabilitation. During wakefulness several features can mask a latent mechanical impairment of the respiratory system, which becomes apparent when behavioral control is suppressed, clinostatism is mandatory, and muscular tone decreases. Sleep-related respiratory disorders occur more often in obese patients and usually the greater the BMI the higher the probability of a respiratory disorder (i.e., obstructive apneas, hypoventilation, desaturations caused by ventilation/perfusion mismatch).

8.2 Work Capacity in Obesity

In general, obese subjects have a decreased work capacity when compared to normals. Studied at a progressive cycle-ergometer test, young obese subjects 26–27 years old have a maximal sustainable work capacity similar to that of normals, but an anaerobic threshold (AT) at work rates significantly lower than those of normals. According to Wasserman (1984), this implies that endurance time at maximal effort is less, and therefore that work capacity is less in obesity. In older obese subjects there is a significant reduction both in AT and maximal sustainable work capacity, which may probably mean a heavier reduction in performance with aging compared to normals. Cycle-ergometer has the advantage that the work-output performed is known with accuracy and more reliable information is learned about cardiovascular function and gas exchange. On the other hand, the differences in constitution compared with normal subjects are probably weakened (Wasserman and Whipp 1975). Lafortuna e coll (2010) found that in obese adolescents cycling imposes a metabolic involvement at the level of the single active muscle greater than walking. In fact, walking is more convenient permitting to attain the same energy expenditure at lower heart rate with lower blood lactic acid concentrations and greater fat oxidation.

AT may be considered an effective gauge of physical fitness in health as well as in cardiorespiratory diseases (Matsumura et al. 1983). In obesity, it decreases in a linear correlation with increasing BMI and/or percent of ideal weight, while in normals it increases with increasing BMI and/or percent of ideal weight (Salvadori et al. 1991) (Fig. 8.1). More recently, Lakoski (2011) referred that BMI is strongly related to cardiorespiratory fitness so that the highest fitness is obtained in lowest BMI.

The term "mechanical efficiency" defines the energy used to perform a given amount of work and during steady-state exercise it may be conceptually compared with the "muscular contraction efficiency" which is based on thermodynamic

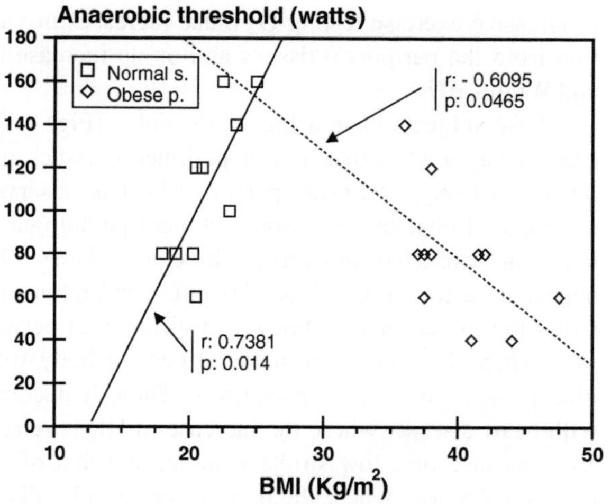

Fig. 8.1 Linear regression lines between anaerobic threshold (AT) and body mass index (BMI) in obese and normal subjects. By permission of Respiration (58; 311–315, 1991)

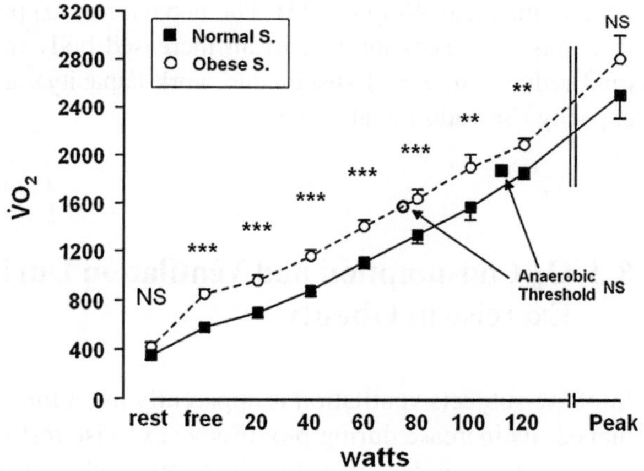

Fig. 8.2 Mean oxygen consumption (VO_2) during an incremental exercise testing in obese and normal subjects; ** $p < 0.01$, *** $p < 0.001$ at a two-tailed analysis of variance and Dunnet's method

considerations (Whipp and Wasserman 1969). Obese subjects use a greater amount of O_2 (VO_2) to accomplish the same external work-rate of normals. During a progressive physical exercise, the increase of VO_2 is the same in obese as well as in normal subjects; this indicates that the gross mechanical efficiency (watts/VO_2) is less in obesity, while the net mechanical efficiency (watts/VO_2-VO_2 free wheeling) is virtually identical between obese and lean subjects (Salvadori et al. 1992) (Fig. 8.2).

Heart Rate (HR) is higher in obese subjects at rest and up to AT, and significantly lower at maximal peak exercise when compared to controls. In normal subjects there are close relationships between VO_2 and HR and their quotient is the "O_2 pulse". It is the volume of O_2 extracted by the peripheral tissues, or the volume of O_2 added to the pulmonary blood per heart beat, and is equal to the product of stroke volume and arterial-mixed venous O_2 difference. During a progressive exercise testing O_2 pulse increases in view of an enhanced O_2 extraction from the peripheral tissues and on an increase in stroke volume (Wasserman and Whipp 1975).

Obese subjects have a higher O_2 pulse (Fig. 8.3) during exercise which indicates a larger O_2 extraction at peripheral tissues per heart beat. Generally, any increase of oxygen uptake per heart beat is observed in athletes, and it may be a beneficial change when some clinical pathologies (like chronotropic incompetence) are excluded, or during β-blockade (Zipes 1992). Therefore, at first it may appear hard to couple higher O_2 pulse and reduced work capacity. Here we can stress the importance of body size differences between obesity and normality. In fact, when O_2 pulse is normalized per Kg body weight, or Kg fat-free mass, in obesity it is lower than in controls. Then, it becomes similar to that of patients with heart disease, where the increase in HR is relatively steep for the increase of VO_2 because of a low stroke volume, and that of patients with chronic obstructive lung disease, where the pulmonary vascular disorder limits the rate of venous return to the left side of the heart, and consequently the left ventricular output (Wasserman and Whipp 1975). The behavior of O_2 pulse per Kg weight or Kg fat-free mass enhances the role of an increased body mass in causing a reduced AT (and reduced maximal sustainable work capacity), and therefore a reduced work capacity (Salvadori et al. 1999a, b).

8.3 O_2 Consumption and Ventilation During Exercise in Obesity

In obese subjects ventilation is apparently not a limiting factor to physical performance. Its increase during progressive exercise testing up to maximal sustainable work capacity is less than that in normal subjects, especially when considering ventilation in relation to Fat-Free Mass (FFM) (Fig. 8.4). Nevertheless, percutaneous oxygen saturation is constantly within normal ranges during the test up to exhaustion, probably due to an optimization of ventilation/perfusion ratio (Salvadori et al. 2008).Ventilation of obese subjects seems appropriate to remove the increased amount of CO_2 produced at every power output, as indicated by end-tidal pressure of CO_2 ($PETCO_2$), being quite similar to controls; on the other hand, the higher VO_2 of the obese subjects seems to be attained even though an increased peripheral tissues extraction, as indicated by a tendency for lower end-tidal pressure of O_2 ($PETO_2$). These data agree with the established strong correlation between VCO_2 and VE during exercise (Hardarson et al. 1998), but at every power output VE is lower than that required to fulfill their increased need for O_2.

A voluntary hyperpnea test in an open circuit while breathing CO_2 enriched air, reproducing ventilations at determined work outputs and maintaining $PETCO_2$ constant, demonstrates slight increases of VO_2 in young obese subjects compared

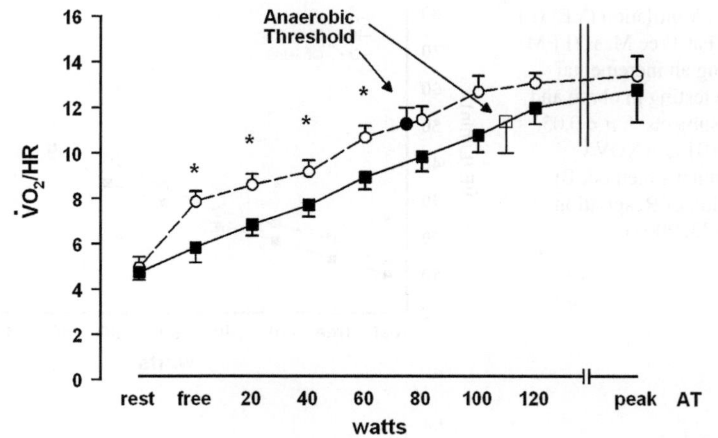

	rest	free	20	40	60	80	100	120	peak	AT
–■– Normal s.	4,74	5,83	6,84	7,72	8,95	9,84	10,79	11,99	12,77	11,26
SEM	0,34	0,67	0,48	0,52	0,56	0,64	0,75	0,72	1,40	1,01
–O– Obese p.	4,94	7,87	8,61	9,18	10,69	11,48	12,70	13,08	13,40	11,00
SEM	0,47	0,45	0,47	0,45	0,51	0,57	0,68	0,44	0,85	0,57

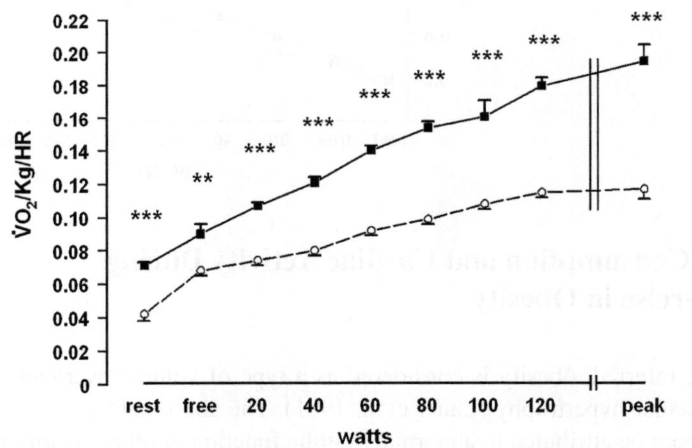

	rest	free	20	40	60	80	100	120	peak
–■– Normal s.	0,071	0,090	0,107	0,121	0,140	0,154	0,161	0,180	0,195
SEM	0,002	0,006	0,002	0,003	0,003	0,004	0,010	0,005	0,010
–O– Obese p.	0,042	0,068	0,074	0,080	0,092	0,099	0,108	0,115	0,117
SEM	0,004	0,003	0,003	0,003	0,003	0,003	0,003	0,003	0,006

Fig. 8.3 Mean O_2 pulse and ratio between O_2 pulse and kilogram body weight during an incremental exercise testing for obese subjects compared with normal subjects; * $p < 0.05$, ** $p < 0.01$, *** $p < 0.001$ at a two-tailed ANOVA with Dunnet's method. By permission of Respiration (66; 25–33, 1999)

to controls. This may indicate that only a small proportion of the increase of VO_2 in obese subjects can be attributed to the increased O_2 cost of breathing during a cycle-ergometer test (Salvadori et al. 1992).

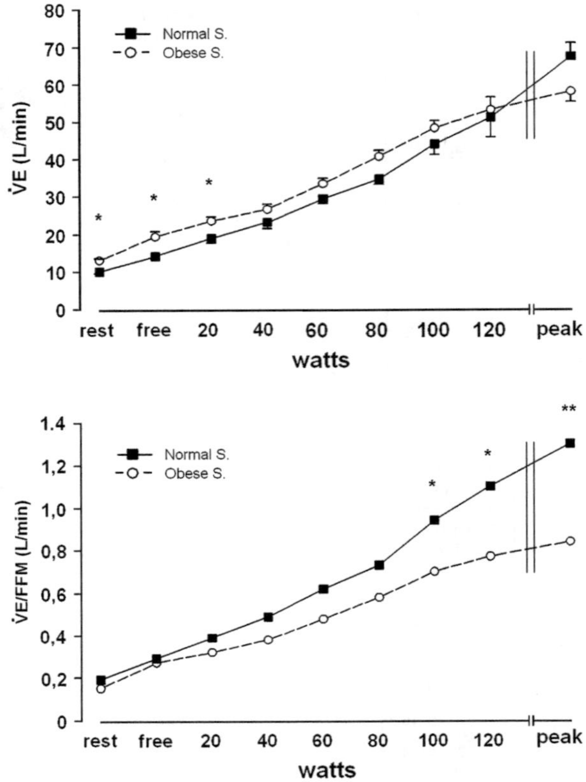

Fig. 8.4 Ventilation (VE) (**a**) and VE/Fat-Free Mass (FFM) (**b**) during an incremental exercise testing in obese and normal subjects; * $p < 0.05$, ** $p < 0.01$ at ANOVA with Dunnet's method. By permission of Respiration (75; 26–33, 2008)

8.4 O_2 Consumption and Cardiac Activity During Exercise in Obesity

As before referred, obesity is considered as a type of volume-overload state with left ventricular hypertrophy (Lauer et al. 1991). The abnormalities of left diastolic filling cannot be attributed to abnormal systolic function or other conditions known to impair diastolic filling and may antedate a contractile impairment, representing a sub-clinical form of cardiomyopathy in the obese subjects (Zarich et al. 1991).

In general, during exercise the increased transport and peripheral unloading of oxygen is assured by an increase of cardiac output (Q) together with a greater tissue extraction. The determinants of Q for a given submaximal VO_2 level are subject to large individual fluctuations, whereas the Q/VO_2 relationship appears to be the same for all individuals in most experimental conditions (Cerretelli and Di Prampero 1987). As shown in Fig. 8.5, in normal subjects an increase in oxygen consumption occurs together with an increase in arterial–venous difference. The same regression in obesity displays a significant lower slope. By the relation between Q and VO_2 we may argue that during a physical effort obese subjects have a reduced Q for the same VO_2 when compared to normal subjects, and

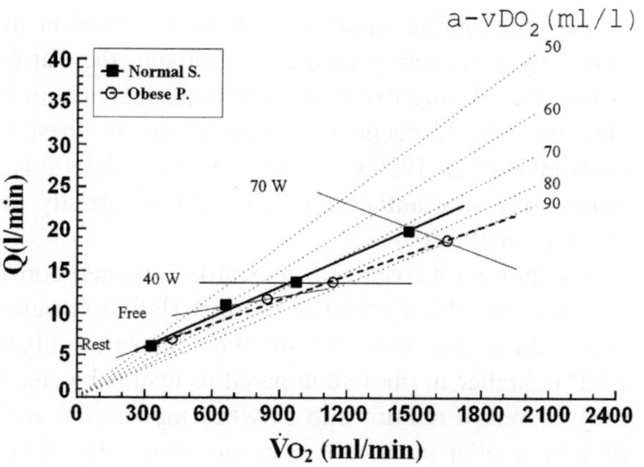

Fig. 8.5 Cardiac output (Q) versus oxygen consumption (VO_2) in normal and obese subjects with increasing workload. Iso—workload, and iso-arterial–venous oxygen difference lines are shown. By permission of Journal of Clinical and Basic Cardiology (2; 229–236, 1999)

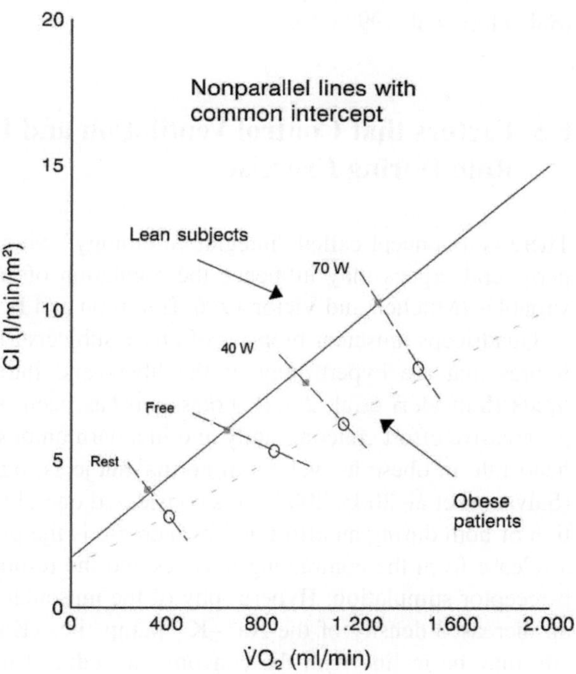

Fig. 8.6 Cardiac index (CI) versus oxygen consumption (VO_2) in normal and obese subjects with increasing workload. By permission of Respiration (66; 25–33, 1999)

they must rely on a greater arterial–venous difference (Salvadori et al. 1991a, b). In normal subjects an increase in oxygen consumption occurs with an increase in cardiac index (CI) which expresses the ratio Q/body surface. This index may minimize the difference in body sizes between obesity and normality, hence allowing a more correct comparison. Similar to the relation Q/VO_2, this regression in obesity displays a significant lower slope. In agreement with this sentence, the different slopes in the linear correlations of CI versus VO_2 would suggest a reduced cardiac performance while proceeding with heavier workloads in obesity (Fig. 8.6).

An unfavorable situation of muscle perfusion in obese subject can be appreciated by partitioning cardiac output into the Fat-Free Mass (FFM). Although it is true that during exercise there is an increase in blood flow to exercising muscles, this should occur to a lesser extent in obese subjects compared to controls (Salvadori et al. 1999a, b). All these considerations, together with those about O_2 pulse, may substantiate the lower AT of obesity and, therefore, the lower work capacity in obesity.

An indirect measure of myocardial oxygen consumption is given by the heart rate systolic blood pressure product (HRP) (Braunwald et al. 1976). This index expresses a less favorable situation in obese subjects; in fact for any workload HRP is higher in obese compared to lean subjects. HRP to total VO_2 ratio, which may express a relationship between myocardial and total body oxygen consumption, is smaller in the obese group, especially at the beginning of exercise going from rest to free pedaling that may mean a period of particular myocardial stress (Salvadori et al. 1999a, b).

8.5 Factors that Control Ventilation and Heart Rate During Exercise

There is a concept called "integrative biology" which states that modifications in peripheral organs may influence the regulation of cardiovascular and respiratory variables (Mitchell and Victor 1996; Hamilton and Booth 2000).

Quadriceps muscular biopsies of obese subjects are characterized by two major figures that are hypertrophy of the fibers and intracellular amounts of neutral lipids (Salvadori et al. 2004). Potassium has been observed to increase during a progressive effort concomitantly and in a harmonious manner with ventilation and heart rate in obese as well as in normal subjects, but significantly less in obesity (Salvadori et al. 2003, 2008). It is considered one of the major drives in the regulation of both during an effort and its increase is the product of the balance between a release from the contracting muscles and the re-uptake by muscular and kidney β-receptor stimulation. Hypertrophy of the muscular fibers has been associated to an increased density of the Na^+–K^+ pump sites (Knochel et al. 1985). This density may be in line with the concomitant reduced increase in plasma potassium, ventilation, and heart rate noted in obesity when compared with controls.

Intracellular amounts of neutral lipids are related to a metabolic condition of insulin resistance (Perseghin et al. 1999) and are accompanied by a reduced increase in lactic acid during a progressive effort in obesity compared to controls. The lower production in lactate anions, probably linked to insulin resistance, with the resultant lower decline in arterial pH during strenuous exercise is associated with a less ventilatory drive in VE above the lactate threshold.

In conclusion, the lower answers to the heavier work outputs of HR and VE may be linked to structural modifications of heart and chest, together with the behaviors of some of the most accepted driving factors that appear modified in agreement with structural modifications of peripheral muscles (hypertrophy of fibers) and metabolism (insulin resistance).

8.6 Prescription of Physical Activity in Obesity

Physical activity is difficult to assess in free-living populations and a gold standard measurement does not exist.

However, for a weight loss intervention to be successful a negative energy balance must be attained. In this view, energy expenditure in the form of physical activity is an important part in the management of obesity for losing weight and reducing associated risk factors for chronic disease. Increased physical activity has only small effects on weight loss in the short term.

Many formulas are recommended, principally in the form of "moderate intensity exercise" typically defined as 55–70 % of maximum heart rate 30–60 min a day, 5–7 days a week (Matus and Klaege 2007). It is equivalent, for example, to brisk walking at a pace of 3.5 miles per hour and is associated with an energy expenditure of 3 kcal/kg/h. Ratings of perceived exertion scores like Borg 15 category scale may be used in absence of access to heart rate monitoring systems. This degree of work is considered sufficient to achieve moderate levels of cardiorespiratory fitness which improve health and function, whether or not resulting in weight loss (LaMonte and Blair 2006). Less is known about advisable heavier intensities of exercise. With regard to this last topic, it has been observed that bouts of exercise beyond AT added to a program of aerobic activity significantly increase the release of GH to a greater extent than aerobic activity alone during physical activity (Salvadori et al. 2010). This may be of some interest considering that GH is known as one of the most active substances able to ameliorate the ratio Fat Mass/Fat-Free Mass. On the other hand, added bouts at heavy intensities promote a greater decrease in Fat Mass but do not seem to guarantee those metabolic improvements (like the reduction in insulin resistance) noted after a period of training at lower, continuous work outputs. Lately, the modality "repeated brief heavy efforts interrupted by brief recoveries" is growing in interest. In fact, recent observations about risk factors in people with metabolic syndrome have shown significant improvements after prolonged as well as after short periods of training in accordance with this model (Gibala et al. 2012).

8.7 Physiological Changes During Sleep and Sleep Disability in Obesity

Sleep requires a lower energy expenditure compared to wakefulness, therefore a reset of ventilatory regulation and respiratory muscle function occurs, so that minute ventilation decreases of about 1 l per minute, $PaCO_2$ can increase 2–6 mm Hg and thoracic volume can decrease accordingly (Phillipson 1978).

In normal subjects there is no relevant consequence, but in many obese patients the break point can be overcome. Nevertheless, two subjects with an apparently similar phenotype (same sex, age, BMI, fat distribution, hip/waist ratio) can be totally different during sleep, one with deep desaturations and sleep fragmentation, the other with almost no changes compared to wakefulness. The reason lies in the concept of "adequate compensation" (Younes 2003), which requires that an individual achieve a sustainable level of ventilation during sleep that is sufficient to prevent arousal (ventilation need not return to the eupneic level).

Arousal is the "emergency exit" of the respiratory drive when the $PaCO_2$ is higher than scheduled and the compensatory output to the pharyngeal and respiratory muscles does not achieve the desired level of ventilation (Berry and Gleeson 1997; Wellman et al. 2011), so that the resumption of muscular tone of wakefulness becomes mandatory. It will occur at a definite level identified as arousal threshold, which is individual specific and state specific.

Sleep is not a stable condition: light sleep, deep sleep, and REM sleep alternate overnight in cycles of about 90 min, each showing a specific regulation of respiratory drive, airway resistances, and muscular function. During REM sleep, particularly during eye movements (or phasic REM sleep), the muscular tone is almost abolished and the only muscles still active are those controlling oculomotion, some muscles of internal ear, and the diaphragm, which carries all the burden of ventilation (Tabachnik et al. 1981). Adequate compensation is therefore a dynamic concept, changing overnight, which can be maintained or overwhelmed in obese patients. The main reasons for inadequate compensation can be summarized as follows:

- *Increased work of breathing.* Morbid obesity, particularly central obesity, adds a load on the respiratory system proportional to the fat accumulation in the abdomen and around the thorax. FRC is reduced, mainly in supine position because of the rostral positioning of the diaphragm (Yap et al. 1995), with reduced lung compliance (Behazin et al. 2010).
- *Increased pharyngeal load.* The fat pads around the pharynx put an additional load on the pharyngeal dilator muscles, so that the upper airway in obese patients collapses at a more positive pharyngeal lumen pressure (Schwab et al. 2003, Wellman et al. 2011).
- *Ventilatory drive.* Usually obese patients have an increased ventilatory drive during wakefulness (Steier et al. 2009), but those who are prone to hypoventilation during sleep show an attenuated ventilatory response both to hypoxia and to hypercapnia (Piper 2010).

8.8 Obstructive Sleep Apnea and Obesity Hypoventilation Syndrome

Obstructive sleep apnea is a common disorder caused by a repetitive collapse of the pharynx during sleep, obesity and male sex being the most predictive factors (Young et al. 1993). Obesity leads to anatomically smaller airway (Isono et al. 1997), lower upper airways dilator muscle activity at sleep onset (Fogel et al. 2005), reduced end-expiratory lung volume with decrement of respiratory muscle activity both during inspiration and expiration, which potentially contribute to the increased propensity for UA collapse in OSA patients at sleep onset (Stadler et al. 2010).

In the literature it is not easy to discriminate OSA with hypercapnia and OHS, the latter being a condition of chronic hypercapnia (daytime $PaCO_2 > 45$ mm Hg) in an obese patient without a lung, cardiac, or chest disorder (i.e., normal respiratory function test or a mild restrictive pattern proportional to the degree of obesity) (Piper 2010).

The link between OSA and OHS is close, since about 90 % of patients with OHS have OSA (Mutlu and Rubinstein 2005), the other 10 % showing a pure hypoventilation. OSA in these patients causes a CO_2 loading at each respiratory event which is not removed completely in the inter-event ventilatory phase (Berger et al. 2002), leading to bicarbonate retention which further blunts the chemo responsiveness of the subject. Daytime hypercapnia and bicarbonate levels >27 mEq/L should be considered as the best homeostatic set-point possible, which anyway carries the burden of a higher risk of acute-on-chronic respiratory insufficiency in case of an acute exacerbation. In about 10 % of OHS patients there is no upper airway obstruction, but a constant hypoventilation with hypoxemia and hypercapnia which has been labeled as sleep hypoventilation (Mutlu and Rubinstein 2005). It is defined as an increase in $PaCO_2$ during sleep by 10 mm Hg above wakefulness or significant oxygen desaturation that is not explained by obstructive apneas or hypopneas.

8.9 Patients Assessment

It is important to recognize early the occurrence of a sleep respiratory disorder in these patients. OHS leads to higher mortality and morbidity (Budweiser et al. 2007; Priou et al. 2010), adding to the well-known cardiovascular morbidity and mortality of OSA (Marin et al. 2005). Sleep history usually unveils unrefreshing sleep, excessive daytime somnolence, witnessed apneas, snoring, repetitive awakenings to avoid, and sometimes choking episodes. This is not different from what a common OSA patient refers, but it is important to distinguish simple OSA and OHS since the diagnostic and treatment protocols can be different. Ambulatory cardiorespiratory monitoring and auto CPAP training is becoming a standard option in many countries for the management of OSA (Masa et al. 2011; Kuna et al. 2011; Braghiroli et al. 2001), but in patients with OHS this protocol has not been validated. OHS should be suspected when serum bicarbonate is increased (Mokhlesi et al. 2007), PaO_2 is <70 mm Hg (or SaO_2 <93 %) in the absence of concomitant diseases. Formal polysomnography including transcutaneous PCO_2 monitoring is the gold standard for the diagnosis, but a cardiorespiratory monitoring can offer an adequate alternative in expert hands.

8.10 Treatment with Positive Pressure in the Airways

An acute episode of respiratory insufficiency is often the first access of an OHS patient to the hospital and requires the immediate institution of a ventilatory support. Usually it is caused by the worsening of another comorbidity (i.e., decompensation of a chronic heart failure, acute respiratory infection) and can require an ICU setting even for noninvasive ventilation (NIV), because of the risk of failure of noninvasive technique and possible intubation if pH remains too low and for the high performance requested to the ventilator by a patient with high respiratory resistances and high respiratory rate (Powers 2008). However in the rehabilitation setting most of the patients are in stable conditions, with compensated respiratory acidosis. Therefore the treatment of sleep-related respiratory disorder can be instituted, together with a weight loss program and exercise training, in the ward or even at home. In Fig. 8.7 a flowchart of the possible progression of this approach is proposed based on the use of more complex and expensive devices when the simpler failed.

8.10.1 Continuous Positive Airway Pressure

CPAP is the gold standard for the treatment of obstructive sleep apnea and consists of a positive pressure applied to the upper airways through a nasal or oro-nasal mask, acting as a pneumatic splint which prevents the pharyngeal collapse during sleep (Montserrat et al. 2010). This is important in OHS patients because the

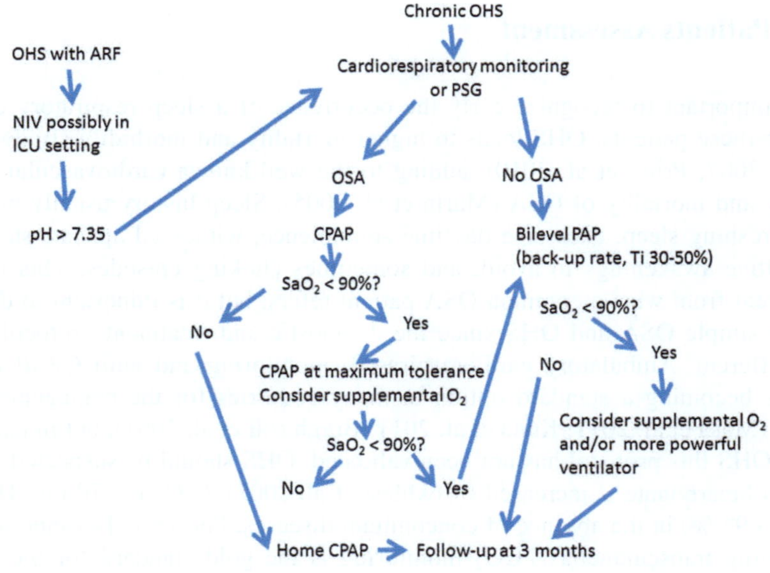

Fig. 8.7 Flowchart of ventilator approach to a patient with obesity hypoventilation syndrome

correction of obstructive respiratory events reduces the CO_2 load. When the pressure applied prevents small airway closure and increases FRC the benefits on gas exchange and the unloading of respiratory muscles further improve oxygenation and decrease the work of breathing.

In OHS patients the titration of CPAP should therefore not be performed with an auto CPAP in a home setting but in a sleep lab trying to obtain at least 90 % SaO_2. CPAP is very effective in improving symptoms, particularly hypersomnolence, awake blood gases, quality of life, and responsiveness to hypoxia and hypercapnia (Piper 2010). Nevertheless, it is not a ventilatory treatment and it requires weeks before the kidneys eliminate the excess of bicarbonate: a follow-up period of 1–3 months is advised before considering the treatment ineffective (Pérez de Llano et al. 2005).

8.10.2 Bilevel PAP

Again the literature is confusing on the application of bilevel PAP in OHS. The device delivers a pressure in inspiration (IPAP) and expiration (EPAP) which can be set independently, according to at least three different criteria.

- CPAP required is higher than maximum patient's tolerance: since the collapsing forces in inspiration are higher than in expiration, IPAP is set to prevent inspiratory events (hypopnea and inspiratory flow limitation), EPAP to prevent expiratory events (apnea and snoring). The difference between IPAP and EPAP can be even small and patient's comfort is the main target;
- Treatment of OSA and hypoventilation: EPAP is set to the pressure which eliminates all upper airway events (apneas, hypopneas, intermittent flow limitation, snoring, respiratory effort-related arousals), IPAP is increased up to the resolution of desaturations or up to the maximum patient's tolerance;
- Pressure support ventilation: EPAP is set to balance intrinsic PEEP, IPAP is the maximum pressure in the airways, the pressure support is the difference between the two.

Considering that OHS is a disease with hypoventilation, bilevel PAP could theoretically be considered the elective treatment of this disorder, but the reality does not fully support this view. Randomized controlled studies failed to demonstrate a superiority compared to CPAP both in patients with mean BMI 41 kg/m^2 (Storre et al. 2006) and in even more obese patients (BMI 52–54 kg/m^2)(Piper et al. 2008), but follow-up was limited to 3 months.

Another problem occurring with bilevel PAP is the possibility of ineffective efforts which can affect sleep architecture and ventilator efficiency; two studies addressed this issue showing that patient-ventilator asynchronies can occur often in OHS patients (Fanfulla et al. 2007; Guo et al. 2007). Positive pressure ventilation improves survival which is lower when supplementary oxygen is needed (Priou et al. 2010). Females are probably less adherent to a regular use of ventilator and should be periodically reinforced (Priou et al. 2010).

8.10.3 Intermittent Positive Pressure or Volume Cycled Ventilators

Volume cycled ventilators deliver a preset volume at pressures dependent on airway resistance; pressure cycled ventilators have a preset pressure support, the volume delivered being dependent on airway resistance. Both ventilators rely on an expiratory valve which opens at the beginning of expiration and allows a more reliable inspiratory time, since the patient cannot exhale until the valve opens. The blower is often more powerful and can deliver higher inspiratory pressures compared to bilevel devices, if necessary. They cannot offer a leak compensation, cause more gastrointestinal inflation, and are more expensive than bilevel devices; in common practice their use is nowadays more and more limited.

8.10.4 Supplemental Oxygen

Oxygen therapy alone is not recommended in OHS since it worsens hypoventilation, causing more respiratory events and worse sleep quality (Masa et al. 2001) and its usage at the beginning of CPAP/NIV treatment has to be considered with caution.

References

Behazin N, Jones SB, Cohen RI, Loring SH (2010) Respiratory restriction and elevated pleural and esophageal pressures in morbid obesity. J Appl Physiol 108:212–218

Berger KI, Ayappa I, Sorkin IB, Norman RG, Rapoport DM, Goldring RM (2002) Postevent ventilation as a function of CO2 load during respiratory events in obstructive sleep apnea. J Appl Physiol 93:917–924

Berry RB, Gleeson K (1997) Respiratory arousal from sleep: mechanisms and significance. Sleep 20:654–675

Braghiroli A, Marrone O, Sanna A, Vianello O, Ferini Strambi L, Manni R, Mondini S, Spaggiari C (2001). Linee guida di procedura diagnostica nella sindrome delle apnee nel sonno dell'adulto. Rassegna di Patologia dell'Apparato Respiratorio 16:278–280

Braunwald E, Ross JJ, Sonnenblick ES (1976) In mechanisms of contraction of the normal and failing heart, Little Brown and Co, Boston, pp 166–200.

Budweiser S, Riedl SG, Jorres RA, Heinemann F, Pfeifer M (2007) Mortality and prognostic factors in patients with obesity–hypoventilation syndrome undergoing noninvasive ventilation. J Intern Med 261:375–383

Cerretelli P, Di Prampero E (1987) Gas exchange in exercise. In: Farhi LE, Tenney SM (eds) Handbook of physiology, the respiratory system. Gas exchange, American Physiology Society, Bethesda, pp 307–309

Fanfulla F, Taurino AE, Lupo ND, Trentin R, D'Ambrosio C, Nava S (2007) Effect of sleep on patient/ventilator asynchrony in patients undergoing chronic non-invasive mechanical ventilation. Respir Med 101:1702–1707

Fogel RB, Trinder J, White DP, Malhotra A, Raneri J, Schory K, Kleverlann D, Pierce RJ (2005) The effect of sleep onset on upper airway muscle activity in patients with sleep apnoea vs. controls. J Physiol 564:549–562

Gibala MJ, Little JP, Macdonald MJ, Hawley JA (2012) Physiological adaptations to low-volume, high-intensity interval training in health and disease. J Physiol 590:1077–1084

Guo YF, Sforza E, Janssens JP (2007) Respiratory patterns during sleep in obesity-hypoventilation patients treated with nocturnal pressure support: a preliminary report. Chest 131:1090–1099

Hamilton MT, Booth FW (2000) Skeletal muscle adaptation to exercise: a century of progress. J Appl Physiol 88:327–331

Hardarson T, Skarphedinsson JO, Sveinsson T (1998) Importance of the lactate anion in control of breathing. J Appl Physiol 84:411–416

Isono S, Remmers JE, Tanaka A, Sho Y, Sato J, Nishino T (1997) Anatomy of pharynx in patients with obstructive sleep apnea and in normal subjects. J Appl Physiol 82:1319–1326

Knochel JP, Blachley JD, Johnson JH, Carter NW (1985) Muscle cell electrical hyperpolarization and reduced exercise hyperkaliemia in physically conditioned dogs. J Clin Invest 75:740–745

Kuna ST, Gurubhagavatula I, Maislin G, Hin S, Hartwig KC, McCloskey S, Hachadoorian R, Hurley S, Gupta R, Staley B, Atwood CW (2011) Noninferiority of functional outcome in ambulatory management of obstructive sleep apnea. Am J Respir Crit Care Med 183:1238–1244

Lafortuna CL, Lazzer S, Agosti F, Busti C, Galli R, Mazzilli G, Sartorio A (2010) Metabolic responses to submaximal treadmill walking and cycle ergometer pedalling in obese adolescents. Scand J Med Sci Sports 20:630–637

Lakoski SG, Barlow CE, Farrell SW, Berry ID, Morrow JR, Haskell WL (2011) Impact of body mass index, physical activity, and other clinical factors on cardiorespiratory fitness (from the Cooper Center Longitudinal Study). Am J Cardiol 108:34–39

LaMonte MJ, Blair SN (2006) Physical activity, cardiorespiratory fitness, and adiposity: contributions to disease risk. Curr Opin Clin Nutr Metab Care 9:540–546

Lauer MS, Anderson KM, Kannel WB, Levy D (1991) The impact of obesity on left ventricular mass and geometry. JAMA 266:231–236

Marin JM, Carrizo SJ, Vicente E, Agusti AG (2005) Long-term cardiovascular outcomes in men with obstructive sleep apnoea-hypopnoea with or without treatment with continuous positive airway pressure: an observational study. Lancet 365:1046–1053

Masa JF, Celli BR, Riesco JA, Hernández M, Sánchez De Cos J, Disdier C (2001) The obesity hypoventilation syndrome can be treated with noninvasive mechanical ventilation. Chest 119:1102–1107

Masa JF, Corral J, Pereira R, Joaquin Duran-Cantolla J, Cabello M, Hernàndez-Blasco L, Monasterio C, Alonso A, Chiner E, Rubio M, Garcia-Ledesma E, Cacelo L, Carpizo R, Sacristan L, Salord N, Carrera M, Sancho-Chust JN, Embid C, Vàsquez-Polo F-J, Negrin MA, Montserrat JM (2011) Effectiveness of home respiratory polygraphy for the diagnosis of sleep apnoea and hypopnoea syndrome. Thorax 66:567–573

Matsumura N, Nishijma H, Koijma S, Hashimoto F, Minami M, Yasuda H (1983) Determination of anaerobic threshold for assessment of functional state in patients with chronic heart failure. Circulation 68:360–367

Matus CD, Klaege K (2007) Exercise and weight management. Prim Care Clin Office Pract 34:109–116

Mitchell JH, Victor RG (1996) Neural control of the cardiovascular system: insights from muscle sympathetic nerve recordings in humans. Med Sci Sports Exerc 28(suppl 10):S60–S69

Mokhlesi B, Tulaimat A, Faibussowitsch I, Wang Y, Evans A (2007) Obesity hypoventilation syndrome: prevalence and predictors in patients with obstructive sleep apnea. Sleep Breat 11:117–124

Montserrat JM, Navajas D, Parra O, Farré R (2010) Continuous positive airway pressure treatment in patients with OSA. Eur Respir Mon 50:244–266

Mutlu GM, Rubinstein I (2005) The saga of obstructive sleep apnea syndrome and daytime hypercapnia: work in progress. Chest 127:698–699

Naimark A, Cherniack RM (1960) Compliance of the respiratory system and its components in health and obesity. J Appl Physiol 15:377–382

Pérez de Llano LA, Golpe R, Piquer MO, Racamonde AV, Caruncho MV, Muinelos OC, Carro CA (2005) Short-term and long-term effects of nasal intermittent positive pressure ventilation in patients with obesity-hypoventilation syndrome. Chest 128:587–594

Perseghin G, Scifo P, De Corbelli F, Pagliato E, Battezzati A, Arcelloni C, Vanzulli A, Testolin G, Pozza G, Del Maschio A, Luzi L (1999) Intramyocellular triglyceride contents as a determinant of in vivo insulin resistance in humans. Diabetes 48:1600–1606

Phillipson EA (1978) Control of breathing during sleep. Am Rev Respir Dis 118:909–939

Piper AJ, Wang D, Yee BJ, Barnes DJ, Grunstein RR (2008) Randomised trial of CPAP vs bilevel support in the treatment of obesity hypoventilation syndrome without severe nocturnal desaturation.Thorax 63:395–401

Piper AJ (2010) Obesity hypoventilation syndrome: therapeutic implications for treatment. Expert Rev Resp Med 4:57–70

Powers MA (2008) The obesity hypoventilation syndrome. Respir Care 53:1723–1730

Priou P, Hamel J-F, Person C, Meslier N, Racineux J-L, Urban T, Gagnadoux F (2010) Long-term outcome of noninvasive positive pressure ventilation for obesity hypoventilation syndrome. Chest 138:84–90

Salvadori A, Fanari P, Cavestri R, Mazza P, Baudo S, Longhini E (1991) Relationship between body mass and tolerance to physical stress in obese patients. Respiration 58:311–315

Salvadori A, Fanari P, Mazza P, Agosti R, Longhini E (1992) Work capacity and cardiopulmonary adaptation of the obese subject during exercise testing. Chest 101:674–679

Salvadori A, Fanari P, Fontana M, Buontempi L, Saezza A, Baudo S, Miserocchi G, Longhini E (1999a) Oxygen uptake and cardiac performance in obese and normal subjects during exercise. Respiration 66:25–33

Salvadori A, Fanari P, Palmulli P, Giacomotti E, Arreghini M, Bolla G, Miserocchi G, Longhini E (1999b) Cardiovascular and adrenergic response to exercise in obese subjects. J Clin and Basic Cardiol 2:229–236

Salvadori A, Fanari P, Giacomotti E, Palmulli P, Bolla G, Tovaglieri I, Luzi l, Longhini E (2003) Kinetics of catecholamines and potassium, and heart rate during exercise testing in obese subjects. Eur J Nutr 42:181–187

Salvadori A, Fanari P, Dworzak F, Azan G, Brunani A, Tovaglieri I, Luzi L, Longhini E (2004) Respiratory and metabolic responses during exercise, and skeletal muscle morphology in obesity. Sport Sci Health 1:47–54

Salvadori A, Fanari P, Tovaglieri I, Giacomotti E, Nibbio F, Belardi F, Longhini E (2008) Ventilation and its control during incremental exercise in obesity. Respiration 75:26–33

Salvadori A, Fanari P, Marzullo P, Codecasa F, Tovaglieri I, Cornacchia M, Walker G, Brunani A, Longhini E (2010) Dinamics of GH secretion during incremental exercise in obesity, before and after a short period of training at different work-loads. Clin Endocrinol 73: 491–496

Schwab RJ, Pasirstein M, Pierson R, Mackley A, Hachadoorian R, Arens R, Maislin G, Pack AI (2003) Identification of upper airway anatomic risk factors for obstructive sleep apnea with volumetric magnetic resonance imaging. Am J Respir Crit Care Med 168:522–530

Stadler DL, McEvoy RD, Bradley J, Paul D, Catcheside PG (2010) Changes in lung volume and diaphragm muscle activity at sleep onset in obese obstructive sleep apnea patients vs. healthy-weight controls. J Appl Physiol 109:1027–1036

Steier J, Jolley CJ, Seymour JM, Roughton M, Polkey MI, Moxham J (2009) Neural respiratory drive in obesity. Thorax 64:719–725

Storre JH, Seuthe B, Fiechter R, Milioglou S, Dreher M, Sorichter S, Windisch W (2006) Average volume-assured pressure support in obesity hypoventilation: A randomized crossover trial. Chest 130:815–821

Tabachnik E, Muller NL, Bryan AC, Levison H (1981) Changes in ventilation and chest wall mechanics during sleep in normal adolescents. J Appl Physiol 51:557–564

Wasserman K, Whipp J (1975) Exercise physiology in health and disease. Am Rev Respir Dis 112:219–246

Wasserman K (1984) The anaerobic threshold measurement to evaluate exercise performance. Am Rev Respir Dis 129(Suppl):S35–S40

Wellman A, Eckert DJ, Jordan AS, Edwards BA, Passaglia CL, Jackson AC, Gautam S, Owens RL, Malhotra A, White DP (2011) A method for measuring and modelling the physiological traits causing obstructive sleep apnea. J Appl Physiol 110:1627–1637
Whipp BJ, Wasserman K (1969) Efficiency of muscular work. J Appl Physiol 26:644–648
Yap JCH, Watson RA, Gilbey S, Pride NB (1995) Effects of posture on respiratory mechanics in obesity. J Appl Physiol 79:1199–1205
Younes M (2003) Contributions of upper airway mechanics and control mechanisms to severity of obstructive apnea. Am J Respir Crit Care Med 148:645–658
Young T, Palta M, Dempsey J, Skatrud J, Weber S, Badr S (1993) The occurrence of sleep-disordered breathing among middle-aged adults. N Engl J Med 328:1230–1235
Zarich SW, Kowalchuk EA, Nesto RW (1991) Left ventricular filling abnormalities in asymptomatic morbid obesity. Am J Cardiol 68:377–381
Zipes DP (1992) Management of cardiac arrhythmias. Pharmacological, electrical, and surgical techniques. In: Braunwald E (ed) Heart disease: a textbook of cardiovascular medicine. Saunders, Philadelphia, pp 644–646

Chapter 9
Psychological Aspects and Rehabilitation Protocols

Gianluca Castelnuovo, Gian Mauro Manzoni, Valentina Villa, Giada Pietrabissa and Enrico Molinari

Abstract Research on psychological dimensions of obesity has evolved considerably in recent years. This chapter aims at underlying the main psychological aspects, such as discrimination and stigma, psychopathological complications, correlations between obesity and personality disorders, eating disorders (particularly Binge Eating Disorders), mood disorders, and stress-anxiety. A brief overview about the connection between obesity and addiction is also provided. Then quality of life and self-esteem in obesity are discussed. The second part of the chapter aims at describing the main psychological rehabilitation protocols for obesity, according to the most evaluated approaches such as cognitive-behavioral, interpersonal, systemic-strategic ones. Particularly, attention is dedicated to the transtheoretical model stages of change (TTM SOC). Some promising treatments are the mindfulness approach as described by Kabat-Zinn and the EMDR-based techniques. Finally, a section is dedicated to the promotion of collaborative rehabilitation protocols with in-patient and out-patient settings using telemonitoring for the continuity of care (the TECNOB project), discussing the psychological components crucial in technology-based remote weight-loss interventions. In conclusion, future directions for research and practice in clinical psychology for obesity rehabilitation are discussed.

G. Castelnuovo (✉) · G. M. Manzoni · V. Villa · G. Pietrabissa · E. Molinari
Psychology Research Laboratory, Istituto Auxologico Italiano IRCCS,
Ospedale San Giuseppe, Verbania, Italy
e-mail: gianluca.castelnuovo@auxologico.it

G. M. Manzoni
e-mail: gm.manzoni@auxologico.it

V. Villa
e-mail: v.villa@auxologico.it

G. Pietrabissa
e-mail: giada.pietrabissa@gmail.com

E. Molinari
e-mail: molinari@auxologico.it

G. Castelnuovo · G. M. Manzoni · G. Pietrabissa · E. Molinari
Department of Psychology, Catholic University of Milan, Milan, Italy

P. Capodaglio et al. (eds.), *Disabling Obesity*, DOI: 10.1007/978-3-642-35972-9_9,

9.1 Psychological Aspects in Obesity

9.1.1 Psychological Aspects in Obesity: An Introduction

Obesity is traditionally defined as a body mass index (BMI) of >30 kg/m^2 and is today considered a public health problem and epidemia (globesity) (Gutierrez-Fisac et al. 2006). According to World Health Organization (WHO), it is estimated globally that more than 700 million adults will develop a condition of obesity by 2015 (Davin and Taylor 2009; WHO 2006).

Consequently, it can be considered one of the most important current health problems affecting millions of people around the world (Ayensa and Calderon 2011). Obesity is associated with early death (Flegal et al. 2005; Whitlock et al. 2009) and is universally recognized as a risk factor for many health complications such as cardiovascular diseases, some types of cancer, osteoarthritis, hypertension, dyslipidemia, hypercholesterolemia, and diabetes (Castelnuovo et al. 2010). Moreover, obesity is a strong risk factor for the development of type 2 diabetes (Klein et al. 2004a, b).

Regarding the etiology of obesity, investigations have shown that inheritance (Price 2002) and environment (Horgen and Brownell 2002) are the main causes of the obesity epidemic (Swencionis and Rendell).

Nevertheless, there is strong consensus in the scientific community that the etiology of obesity is multifactorial: a genetic base interacts in an intricate modality with individual and environmental factors that lead to the expression of heterogeneous conditions and results (Marcus and Wildes 2009).

Moreover, according to Dombrowski et al. (2012), "Behavioral factors, i.e., poor diet and physical inactivity are among the main proximal causes linked to obesity (Haslam and James 2005), obesity-related morbidity (Alley and Chang 2007) and mortality (Flegal et al. 2005)" (Dombrowski et al. 2012, p. 7). As noted by (Davin and Taylor 2009), "When looking across the obesity literature, the impact of psychological variables in the development and treatment of obesity is certainly acknowledged but overshadowed by the physical and medical components of obesity which dominates the literature. Given the failures of weight loss methods, even surgical methods, an understanding of the psychological variables that impact success of surgical and nonsurgical weight loss is imperative, as the field continues to strive to reduce the public health impact of obesity" (Davin and Taylor 2009, p. 717).

The literature on the psychosocial aspects of obesity has a long history: for example Stanley Hall, the first person to earn a Ph.D. in Psychology in the United States, started research on eating behaviors and obesity in the nineteenth century (Lepore 2011; Swencionis and Rendell 2012).

Then research on psychological aspects of obesity improved in the 1950s and evolved considerably in recent years. As noted by Fabricatore and Wadden (2004), the earliest literature was more theoretical and focusing on the manifestation of underlying psychopathology and suboptimal development. "The first empirical studies of the matter attempted to find commonalities in the psychological profiles of obese persons. No "obese personality" was ever identified, however. Comparisons of overweight and average weight individuals yielded inconsistent results that depended on the characteristics of the samples. Some studies showed that obese persons suffered greater levels of psychopathology, whereas others demonstrated an apparent protective effect of excess body weight on psychological distress. These studies, collectively, have been criticized for methodological shortcomings (Friedman and Brownell 1995). The best evidence (from nationally representative samples of the U.S. population) suggests that obese women but not obese men have a slightly higher rate of depression and suicidal ideation than their normal-weight counterparts" (Carpenter et al. 2000; Istvan et al. 1992) (Fabricatore and Wadden 2004, p. 332).

The research on psychosocial correlates of obesity has evolved into a second phase in which scientists aimed at identifying risk factors for psychopathology for obese population (for example, a relevant result was that female gender is one important risk factor) (Fabricatore and Wadden 2004).

The last decade has been the beginning of the third step of research into the association between obesity and psychosocial features.

The focus is now more on depression (versus other psychopathological conditions) and on discovering the temporal and causal relationship between psychological distress and obesity.

9.1.2 Discrimination and Stigma

The role of genetic contributions toward the condition of obesity has been largely acknowledged, unfortunately, social prejudice and discrimination still persist toward afflicted individuals and obesity in general. In western cultures, the promotion of being thin, is not only dysfunctional and dangerous but intensifies and polarize the issue of weight, leading to further ignorance.

Subjects with a condition of obesity are strongly stigmatized and they have to daily cope with growing modalities of discrimination and prejudice about the overweight. This topic has been largely studied and underlined by Puhl and Brownell (2001, 2003a, b, 2006; Puhl and Heuer 2009; Puhl and Latner 2007; Puhl et al. 2005, 2007, 2008). According to a recent Puhl's review, "These stereotypes are prevalent and are rarely challenged in Western society, leaving overweight and obese persons vulnerable to social injustice, unfair treatment, and impaired quality of life as a result of substantial disadvantages and stigma… In recent years, attention to weight bias has increased, with a growing recognition of the pervasiveness of weight bias and stigma, and its potential harmful consequences for obese persons." (Puhl and Heuer 2009, p. 941). The stigma is also present in extremely obese individuals who presented for bariatric surgery (Sarwer et al. 2008).

Discrimination may also affect employment settings. Roehling et al. (2008) demonstrated that obese employees and applicants were considered with a more negative evaluation and obtained more negative outcomes with regard to employment in comparison with non-overweight persons in the same working conditions and places. This unpleasant stigma is widespread also in the health care settings: physicians, psychologists, dietitians, nurses, and other health care professionals keep negative attitudes, beliefs, and biases toward obese patients (Capodaglio et al. 2010; Puhl and Brownell 2001) and dedicate them inadequate time even if the scientific community recognizes the importance of providing them with information and other tools to improve healthy lifestyles (Capodaglio et al. 2010; Foster et al. 2003; Puhl and Heuer 2009).

An exhaustive summary of the key findings about the relationship between stigma and obesity is reported by (Puhl and Heuer 2009). The significative psychological consequences of the weight biases are well reported: "Weight bias increases vulnerability to depression, low self-esteem, and poor body image (moderate evidence), Weight bias contributes to maladaptive eating behaviors among obese individuals (strong evidence), Weight bias contributes to less participation/avoidance of physical activity (limited evidence)" (Puhl and Heuer 2009, p. 948).

The future scenario does not appear so promising: "Unfortunately, it does not appear that the increasing prevalence of obesity has attenuated negative societal attitudes toward obese people… Without sufficient attention to this issue in the obesity field and in larger society, it is likely that weight bias will remain both a social injustice and a public health issue, impairing the quality of life for both present and future generations of obese individuals" (Puhl and Heuer 2009, p. 960).

9.1.3 Psychopathology

Current knowledge gives no clear link between psychopathology and obesity and whether the different psychological disorders are a cause or a consequence of obesity has not yet been explained either.

In fact, the first studies about the psychosocial status of obese individuals in the general population provided not clear results (Fagiolini et al. 2005, 2008). In some cases researchers found that obesity was connected with a greater emotional distress, whereas other studies noted that obese people had a less psychological level of disorder (Marcus and Wildes 2009). No clinically significant results were found in order to create different psychological profiles between obese and non-obese individuals.

Even if recent studies failed to find consistent results, epidemiologic investigations have reported positive correlations between obesity and personality disorders, eating disorders, mood disorders, and anxiety disorders (Hudson et al. 2007; Petry et al. 2008; Pickering et al. 2007; Scott et al. 2008). These findings are particularly significant in women (Barry et al. 2008; McIntyre et al. 2006) and subjects affected by severe obesity (such as BMI $\geq$ 35) (Hudson et al. 2007; Marcus and Wildes 2009; Scott et al. 2008).

According to Ayensa it is possible to conclude that "certain psychological disorders seem to be more common in obese people, such as depression, anxiety, substance abuse, etc., although the direction of the explanatory relation is not clear; moreover, the opposite result has also been found. The evidence of likely comorbid psychopathology means that the practitioner must consider its assessment in obesity cases, and include a plan for a psychiatric and/or psychological therapeutic intervention" (p. 253, Ayensa and Calderon).

9.1.3.1 Stress and Anxiety

There is important evidence that the effects of chronic stress on central functioning may lead to a condition of obesity in the useless eater. The combination of ongoing stress and the wide availability of poor quality foods promote the development of overeating and obesity. In particular some people may enhance overeating behaviors in order to modulate activity of the chronic stress network and reduce negative affects (Marcus and Wildes 2009).

While links between obesity and physical activity are well known, the currently available research literature gives not yet significant evidence for the relationship between obesity and psychological distress. According to Hill (2005) "Some obese (and lean) people have serious psychological problems; others have mild problems, and some very few at all. The challenge is to determine which people suffer psychologically from their obesity, and how they display their distress" (Hill 2005, p. 27).

Greeno and Wing reviewed many studies and noted significant individual differences in stress-induced eating, concluding that restrained eating behaviors can predict stress-induced eating in women. Anyhow they underlined the absence of a direct demonstration of the link between stress and weight gain or development of obesity (Greeno and Wing 1994).

Van Strien et al. examined the difference between low and high emotional eaters in gaining weight as a response to negative life events: only male emotional eaters who severally experienced negativity have shown greater increases in BMI over a 6-month period (Van Strien et al. 1986).

Recent articles and meta-analytic reviews confirm that psychosocial stress is positively linked with the development of adiposity in prospective studies (Wardle 2007; Wardle et al. 2010; Wardle and Cooke 2005) and major life-events or periods of acute stress can play a greater part in the etiology of obesity for men than women (Torres and Nowson 2007). "Men consistently show stronger physiological responses to stress than women, including greater cardiovascular and neuroendocrine activation (Stoney et al. 1987), and higher cortisol levels after exposure to acute real-life psychological stress (e.g., examinations) or controlled laboratory experiments (Kirschbaum et al. 1992; Stroud et al. 2002)" (Wardle et al. 2010, pp. 775–776). Psychosocial stress promotes weight gain, although with a modest effect size. In further investigations, then, psychobiological mechanisms that lead stress to create "obesogenic" effects should be better investigated and understood (Wardle et al. 2010).

9.1.3.2 Obesity as Addiction?

Recent researches have also revealed differences between obese and lean individuals with respect to neural activation in response to tasting and consuming food (Del Parigi et al. 2002a, b; Gautier et al. 2000, 2001).

Some recent theories have underlined the role of the neural reward system in the development and maintenance of obesity. Specifically these researches have hypothesized that a "dysfunction of brain reward circuitry in response to food cues may predispose some individuals to obesity via an increased likelihood of overeating, particularly excessive consumption of palatable foods" (Marcus and Wildes 2009, p. 744). The hypothesis that some kind of obesity may be considered as a kind of food "addiction" has obtained growing success (Gearhardt and Corbin 2011; Gearhardt et al. 2011a, b, c, d, 2012; Wang et al. 2001, 2002, 2004, 2009).

9.1.3.3 Depression

Obesity is differently associated with depression: the percentage of risk for depression in women has increased by 37 %, whereas the same rate of men is currently less at risk. Similar associations were found between obesity and suicide attempts, where gender still represents a determinant variable (obese women have an increased risk whereas obese men have a decreased risk). Moreover, anxiety and depression have currently been detected more in obese people seeking clinical assistance rather than in a community sample.

Obese women, especially with a high socioeconomic background (or status), have more relevant risk of anxiety and depression than obese men. (Hill 2005; Stunkard et al. 2003). Istvan et al. studying a large sample of adults (age 25–74), did not find a relationship between BMI and depression among men, discovering, instead, women with highest BMI having more probability (38 %) to report clinically significant depressive symptoms (Istvan et al. 1992).

Carpenter et al. (2000) studied more than 40,000 people and found that the relationship between obesity and depression was gender dependent. As noted by Fabricatore about this important study, "men with a BMI $\geq$ 30 kg/m^2 were significantly less likely to report a history of major depression…, suicidal ideation, or suicide attempts in the past year than average-weight men (BMI = 20.8–29.9 kg/m^2). In contrast, underweight men (BMI $\leq$ 20.8 kg/m^2) experienced a 25 % increased risk for depression, 81 % increased risk for suicidal ideation, and 77 % increased risk for suicide attempts compared with average-weight men. A different pattern was seen for women. The 1-year prevalence of major depression was 37 % higher among obese females than in their average weight peers. Women with a

BMI ≥ 30 kg/m^2 also were 20 % more likely to report suicidal ideation and 23 % more likely to have made a suicide attempt in the past year. There was no association with depression or suicide for underweight women." (Fabricatore and Wadden 2004, p. 333).

9.1.3.4 Binge-Eating Disorder

The typical eating disorder, added to DSM-IV in the research section, connected with obesity is binge-eating disorder (BED) (American Psychiatric Association 2000; Hill 2005). Although binge-eating occurs not only in overweight conditions but at all body weights, there has been an increase of interest in obese binge-eaters. According to Hill, "it is apparent that BED is more common in the obese than in normal-weight individuals. In US weight loss clinics, 20–40 % of patients are reported to have BED, although the use of a strict diagnostic interview reduces this to well below 20 %. In community samples, BED is much less common, apparent in 1–3 % of respondents. Overall, the prevalence of BED in any group increases with increasing obesity" (Hill 2005, p. 27). About the key features of BED with obesity, Hill noted that "comparing samples of obese people with and without BED shows several important differences, those with BED having:

- An Earlier Onset of Obesity
- more severe obesity
- greater weight fluctuations and weight distress
- greater likelihood of past treatment for psychological disorders
- more impairment in work/social functioning
- more psychological distress and self-esteem problems" (Hill 2005, p. 28).

9.1.3.5 Quality of Life

Obesity is traditionally associated with perception of disability, poorer control of symptoms and poorer quality of life, weight-based stigmatization and significant psychosocial impairment (Lavoie et al. 2006). Many empirical studies have shown that obese people report more physical impairments and limitations in comparison with normal weight individuals (Doll et al. 2000; Fine et al. 1999; Ni Mhurchu et al. 2004). Other investigations have shown that obese individuals do not simply report impairments in the area of physical health, but also perceive significant reductions of psychosocial functioning, psychological well-being, and in general, in the mental health field (Fontaine et al. 1996; Kolotkin et al. 2001; Kushner and Foster 2000).

After the biopsychosocial paradigm (Engel 1977) has been applied to the area of obesity, it is relevant to consider, measure, and monitor the "health-related quality of life" (HRQoL) for obese patients. HRQoL is a multidimensional concept that includes domains related to physical, mental, emotional, and social functioning and is referred to the subjective perception of effects of diseases (and subsequent treatments) (Testa and Simonson 1996). The traditional medical model centered on the disease has gradually moved to a more modern patient-centered model and the improvement of quality of life in patients has been considered and added to the clinical goals, traditionally oriented only towards bio-physiological dimensions.

Evaluation of the quality of life has become a focus of interest not only in population studies, but also in clinical medicine (Blissmer et al. 2006). Population research is based on the application of generic instruments such as SF-36, which evaluates both physical and mental health of the individual (Vasiljevic et al. 2008; Ware et al. 1993). In the field of obesity the perceived impact of overweight on health can be evaluated by the use of generic as well as disease-specific assessment scales (Doll et al. 2000; Kolotkin and Crosby 2002). For a specific and detailed review of the main obesity-specific HRQoL assessment scales please see (Manzoni and Castelnuovo 2011).

According to Hill, "Further studies of large representative samples using established measures of well-being (such as the SF-36 self-report questionnaire) have helped to separate physical functioning from psychological health, while also showing their interdependence. Age is an important moderator of the relationship between obesity and well-being. Overweight and obese young women in their late teens and early 20 s score significantly lower in physical functioning, vitality and general health, but show few differences on any of the main psychological health measures. The pattern for middle-aged women aged 45–49 years is different: there are proportionately more overweight and obese women in this age group, and these older obese women score significantly lower than those of average weight on all the physical and psychological health scales.

Most of the available evidence shows that the greatest deficits are in people who are severely or morbidly obese (BMI > 40). Scores on psychological health scales for people who are moderately obese (BMI 30–40) are similar to those of underweight people (BMI < 20), whereas the scores for overweight women are similar to those of normal-weight individuals" (Hill 2005, p. 30).

9.1.3.6 Self-Esteem

Obesity and self-esteem are related in a way that changes according to the period of life considered. For example, adults with obesity have a modest reduction in self-esteem, sometimes limited only to those with morbid pathology. In pre-adolescents, self-esteem is affected by obesity only with limited impact, while adolescents' self-perceived overweight is more strongly associated with reduced self-esteem, and this correlation is more significant for females than for males.

As noted by Hill, body esteem is the domain of self-worth that can be most affected by obesity, especially for obese adolescent and young women, producing very high levels of body dissatisfaction. Fortunately, psychotherapy focused on improving body image in obese women has shown discrete success in reducing psychopathological symptoms, but has very low impact on body weight reduction (Hill 2005).

9.2 Psychological Rehabilitation Protocols for Obesity

9.2.1 Psychological Rehabilitation Protocols for Obesity: An Overview

According to the Cochrane guidelines (Shaw et al. 2005), when a psychological therapy, and particularly cognitive-behavior therapy (CBT), is combined with a diet/exercise plan, the resulting treatment will increase weight loss, if compared with the only diet/exercise intervention. In fact, psychological treatments commonly used for clinical trials, as well as in clinics and psycho-educational approaches, are widespread and usually composed of imparting diet, exercise, behavioral information, and different forms of follow-up sessions (Swencionis and Rendell 2012; Wing 2002). Behavior therapy is another spread approach in obesity treatment: it is characterized by self-monitoring (for example, using diaries), stimulus control (for example, restricting quantities of food), and behavioral modification (for example, chewing slowly, taking time to really taste and enjoy the food, maximizing the pleasure from it) (Foster et al. 2005; Swencionis and Rendell 2012; Wing 2002).

Cognitive-behavioral approach represents the gold standard for the treatment of obesity. It is focused on dysfunctional behaviors and cognitive processes, such as unrealistic weight goals and body image perceptions (Cooper et al. 2004). In addition, hypnosis could supplementarily be applied, toward particular subjects, enhancing patients to correct dysfunctional thoughts, attitudes, and beliefs using repeated self-statements (Wylie-Rosett et al. 2007; Xu and Cardena 2008).

Other important psychotherapies used for the treatment of obesity are the interpersonal approach (Brambilla et al. 2009; Fabricatore and Wadden 2004; Tanofsky-Kraff et al. 2007), the systemic and strategic treatments (Castelnuovo et al. (2011a, b); Nowicka and Flodmark 2010; Peterson 2005), the psychodynamic approach (Becker 1960; Beutel et al. 2006; Fernandez-Paredes and Sumano-Avendano 1986; Ingram 1976; Rand and Stunkard 1977, 1978, 1983) and the feminist-theories-based treatment (Allan 1994; Orbach 2006; Rgon 1995).

Another interesting treatment used in psychological rehabilitation of obesity is the mindfulness approach as described by Kabat-Zinn (1994; Ludwig and Kabat-Zinn 2008). Different researches have reported that mindfulness training could be effective in reducing episodes of binge and other dysfunctional eatings using meditation to control food cravings (Alberts et al. 2010; Dalen et al. 2010; Kabat-Zinn 1994; Lillis et al. 2009; Ludwig and Kabat-Zinn 2008; Singh et al. 2008; Swencionis and Rendell 2012).

A promising treatment could also be the EMDR approach (Bloomgarden and Calogero 2008; Hudson et al. 1998), but needs to be more investigated.

Even if clinicians can use different psychological approaches in treating obesity (such as behavioral, cognitive-behavioral, interpersonal, systemic-strategic, psychodynamic, etc.), interventions that aim at changing dysfunctional behaviors might be carried out in a multidisciplinary context (with a clinical team composed of psychologists, dieticians, psychiatrists, endocrinologists or nutritionists, physiotherapists, etc.) and might benefit from including specific instructions about changing diet, asking participants to self-monitor their dietary intake, and motivating patients to maintain initially achieved goals, anticipating possible future relapses, and learning strategies to cope with difficult moments or situations Dombrowski et al. (2011, 2012) .

Moreover, one of the greatest challenges in the management of obese patients is working to work on the significant disparity between real and expected weight losses. For clinicians a 10 % weight loss is generally considered a good success, due to an important associated improvement in comorbidities, but patients typically look for a more or less 30 % reduction in body weight (Foster et al. 1997; Jeffery et al. 1998; O'Neil et al. 2000; Wadden et al. 2000). Specific recommendations are needed in order to help patients in accepting more modest and realistic weight loss outcomes as successful results (Foster et al. 2005; Nonas and Foster 2005).

Moreover, moderate weight loss (5–10 % of initial weight) can lead to some positive psychological changes such as improvements in body satisfaction, self-esteem, and mood, above all if physical activity component is included in the rehabilitation protocol (Hill 2005). Unfortunately, only short-term studies have confirmed these findings. The evidence for longer term psychological benefits and for the long-term effectiveness of most weight-reduction protocols is not so clear (Hill 2005).

For clinicians it is also relevant to evaluate the obese individual's readiness for weight loss during the therapeutic process. As noted by Hill "attempts to identify psychological predictors of weight loss have not been very successful… the presence of depression, anxiety or binge-eating is associated with poorer weight loss, although findings are inconsistent. Similarly, summary measures of readiness to change or motivation to lose weight have generally failed to predict outcome. In contrast, self-efficacy (a person's evaluation of whether he or she can perform the behaviors required for weight loss) is a modest but consistent predictor of success. In terms of setting goals, patients are often more ambitious, and therefore less realistic, in the amount of weight they want to lose compared with the moderate loss advised by informed health professionals. The psychological consequences of a failure to achieve goals need to be recognized. Overall, goals for effective weight loss and maintenance should be specific, attainable and forgiving (i.e. less than perfect) " (Hill 2005, p. 29).

Methods to assess patients' readiness to change and motivation to address weight management behaviors are typically related to Prochaska and DiClemente's stages of change: precontemplation, contemplation, preparation, action, and maintenance (Bolognesi et al. 2006; Boudreaux et al. 2003). The TTM SOC should include dietary strategies based on Prochaska and DiClemente's stages of change, enhances dietary habits (such as reduction in fatty food and daily calories and consumption), and physical exercise strategies with the goal to get a sustainable weight loss and the desired behavioral change (Prochaska and DiClemente 1986; Rao et al. 2011; Tuah et al. 2011).

TTM SOC produced important effects on different outcome measures, such as change in physical activity, modification in dietary intake, and progression through the process theorized by Prochaska and DiClemente. Feedback reports, anthropometric measurements, and counseling are also relevant.

Below, is reported a typical scheme of questions provided to clinicians by the (National Heart Lung and Blood Institute 2000) in order to check the readiness to lose weight in obese patients.

Questions to ask in evaluating an obese person's readiness to lose weight

- Has the person sought weight loss on his/her own initiative?
- What events have led him/her to seek weight loss now?
- What are his/her typical stress levels and mood?
- Does he/she have an eating disorder (e.g. binge-eating disorder), in addition to obesity?
- Does the person understand the requirements of the treatment and believe that he/she can fulfill them?
- How much weight does the patient expect to lose?
- What other benefits does the patient anticipate?

(From National Heart Lung and Blood Institute 2000, and reported also in Hill 2005).

9.2.2 *Promoting Collaborative Rehabilitation Protocols with In-Patient and Out-Patient Settings Using Telemonitoring for the Continuity of Care*

The most successful strategy for the management and rehabilitation of obesity is a collaborative approach, well-defined as a "strategy or set of strategies to help patients achieve and/or maintain a healthy weight that involve collaboration among healthcare professionals in at least two different disciplines (e.g., physicians and dieticians) for the delivery of weight management interventions. We considered interventions that do not significantly disrupt the usual processes of health care in busy practices and that included at least 1 meaningful outcome such as change in dietary behavior or change in weight" (Rao et al. 2011, p. 1190). Unfortunately, only a restricted number of articles report real and practical collaborative approaches, whereas the scientific evidence shows that strategies based on central planning and training, using a "chronic care model" logic, obtain better results (Rao et al. 2011).

The use of Internet-based technologies can improve many steps of the collaborative approach, such as psychological and medical education in food management, physical activity, weight loss reduction and maintenance, functional continuous self-monitoring of health behaviors, individual realistic goals setting and motivation, cognitive and behavioral change, as well as peer social support (Castelnuovo et al. 2003, 2010, 2011a, b, c, d; Castelnuovo and Simpson 2011;

Manzoni et al. 2011; Rao et al. 2011; Riva et al. 2006). This strategy using telemonitoring for the continuity of care, above all in the follow-up and monitoring steps, is essential in reducing costs for the management of chronic pathologies such as obesity (Ekeland et al. 2010, 2011; Manzoni et al. 2011).

9.2.2.1 The TECNOB Telemonitoring Approach

An operative example of collaborative approach is represented by the TECNOB clinical program (TEChNology for OBesity) (Castelnuovo 2008; Castelnuovo et al. 2003, 2010, 2011a, b). It has a total duration of 13 months and consists of two stepped down phases: in-patient (1 month) and out-patient (the following 12 months). During the in-patient phase, participants undergo an intensive 4-week hospital-based and medically managed program for weight reduction and rehabilitation. All patients are placed on a hypocaloric nutritionally balanced diet tailored to the individual after consultation with a dietician (energy intake around 80 % of the basal energy expenditure estimated according to the Harris-Benedict equation and a macronutrient composition of about 16 % proteins, 25 % fat, and 59 % carbohydrates). Furthermore, they receive nutritional counseling provided by a dietician, psychological support provided by a clinical psychologist, and have physical activity training provided by a physiotherapist.

Nutritional rehabilitation program is aimed to improve and promote change in eating habits and consists of both individual (dietary assessment, evaluation of nutrient intake and adequacy, nutritional status, anthropometric, eating patterns, history of overweight, readiness to adopt change) and group sessions (45 min each twice a week), including: information about obesity and related health risks, setting of realistic goals for weight loss, healthy eating in general, general nutrition and core food groups, weight management and behavioral change strategies for preventing relapse).

Psychological support is provided once a week both individually and in group settings. Individual sessions, lasting 45 min each, are mainly based on the cognitive-behavioral approach described by Cooper and Fairburn (Cooper 1990) and emphasize the techniques of self-monitoring, goal setting, time management, prompting and cueing, problem solving, cognitive restructuring, stress management, and relapse prevention. Group sessions ("closed" groups of 5/6 persons), lasting 1 h each, focus on issues such as motivation, assertiveness, self-esteem, self-efficacy, and coping.

Physical activity takes place once a day except for weekends and consists of group programs (20 subjects) based on postural gymnastics, aerobic activity, and walks in the open air (or outdoor). In particular, patients with specific orthopedic complications carry out individual activities planned by physiotherapists and articulated in programs of physical therapy, such as assisted passive and active mobilization and isokinetic exercises (Castelnuovo et al. 2010).

The Internet-based tools show promising results in promoting weight reduction in obese patients but further researches are needed to determine its long-term efficacy and effectiveness from clinical, organizational, and economic points of view. (Manzoni et al. 2008, 2011; Rao et al. 2011).

Even if positive feedbacks have been achieved about the usefulness of mobile phone devices in promoting healthy habits and weight loss behaviors, until now the collected data do not support the use of Internet interventions for weight maintenance (Rao et al. 2011). Indeed, there is a general lack of knowledge regarding understanding of the real costs of telemedicine. Telemedicine seemed to be cost-effective, but few studies draw firm conclusions to this only with regard to chronic disease management. Telemonitoring could reduce travel time and hospital admissions. Other reviewers did not find good evidence about cost-effectiveness. Another review highlighted the need to consider not only costs to health services of interventions, but also costs to service users and their social networks (Ekeland et al. 2010, 2011; Khaylis et al. 2010; Rojas and Gagnon 2008).

In recent years economic analysis has risen in the field of telemonitoring. The majority of these studies come from the United States. Telemedicine resulted to be a cost-effective alternative in 91 % of the studies.

The main telemedicine benefits include:

- decreased hospital utilization,
- improved patient compliance with treatment plans,
- improved patient satisfaction with health services, and
- improved quality of life.

Telemedicine also improved cognitive status, cognitive level, and self-rated health status. Studies also found disadvantages related to telemedicine, such as technical problems and reluctance from patients, caregivers, nurses, and physicians.

9.2.2.2 Psychological Components Crucial in Technology-Based Remote Weight-Loss Interventions

Five psychological components seem to be considered crucial in technology-based weight-loss interventions that are successful in facilitating weight loss (Khaylis et al. 2010).

(1) Self-monitoring. This refers to the process in which individuals regulate and keep track of their own behaviors. Technology can simplify the monitoring process, recording one's progress of food intake and physical activity using online devices. The reason these technologies are likely to be effective is because portable body monitors, pedometers, and handheld PDAs are mobile and, therefore, can be easily used, resulting in continuous self-monitoring. Also, these devices are more convenient for individuals without access to a high-speed Internet connection.

(2) Counselor Feedback and Communication. Feedback from a counselor regarding goals, progress, and results can encourage, motivate, and assist patients in successfully completing a weight-loss program. A functional approach is to

provide online weight-loss interventions with brief weekly or monthly counselor or psychologist visits. Participants typically submit their weekly food and exercise journals online and receive personalized feedback, reinforcement, and recommendations from a counselor over e-mail.

(3) Social support. A group treatment format is typically the preferred delivery of behavioral weight-loss interventions. Not only is this a cost-effective method of delivering treatment to a larger number of people, but also group treatments leverage social support, an important facilitator of behavior change. Group support can foster motivation, encouragement, and commonality. To facilitate communication among participants are useful electronic message boards, forums, "real time" chat rooms, or online meetings.

(4) Structured program. Technology-based weight-loss programs have been structured interventions that incorporated principles of behavior therapy and change. They consisted of structured weekly lessons on various topics, including nutrition, exercise, stimulus control, self-regulation strategies, and goal-setting.

(5) Individually tailored program. Interventions that were individually tailored to participant goals had higher rates of adherence and weight loss. In one case, participants met with a health coach prior to receiving the intervention and selected four high-priority behavior change goals that were subsequently monitored and achieved through behavior skills training. Another study delivered automated, real-time SMS text messages that were specific for each participant's barrier to exercise at that moment.

9.2.3 Future Directions for Research and Practice in Clinical Psychology for Obesity Rehabilitation

It is necessary to collect more scientific evidence to support collaborative approaches for weight management, considering Internet and mobile-based tools in remote settings too. The up-to-date literature of patients should be taken into account in order to select the best "tailored" intervention for each patient. A Guide to Selecting Treatment (see Fig. 9.1) has been reported by NIH Guidelines (NIH/NHLBI/NAASO October 2000). As indicated, behavioral and psychological treatments have to be included in each degree of obesity, from an overweight BMI condition with comorbidities, where a traditional rehabilitation protocol with psychological support, diet, and physical activity is enough, to a moderate and severe obesity, both with and without comorbidities, where the introduction of pharmacotherapy and bariatric surgery has to be considered.

In each situation psychological support is necessary to motivate patients to improve to a more healthy lifestyle or to accept more intrusive medical treatments (such as drugs and weight-loss surgery), helping them o cope with chronic conditions where usually a real target is not a full and complete healing, but a functional management of obesity reducing complications (such as type 2 diabetes, hypertension, and cardiovascular disease) (Nguyen and Lau 2012).

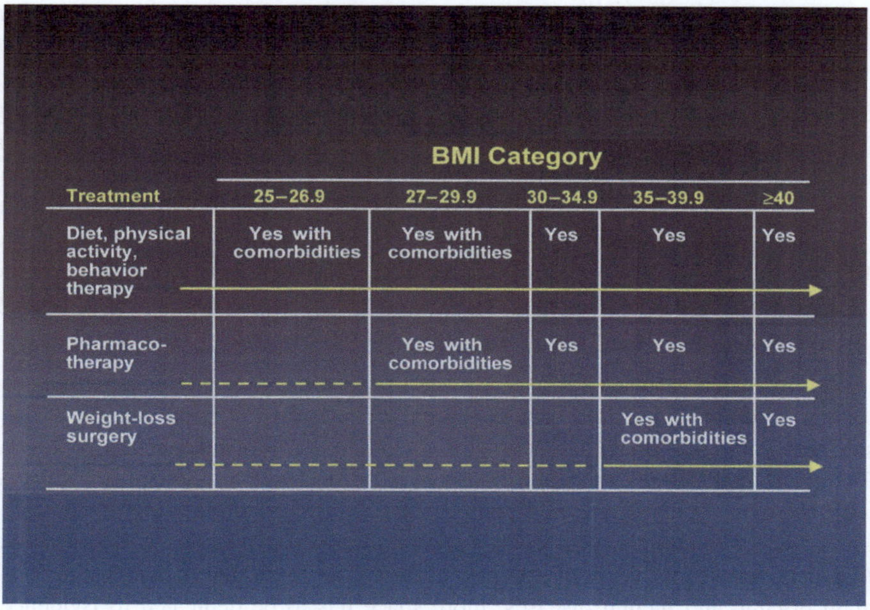

Fig. 9.1 A guide to selecting treatment according to NIH guidelines (NIH/NHLBI/NAASO October 2000) (where "Yes" alone means that the treatment is indicated regardless of the presence or absence and the solid arrow indicates the point at which therapy is initiated)

Detailed recommendations have been reported by (Rao et al. 2011) for future directions for psychological research and practice in obesity rehabilitation. Particularly, "discussions of weight should be performed in a nonjudgmental, respectful, and unhurried manner" (Rao et al. 2011, p. 1200). Then "readiness and self-efficacy to change behaviors should be assessed before weight loss strategies are initiated, and this information should be factored into decisions about what type of approach to use" (Rao et al. 2011, p. 1200). Moreover, collaborative approaches involving physicians, psychologists, nurses, and other clinicians have to be considered using central planning and training modalities. In the end more research has to be dedicated to Internet and other technologies in order to check their possible applications: now, insufficient evidence cannot allow clinicians to make recommendations about their effective use in clinical practice (Rao et al. 2011).

Generally, future directions in obesity and weight-reduction research have to consider the following priority areas:

1. There is a need for larger studies, both those that include technologically based interventions and those that do not, that enroll a diverse spectrum of overweight and obese patients in terms of sex, race, and socioeconomic status. Latino subjects and men, in particular, are underrepresented in obesity studies to date. There is also a need to investigate the specific features of technologically based interventions (e.g., content, format, device) that make such interventions successful in promoting weight loss.

2. Because attrition rates from technology-based studies are very high, there is a need to develop effective strategies to keep patients engaged in using technology tools for the long term.
3. Further evaluation of collaborative approaches (e.g., approaches involving centralized planning, approaches involving nurses in intervention delivery) in general is needed. In particular, larger studies of longer duration are needed to evaluate the effectiveness of the chronic care model as a framework for weight management interventions.
4. Use of electronic health records is increasing, and there is a need to explore the use of these valuable tools, not only for identification and assessment of obesity but also for the delivery of obesity interventions." (Rao et al. 2011, p. 1200).

In conclusion, Clinical Psychology has to face the "globesity" challenge, providing more evidence-based protocols for the psychological rehabilitation in obesity both in the traditional and in new remote clinical settings (Castelnuovo 2010a; Pietrabissa et al. 2012). "Clinical Psychology has to forge a more fruitful alliance with Medicine, "accompanying" it in all the clinical acts and adopting a scientific stance. Guidelines, protocols, and investigations using an Evidence-Based Approach are requested in all the psychological areas related to the treatment of main organic and mental diseases: more space has to be dedicated to Evidence-Based Practice in Clinical Psychology and Empirically Supported Psychological Treatments" (Castelnuovo 2010b, p. 4), above all in the obesity field.

References

Alberts HJ, Mulkens S, Smeets M, Thewissen R (2010) Coping with food cravings. investigating the potential of a mindfulness-based intervention. Appetite 55(1):160–163

Allan JD (1994) A biomedical and feminist perspective on women's experiences with weight management. West J Nurs Res 16(5):524–543

Alley DE, Chang VW (2007) The changing relationship of obesity and disability, 1988-2004. [Research Support, N.I.H., Extramural Research Support, Non-U.S. Gov't]. JAMA 298(17):2020–2027. doi: 10.1001/jama.298.17.2020

American Psychiatric Association (2000). Diagnostic and statistical manual of mental disorders, 4th, text revision: ed. Washington

Ayensa JI, Calderon MJ (2011) Psychopathological comorbidity of obesity. An Sist Sanit Navar 34(2):253–261

Barry D, Pietrzak RH, Petry NM (2008) Gender differences in associations between body mass index and DSM-IV mood and anxiety disorders: results from the National Epidemiologic Survey on Alcohol and Related Conditions. Ann Epidemiol 18(6):458–466

Becker BJ (1960) The obese patient in group psychoanalysis. Am J Psychother 14:322–337

Beutel ME, Dippel A, Szczepanski M, Thiede R, Wiltink J (2006) Mid-term effectiveness of behavioral and psychodynamic inpatient treatments of severe obesity based on a randomized study. Psychother Psychosom 75(6):337–345

Blissmer B, Riebe D, Dye G, Ruggiero L, Greene G, Caldwell M (2006) Health-related quality of life following a clinical weight loss intervention among overweight and obese adults: intervention and 24 month follow-up effects. Heal Qual Life Outcomes 4(1):43

Bloomgarden A, Calogero RM (2008) A randomized experimental test of the efficacy of EMDR treatment on negative body image in eating disorder inpatients. Eat Disord 16(5):418–427

Bolognesi M, Nigg CR, Massarini M, Lippke S (2006) Reducing obesity indicators through brief physical activity counseling (PACE) in Italian primary care settings. Ann Behav Med 31(2):179–185

Boudreaux ED, Wood KB, Mehan D, Scarinci I, Taylor CL, Brantley PJ (2003) Congruence of readiness to change, self-efficacy, and decisional balance for physical activity and dietary fat reduction. Am J Health Promot 17(5):329–336

Brambilla F, Samek L, Company M, Lovo F, Cioni L, Mellado C (2009) Multivariate therapeutic approach to binge-eating disorder: combined nutritional, psychological and pharmacological treatment. Int Clin Psychopharmacol 24(6):312–317

Capodaglio P, Castelnuovo G, Brunani A, Vismara L, Villa V, Capodaglio EM (2010) Functional limitations and occupational issues in obesity: a review. Int J Occup Saf Ergon 16(4):507–523

Carpenter KM, Hasin DS, Allison DB, Faith MS (2000) Relationships between obesity and DSM-IV major depressive disorder, suicide ideation, and suicide attempts: results from a general population study. [Article]. Am J Public Health 90(2):251–257

Castelnuovo G (2008) Move the healthcare where it really needs: TECNOB project to improve outpatient obesity treatment. E-letter in response to Wister, Loewen, Kennedy-Symonds, McGowan, McCoy, Singer, "One-year follow-up of a therapeutic lifestyle intervention targeting cardiovascular disease risk". CMAJ 2007(177):859–865

Castelnuovo G (2010a) Empirically supported treatments in psychotherapy: towards an evidence-based or evidence-biased psychology in clinical settings? Front Psychol 1:27

Castelnuovo G (2010b) No medicine without psychology: the key role of psychological contribution in clinical settings. Front Psychol 1:4

Castelnuovo G, Gaggioli A, Mantovani F, Riva G (2003) From psychotherapy to e-therapy: the integration of traditional techniques and new communication tools in clinical settings. Cyberpsychol Behav 6(4):375–382

Castelnuovo G, Manzoni GM, Cuzziol P, Cesa GL, Tuzzi C, Villa V et al (2010) TECNOB: study design of a randomized controlled trial of a multidisciplinary telecare intervention for obese patients with Type-2 Diabetes. BMC Public Heal 10:204

Castelnuovo G, Simpson S (2011) Ebesity—e-health for obesity—new technologies for the treatment of obesity in clinical psychology and medicine. Clin Pract Epidemiol Ment Heal 7:5–8

Castelnuovo G, Manzoni GM, Corti S, Cuzziol P, Villa V (2011a) & Molinari, E. The TECNOB project. Telemedicine techniques and applications, Clin Psychol Med Treat Obes Outpatient Settings

Castelnuovo G, Manzoni GM, Cuzziol P, Cesa GL, Corti S, Tuzzi C et al (2011b) TECNOB study: Ad interim results of a randomized controlled trial of a multidisciplinary telecare intervention for obese patients with type-2 Diabetes. Clin Pract Epidemiol Ment Heal 7:44–50

Castelnuovo G, Manzoni GM, Villa V, Cesa GL, Molinari E (2011c) Brief strategic therapy vs cognitive behavioral therapy for the inpatient and telephone-based outpatient treatment of binge eating disorder: The STRATOB randomized controlled clinical trial. Clin Pract Epidemiol Ment Heal 7:29–37

Castelnuovo G, Manzoni GM, Villa V, Cesa GL, Pietrabissa G, Molinari E (2011d) The STRATOB study: design of a randomized controlled clinical trial of cognitive behavioral therapy and brief strategic therapy with telecare in patients with obesity and binge-eating disorder referred to residential nutritional rehabilitation. Trials 12:114

Cooper H (1990) Meta-analysis and the integrative research review. Research methods in personality and social psychology, pp 142–163

Cooper Z, Fairburn CG, Hawker DM, Press MF (2004) Cognitive-behavioral treatment of obesity: a clinician's guide. The Guilford, New York

Dalen J, Smith BW, Shelley BM, Sloan AL, Leahigh L, Begay D (2010) Pilot study: Mindful Eating and Living (MEAL): weight, eating behavior, and psychological outcomes associated with a mindfulness-based intervention for people with obesity. Complem Ther Med 18(6):260–264

Davin SA, Taylor NM (2009) Comprehensive review of obesity and psychological considerations for treatment. Psychol Heal Med 14(6):716–725

Del Parigi A, Chen K, Gautier JF, Salbe AD, Pratley RE, Ravussin E et al (2002a) Sex differences in the human brain's response to hunger and satiation. Am J Clin Nutr 75(6):1017–1022

Del Parigi A, Gautier JF, Chen K, Salbe AD, Ravussin E, Reiman E et al (2002b) Neuroimaging and obesity: mapping the brain responses to hunger and satiation in humans using positron emission tomography. Ann N Y Acad Sci 967:389–397

Doll HA, Petersen SE, Stewart-Brown SL (2000) Obesity and physical and emotional well-being: associations between body mass index, chronic illness, and the physical and mental components of the SF-36 questionnaire. Obes Res 8(2):160–170

Dombrowski SU, Avenell A, Sniehott FF (2011) Behavioural interventions for obese adults with additional risk factors for morbidity: systematic review of effects on behaviour, weight and disease risk factors. Obes Facts 3(6):377–396

Dombrowski SU, Sniehotta FF, Avenell A, Johnston M, MacLennan G, AraÃºjo-Soares V (2012) Identifying active ingredients in complex behavioural interventions for obese adults with obesity-related co-morbidities or additional risk factors for co-morbidities: a systematic review. Heal Psychol Rev 6(1):7–32

Ekeland AG, Bowes A, Flottorp S (2010) Effectiveness of telemedicine: a systematic review of reviews. Int J Med Inform 79(11):736–771

Ekeland AG, Bowes A, Flottorp S (2011) Methodologies for assessing telemedicine: a systematic review of reviews. Int J Med Inform 81(1):1–11

Engel GL (1977) The need for a new medical model: a challenge for biomedicine. Science 196(4286):129–136

Fabricatore AN, Wadden TA (2004) Psychological aspects of obesity. Clin Dermatol 22(4):332–337

Fagiolini A, Frank E, Scott JA, Turkin S, Kupfer DJ (2005) Metabolic syndrome in bipolar disorder: findings from the Bipolar Disorder Center for Pennsylvanians. Bipolar Disord 7(5):424–430

Fagiolini A, Frank E, Turkin S, Houck PR, Soreca I, Kupfer DJ (2008) Metabolic syndrome in patients with bipolar disorder. J Clin Psychiatry 69(4):678–679

Fernandez-Paredes F, Sumano-Avendano E (1986) Obesity in childhood and adolescence: evaluation after 1 year of integrated treatment. Bol Med Hosp Infant Mex 43(9):555–557

Fine JT, Colditz GA, Coakley EH, Moseley G, Manson JE, Willett WC et al (1999) A prospective study of weight change and health-related quality of life in women. JAMA 282(22):2136–2142

Flegal KM, Graubard BI, Williamson DF, Gail MH (2005) Excess deaths associated with underweight, overweight, and obesity. JAMA 293(15):1861–1867

Fontaine KR, Cheskin LJ, Barofsky I (1996) Health-related quality of life in obese persons seeking treatment. J Fam Pract 43(3):265–270

Foster GD, Makris AP, Bailer BA (2005) Behavioral treatment of obesity. Am J Clin Nutr 82(1 Suppl):230S–235S

Foster GD, Wadden TA, Makris AP, Davidson D, Sanderson RS, Allison DB et al (2003) Primary care physicians' attitudes about obesity and its treatment. Obes Res 11(10):1168–1177

Foster GD, Wadden TA, Vogt RA, Brewer G (1997) What is a reasonable weight loss? Patients' expectations and evaluations of obesity treatment outcomes. J Consult Clin Psychol 65(1):79–85

Friedman MA, Brownell KD (1995) Psychological correlates of obesity: moving to the next research generation. Psychol Bull 117(1):3–20

Gautier JF, Chen K, Salbe AD, Bandy D, Pratley RE, Heiman M et al (2000) Differential brain responses to satiation in obese and lean men. Diabetes 49(5):838–846

Gautier JF, Del Parigi A, Chen K, Salbe AD, Bandy D, Pratley RE et al (2001) Effect of satiation on brain activity in obese and lean women. Obes Res 9(11):676–684

Gearhardt AN, Corbin WR (2011) The role of food addiction in clinical research. Curr Pharm Des 17(12):1140–1142
Gearhardt AN, Davis C, Kuschner R, Brownell KD (2011a) The addiction potential of hyperpalatable foods. Curr Drug Abuse Rev 4(3):140–145
Gearhardt AN, Grilo CM, DiLeone RJ, Brownell KD, Potenza MN (2011b) Can food be addictive? Public health and policy implications. Addiction 106(7):1208–1212
Gearhardt AN, White MA, Potenza MN (2011c) Binge eating disorder and food addiction. Curr Drug Abuse Rev 4(3):201–207
Gearhardt AN, Yokum S, Orr PT, Stice E, Corbin WR, Brownell KD (2011d) Neural correlates of food addiction. Arch Gen Psychiatry 68(8):808–816
Gearhardt AN, White MA, Masheb RM, Morgan PT, Crosby RD, Grilo CM (2012) An examination of the food addiction construct in obese patients with binge eating disorder. Int J Eat Disord 45(5):657–663
Greeno CG, Wing RR (1994) Stress-induced eating. Psychol Bull 115(3):444–464
Gutierrez-Fisac JL, Angel Royo-Bordonada M, Rodriguez-Artalejo F (2006) [Health-risks associated with Western diet and sedentariness: the obesity epidemia]. Gac Sanit 20(Suppl 1):48–54
Haslam DW, James WP (2005) Obesity. Lancet 366(9492):1197–1209. doi: 10.1016/S0140-6736(05)67483-1
Hill AJ (2005) Psychological aspects of obesity. Psychiatry 4(4):26–30
Horgen K, Brownell K (2002) Confronting the toxic environment: environmental and public health actions in a world crisis. In: Wadden T, Stunkard A (eds) Handbook of obesity treatment. The Guilford Press, New York
Hudson JI, Chase EA, Pope HG Jr (1998) Eye movement desensitization and reprocessing in eating disorders: caution against premature acceptance. Int J Eat Disord 23(1):1–5
Hudson JI, Hiripi E, Pope HG Jr, Kessler RC (2007) The prevalence and correlates of eating disorders in the National Comorbidity Survey Replication. Biol Psychiatry 61(3):348–358
Ingram DH (1976) Psychoanalytic treatment of the obese person. Part III Am J Psychoanal 36(3):227–235
Istvan J, Zavela K, Weidner G (1992) Body weight and psychological distress in NHANES I. Int J Obes Relat Metab Disord 16(12):999–1003
Jeffery RW, Wing RR, Mayer RR (1998) Are smaller weight losses or more achievable weight loss goals better in the long term for obese patients? J Consult Clin Psychol 66(4):641–645
Kabat-Zinn J (1994) Wherever you go, there you are: mindfulness meditation in everyday life, 1st edn. Hyperion, New York
Khaylis A, Yiaslas T, Bergstrom J, Gore-Felton C (2010) A review of efficacious technology-based weight-loss interventions: five key components. Telemed J E Health 16(9):931–938
Kirschbaum C, Wust S, Hellhammer D (1992) Consistent sex differences in cortisol responses to psychological stress. Psychosom Med 54(6):648–657
Klein S, Sheard NF, Pi-Sunyer X, Daly A, Wylie-Rosett J, Kulkarni K et al (2004a) Weight management through lifestyle modification for the prevention and management of type 2 diabetes: rationale and strategies. A statement of the American Diabetes Association, the North American Association for the Study of Obesity, and the American Society for Clinical Nutrition. Am J Clin Nutr 80(2):257–263
Klein S, Sheard NF, Pi-Sunyer X, Daly A, Wylie-Rosett J, Kulkarni K et al (2004b) Weight management through lifestyle modification for the prevention and management of type 2 diabetes: rationale and strategies: a statement of the American Diabetes Association, the North American Association for the Study of Obesity, and the American Society for Clinical Nutrition. Diabetes Care 27(8):2067–2073
Kolotkin RL, Crosby RD (2002) Psychometric evaluation of the impact of weight on quality of life-lite questionnaire (IWQOL-lite) in a community sample. Qual Life Res 11(2):157–171
Kolotkin RL, Meter K, Williams GR (2001) Quality of life and obesity. Obes Rev 2(4):219–229
Kushner RF, Foster GD (2000) Obesity and quality of life. Nutrition 16(10):947–952

Lavoie KL, Bacon SL, Labrecque M, Cartier A, Ditto B (2006) Higher BMI is associated with worse asthma control and quality of life but not asthma severity. Respir Med 100(4):648–657
Lepore J (2011) American chronicles: Twilight. The New Yorker, 14 Mar 2011
Lillis J, Hayes SC, Bunting K, Masuda A (2009) Teaching acceptance and mindfulness to improve the lives of the obese: a preliminary test of a theoretical model. Ann Behav Med 37(1):58–69
Ludwig DS, Kabat-Zinn J (2008) Mindfulness in medicine. JAMA 300(11):1350–1352
Manzoni GM, Castelnuovo G (2011) Scale per la misurazione della qualità di vita nell'obesità. In: Molinari E, Castelnuovo G (eds) Clinica psicologica dell'obesità. Springer, Milan, pp 59–72
Manzoni GM, Castelnuovo G, Molinari E (2008) Weight loss with a low-carbohydrate, Mediterranean, or low-fat diet. N Engl J Med 359(20):2170; author reply 2171–2172
Manzoni GM, Pagnini F, Corti S, Molinari E, Castelnuovo G (2011) Internet-based behavioral interventions for obesity: an updated systematic review. Clin Pract Epidemiol Ment Heal 7:19–28
Marcus MD, Wildes JE (2009) Obesity: is it a mental disorder? Int J Eat Disord 42(8):739–753
McIntyre RS, Konarski JZ, Wilkins K, Soczynska JK, Kennedy SH (2006) Obesity in bipolar disorder and major depressive disorder: results from a national community health survey on mental health and well-being. Can J Psychiatry 51(5):274–280
National Heart Lung and Blood Institute (2000) The practical guide to the identification, evaluation and treatment of overweight and obesity in adults: NIH Publication
Nguyen T, Lau DC (2012) The obesity epidemic and its impact on hypertension. Can J Cardiol 28(3):326–333
Ni Mhurchu C, Bennett D, Lin R, Hackett M, Jull A, Rodgers A (2004) Obesity and health-related quality of life: results from a weight loss trial. N Z Med J 117(1207):U1211
NIH/NHLBI/NAASO (2000) The practical guide: Identification, evaluation, and treatment of overweight and obesity in adults. no. 00–4084
Nonas CA, Foster GD (2005) Setting achievable goals for weight loss. J Am Diet Assoc 105(5 Suppl 1):S118–S123
Nowicka P, Flodmark CE (2010) Family therapy as a model for treating childhood obesity: useful tools for clinicians. Clin Child Psychol Psychiatry 16(1):129–145
O'Neil PM, Smith CF, Foster GD, Anderson DA (2000) The perceived relative worth of reaching and maintaining goal weight. Int J Obes Relat Metab Disord 24(8):1069–1076
Orbach S (2006) Fat is a feminist issue. Arrow Books, New York
Peterson Y (2005) Family therapy treatment: working with obese children and their families with small steps and realistic goals. Acta Paediatr Suppl 94(448):42–44
Petry NM, Barry D, Pietrzak RH, Wagner JA (2008) Overweight and obesity are associated with psychiatric disorders: results from the National Epidemiologic Survey on Alcohol and related conditions. Psychosom Med 70(3):288–297
Pickering RP, Grant BF, Chou SP, Compton WM (2007) Are overweight, obesity, and extreme obesity associated with psychopathology? Results from the national epidemiologic survey on alcohol and related conditions. J Clin Psychiatry 68(7):998–1009
Pietrabissa G, Manzoni GM, Corti S, Vegliante N, Molinari E, Castelnuovo G (2012) Addressing motivation in globesity treatment: a new challenge for clinical psychology. Front Psychol 3:317
Price R (2002) Genetics and common obesities: background, current status, strategies, and future prospects. In: Wadden T, Stunkard A (eds) Handbook of obesity treatment. The Guilford Press, New York
Prochaska J, DiClemente C (1986) Toward a comprehensive model of change. In: Miller W, Heather N (eds) Treating addictive behavior: processes of change. Plenum, New York
Puhl R, Brownell KD (2001) Bias, discrimination, and obesity. Obes Res 9(12):788–805
Puhl R, Brownell KD (2003a) Ways of coping with obesity stigma: review and conceptual analysis. Eat Behav 4(1):53–78

Puhl RM, Brownell KD (2003b) Psychosocial origins of obesity stigma: toward changing a powerful and pervasive bias. Obes Rev 4(4):213–227
Puhl RM, Brownell KD (2006) Confronting and coping with weight stigma: an investigation of overweight and obese adults. Obesity (Silver Spring) 14(10):1802–1815
Puhl RM, Heuer CA (2009) The stigma of obesity: a review and update. Obesity (Silver Spring) 17(5):941–964
Puhl RM, Latner JD (2007) Stigma, obesity, and the health of the nation's children. Psychol Bull 133(4):557–580
Puhl RM, Moss-Racusin CA, Schwartz MB (2007) Internalization of weight bias: implications for binge eating and emotional well-being. Obesity (Silver Spring) 15(1):19–23
Puhl RM, Moss-Racusin CA, Schwartz MB, Brownell KD (2008) Weight stigmatization and bias reduction: perspectives of overweight and obese adults. Health Educ Res 23(2):347–358
Puhl RM, Schwartz MB, Brownell KD (2005) Impact of perceived consensus on stereotypes about obese people: a new approach for reducing bias. Health Psychol 24(5):517–525
Rand C, Stunkard AJ (1978) Obesity and psychoanalysis. Am J Psychiatry 135(5):547–551
Rand CS, Stunkard AJ (1977) Psychoanalysis and obesity. J Am Acad Psychoanal 5(4):459–497
Rand CS, Stunkard AJ (1983) Obesity and psychoanalysis: treatment and four-year follow-up. Am J Psychiatry 140(9):1140–1144
Rao G, Burke LE, Spring BJ, Ewing LJ, Turk M, Lichtenstein AH et al (2011) New and emerging weight management strategies for busy ambulatory settings: a scientific statement from the American Heart Association endorsed by the Society of Behavioral Medicine. Circulation 124(10):1182–1203
Rgon JC (1995) Feminist research: strengths and challenges. Contemp Nurse 4(4):180–186
Riva G, Castelnuovo G, Mantovani F (2006) Transformation of flow in rehabilitation: the role of advanced communication technologies. Behav Res Methods 38(2):237–244
Roehling M, Pilcher S, Oswald F, Bruce T (2008) The effects of weight bias on job-related outcomes: a meta-analysis of experimental studies. Paper presented at the Academy of Management Annual Meeting,
Rojas SV, Gagnon MP (2008) A systematic review of the key indicators for assessing telehomecare cost-effectiveness. Telemed J E Health 14(9):896–904
Sarwer DB, Fabricatore AN, Eisenberg MH, Sywulak LA, Wadden TA (2008) Self-reported stigmatization among candidates for bariatric surgery. Obesity (Silver Spring) 16(Suppl 2):S75–S79
Scott KM, Bruffaerts R, Simon GE, Alonso J, Angermeyer M, de Girolamo G et al (2008) Obesity and mental disorders in the general population: results from the world mental health surveys. Int J Obes (Lond) 32(1):192–200
Shaw K, O'Rourke P, Del Mar C, Kenardy J (2005) Psychological interventions for overweight or obesity. Cochrane Datab Syst Rev(2):CD003818
Singh NN, Lancioni GE, Singh AN, Winton AS, Singh J, McAleavey KM et al (2008) A mindfulness-based health wellness program for an adolescent with Prader-Willi syndrome. Behav Modif 32(2):167–181
Stoney CM, Davis MC, Matthews KA (1987) Sex differences in physiological responses to stress and in coronary heart disease: a causal link? Psychophysiology 24(2):127–131
Stroud LR, Salovey P, Epel ES (2002) Sex differences in stress responses: social rejection versus achievement stress. Biol Psychiatry 52(4):318–327
Stunkard AJ, Faith MS, Allison KC (2003) Depression and obesity. Biol Psychiatry 54(3):330–337
Swencionis C, Rendell SL (2012) The psychology of obesity. Abdom Imag
Tanofsky-Kraff M, Wilfley DE, Young JF, Mufson L, Yanovski SZ, Glasofer DR et al (2007) Preventing excessive weight gain in adolescents: interpersonal psychotherapy for binge eating. Obesity (Silver Spring) 15(6):1345–1355
Testa MA, Simonson DC (1996) Assesment of quality-of-life outcomes. N Engl J Med 334(13):835–840
Torres SJ, Nowson CA (2007) Relationship between stress, eating behavior, and obesity. Nutrition 23(11–12):887–894

Tuah NA, Amiel C, Qureshi S, Car J, Kaur B, Majeed A (2011) Transtheoretical model for dietary and physical exercise modification in weight loss management for overweight and obese adults. Cochrane Datab Syst Rev(10):CD008066
Van Strien T, Rookus MA, Bergers GP, Frijters JE, Defares PB (1986) Life events, emotional eating and change in body mass index. Int J Obes 10(1):29–35
Vasiljevic N, Ralevic S, Marinkovic J, Kocev N, Maksimovic M, Milosevic GS et al (2008) The assessment of health-related quality of life in relation to the body mass index value in the urban population of Belgrade. Heal Qual Life Outcomes 6:106
Wadden TA, Anderson DA, Foster GD, Bennett A, Steinberg C, Sarwer DB (2000) Obese women's perceptions of their physicians' weight management attitudes and practices. Arch Fam Med 9(9):854–860
Wang GJ, Volkow ND, Fowler JS (2002) The role of dopamine in motivation for food in humans: implications for obesity. Expert Opin Ther Targets 6(5):601–609
Wang GJ, Volkow ND, Logan J, Pappas NR, Wong CT, Zhu W et al (2001) Brain dopamine and obesity. Lancet 357(9253):354–357
Wang GJ, Volkow ND, Thanos PK, Fowler JS (2004) Similarity between obesity and drug addiction as assessed by neurofunctional imaging: a concept review. J Addict Dis 23(3):39–53
Wang GJ, Volkow ND, Thanos PK, Fowler JS (2009) Imaging of brain dopamine pathways: implications for understanding obesity. J Addict Med 3(1):8–18
Wardle J (2007) Eating behaviour and obesity. Obes Rev 8(Suppl 1):73–75
Wardle J, Chida Y, Gibson EL, Whitaker KL, Steptoe A (2010) Stress and adiposity: a meta-analysis of longitudinal studies. Obesity (Silver Spring) 19(4):771–778
Wardle J, Cooke L (2005) The impact of obesity on psychological well-being. Best Pract Res Clin Endocrinol Metab 19(3):421–440
Ware J, Snow K, Kosinki M (1993) SF-36 health survey: manual and interpretation guide. New England Med Cent, Boston
Whitlock G, Lewington S, Sherliker P, Clarke R, Emberson J, Halsey J et al (2009) Body-mass index and cause-specific mortality in 900 000 adults: collaborative analyses of 57 prospective studies. Lancet 373(9669):1083–1096
WHO (2006) Definition and diagnosis of diabetes mellitus and intermediate hyperglycemia. WHO, Geneva
Wing RR (2002) Behavioral weight control. In: Wadden TA, Stunkard AJ (eds) Handbook of obesity treatment. Guilford Press, New York, pp 301–316
Wylie-Rosett J, Swencionis C, Caban A, Friedler AJ, Schaffer N (2007) The complete weight loss workbook, 2nd edn. American Diabetes Association, Alexandria
Xu Y, Cardena E (2008) Hypnosis as an adjunct therapy in the management of diabetes. Int J Clin Exp Hypn 56(1):63–72

Chapter 10
Psychiatric Aspects

Massimo Cuzzolaro

Abstract Obesity is currently considered as a medical disease but not as a mental disease per se. However, the recommended approach to assessment, treatment, and rehabilitation of patients suffering from obesity is multidimensional and interdisciplinary. In the last two decades, the clinical picture called Binge Eating Disorder has received an increasing attention from both researchers and clinicians and has contributed to the development of a bio-psycho-social approach to obesity. Night eating and emotional eating are two other patterns of disordered eating that deserve attention. The prevalence of obesity in people diagnosed as having a mental illness is very high. Most psychotropic drugs are associated with the potential to induce obesity and obesity-related disorders but iatrogenic weight gain is only a component of the obesity-mental illness association that is a double factor of social stigma and disability. Obesity and mental disorder each accounts for substantial burden of disease, social prejudice, and discrimination and internalized stigma. A multifaceted relationship links psychopathology, obesity, social discrimination, and disability with circular connections.

10.1 Is Obesity a Disease? Should Obesity Per Se be Considered a Mental Disorder?

To identify obesity solely on the basis of a body mass index (BMI) or percentage of body fat in excess of some threshold and to label it a disease with no discussion of the appropriateness of the term leads to a lot of conceptual problems (Heshka and Allison 2001). Nevertheless, at present obesity is generally considered as a medical disease by a number of authoritative bodies, with some ambiguity.

The World Health Organization (WHO) reports on obesity typically refer to this condition as a disease but there is no discussion about the correctness of the term.

M. Cuzzolaro (✉)
University of Roma Sapienza Eating and Weight Disorders, Studies on Anorexia Bulimia Obesity, Roma, Italy
e-mail: massimo.cuzzolaro@alice.it

P. Capodaglio et al. (eds.), *Disabling Obesity*, DOI: 10.1007/978-3-642-35972-9_10,

In actual fact obesity is listed in the WHO International Classification of Diseases (ICD-10, 10th edition), but it is noteworthy that the full name of ICD is the *International Statistical Classification of Diseases and Related Health Problems* because it includes conditions which are not diseases per se but represent risk factors to health.

Consequently, should obesity be considered a disease or a risk factor?

In 2008 a white paper on evidence and arguments commissioned by the Council of the Obesity Society recognized that it is reasonable to call obesity a disease on the basis of a speculative utilitarian analysis: "considering obesity a disease is likely to have far more positive than negative consequences and benefit the greater good by soliciting more resources into research, prevention, and treatment of obesity" (Allison et al. 2008).

In the same year, The Obesity Society Council published its resolution: "Obesity is a complex condition with numerous causes, many of which are largely beyond an individual's control… obesity causes much suffering, ill health, and earlier mortality, and persons with obesity are subject to enormous societal stigma and discrimination. Whether obesity should be declared a disease is a topic on which thoughtful arguments have been expressed on both sides of the issue… After extensive dialogue and careful consideration, the Council concludes that the official position of The Obesity Society is that obesity should be declared a disease. The Obesity Society will henceforth refer to obesity as a disease and encourages others to do so" (Obesity Society Council 2008).

Three years later the American Association of Clinical Endocrinologists (AACE) stated that "there is significant clinical evidence to declare obesity as a disease state. The Association believes that the declaration will help lead the way for more effective therapies and treatments to help the 34 % of Americans currently suffering with obesity" (AACE 2011).

Another related and recurrent question is the following: should obesity be considered a mental disorder?

Considering obesity as a mental or behavioral disorder raises many problems. In 2007 Devlin wondered if there is a place for obesity in the future DSM-5, Diagnostic and Statistical Manual of Mental Disorders, 5th edition (Devlin 2007). He reviewed three possible models of non-homeostatic eating that result in obesity: the first model weighs up the *form* of overeating to classify obesity as an eating disorder; the second model stresses the *consequences* of non-homeostatic eating and considers obesity as an addictive disorder (food addiction); the third model focuses on the *function* of food and concentrates on difficulties in the area of affect regulation and stress response. The author concluded that to devise diagnostic criteria based on the above models raises multiple difficulties because the phenomena central to each model are basically dimensional. The neurobiological relationships among eating behavior, reward systems, and affect regulation systems represent a very interesting topic (Orford 2001; Riva et al. 2006; Acosta et al. 2008; Avena 2010; Gearhardt et al. 2011) but a detailed understanding of the connections is still lacking.

At present, the recommended approach to assessment and treatment of obesity is multidimensional and interdisciplinary: it is essential to consider together the overall physical condition, medical complications, disabilities, psychiatric comorbidity, self-regulation, motivation, self-efficacy, family, social resources, the role of stigma, prejudice, discrimination, environment, and available services (Donini et al. 2010; Karasu 2012).

10.2 Binge Eating Disorder

Binge eating disorder (BED) is not yet an approved DSM-IV-TR (American Psychiatric Association 2000) or ICD-10 (World Health Organization 1992) diagnostic category but it is already accepted as an eating disorder (ED) in actual practice and will be included in DSM-5 as a distinct syndrome.

BED has connected the medical area of obesity with the psychiatric field of eating disorders. In the last 20 years, this bridge attracted more attention to the psychological and psychiatric aspects of obesity and contributed to the development of a multidimensional team approach to the assessment and treatment of eating and weight disorders (Cuzzolaro and Vetrone 2009).

The core feature of BED is the presence of recurrent episodes of overeating with loss of control and no regular use of inappropriate weight-loss behaviors like self-induced vomiting. The frequency cut-point for diagnosis is 2 days per week for 6 months in DSM-IV but probably will be once per week for 3 months in DSM-5 that should appear in May 2013.

The results of the National Comorbidity Survey Replication, a face-to-face household study conducted in a large representative sample ($n = 9282$), indicated the following lifetime community prevalence rates of DSM-IV-BED in USA: 3.5 % in women and 2.0 % in men; F to M ratio was 1.75 to 1.0. Furthermore, lifetime BED was significantly associated with current obesity class III (BMI ≥ 40 kg/m^2) (Hudson et al. 2007).

Subjects with obesity and BED often are trapped in a cycle of attempting to diet, then losing control, binge eating, and gaining even more weight (Herman and Polivy 1990). However, it remains uncertain which comes first: bingeing or dieting?

There is a significant relationship between BED and type 2 diabetes. BED is likely a risk factor for an accelerated weight gain, which often involves an increase in insulin resistance (Herpertz et al. 2006).

The presence of BED is a useful marker of comorbid psychopathology especially depression, anxiety disorders, and substance abuse disorders (Stunkard and Allison 2003; Guerdjikova et al. 2007). In obesity surgery candidates, BED is associated with an increased prevalence of axis I psychopathology (in particular, current and lifetime mood and anxiety disorders), beyond the already elevated rate observed with obesity class III (Jones-Corneille et al. 2012).

Studies on BED treatment are burdened by several limitations as selection biases (e.g., mostly women and overweight), small samples, high drop-out rates, and placebo response (Ramacciotti et al. accepted for publication).

Psychological treatments may be useful, in particular cognitive-behavioral therapy (CBT), dialectical behavior therapy, and interpersonal psychotherapy (Molinari et al. 2005; Castelnuovo et al. 2011; Tasca et al. 2012). For example, a manualized CBT approach offered one day a week during an average of 7 months produces benefits that are durable up to 3.5 years post treatment (Vanderlinden et al. 2012).

A number of psychotropic drugs (antidepressants, in particular SSRIs, selective serotonin reuptake inhibitors, like fluoxetine and sertraline; anticonvulsants/mood stabilizers like topiramate and zonisamide) may have modest beneficial effects. For example, zonisamide augmentation to individual cognitive behavior therapy can improve the treatment of binge eating disorder patients, reducing body weight and the number of binge eating episodes. These results were maintained 1 year after the end of treatment (Ricca et al. 2009).

Loss of control is central to psychological disturbance associated with BED. Feelings of loss of control could drive binge eaters to seek weight loss surgery more often than non-BED obese subjects in an attempt to gain control over both compulsive eating behavior and body shape (Colles et al. 2008).

The prevalence of BED among bariatric surgery candidates is high but this syndrome should not be considered as an absolute contraindication. Adequate pre- and post-surgical treatment (e.g., CBT-group therapy with or without SSRIs) may help BED patients to obtain positive outcomes from weight loss surgery (Ashton et al. 2009, 2011).

Binge eating or loss of control eating may be observed in children and adolescents as well, usually associated with overweight/obesity (Marcus and Kalarchian 2003; Tanofsky-Kraff et al. 2008; Cebolla et al. 2012). Binge eating appears independently associated with emotional eating, depression, and anxiety and positively related to parental problems (e.g., underinvolvement, arguments, and depression of family members), and social problems (d'Autume et al. 2012; Hartmann et al. 2012). As regards the last point, the interpersonal model proposes that social difficulties lead to negative affect which, in turn, precipitates binge eating episodes (Elliott et al. 2010).

10.3 Night Eating Syndrome and Emotional Eating

Two other patterns of disordered eating are frequently associated with obesity: the so-called night eating syndrome (NES) and emotional eating (EE). Albert Stunkard is one of the researchers who supposed that night eating should be classified as a distinct eating disorder (Stunkard et al. 1955, 2008; Stunkard and Lu 2010). There are no approved DSM or ICD criteria for NES that could be tentatively defined by two core criteria: evening hyperphagia (ingestion of at least 25 % of daily calories after supper) and/or awakenings with ingestions (at least three times a week). According to these criteria, NES affects men and women and its prevalence is about 1.1–1.5 % of the general population, 6–16 % of patients in weight reduction programs, and 8–42 % of candidates for bariatric surgery.

NES is frequently associated with depressed mood that worsens in the evening and is gaining large recognition for its role in the development and maintenance of obesity (Gallant et al. 2012; Milano et al. 2012). However, the relationship between this syndrome and obesity remains unclear. Circadian genes may play a role (Gallant et al. 2012). Limited evidence suggests that psychological interventions, some psychotropic drugs (e.g., selective serotonin re-uptake inhibitors, topiramate, agomelatin, etc.) and bariatric surgery may diminish symptoms.

It is well known that eating usually moderates distress. In some people there is a very strong relationship between eating, energy intake, and emotions, above all depression, anger, anxiety (Goldbacher et al. 2012), and boredom (Koball et al. 2012).

EE may be observed in adults, adolescents, and children (Michels et al. 2012; Vannucci et al. 2012).

Several studies suggest that the hormone ghrelin might play a key role in the relationship between acute and chronic stress and food intake (Schellekens et al. 2012). At present, ghrelin is the only known orexigenic hormone from the periphery to stimulate food intake and its plasma levels are enhanced under conditions of physiological stress.

10.4 Obesity, Anxiety, and Depression. The Hen or the Egg?

Anxiety and depression often go together with BED- and non-BED-obesity.

Which came first? Some studies indicate that childhood–adolescence obesity is a strong and independent risk factor for adult depression. For example, a large cohort study of 91.798 female registered nurses followed longitudinally for 12 years found that women in the two highest categories of body shape at age 10 had both higher prevalence (OR = 2.59, 95 % CI 1.46–4.61) and incidence (OR = 2.01, 95 % CI 1.08–3.71) of depression than lean women of the same age. Similar results were obtained for BMI at age 18 (Sanchez-Villegas et al. 2012).

However, a number of studies suggest a bidirectional relation between obesity and depression. The Nurses' Health Study prospectively followed from 1996 to 2006, 65.955 women aged 54–79 years. Information about body weight, depression status, and various covariates were updated every 2 years. Depression at the baseline period was associated with an increased risk of obesity at the follow-up period also in baseline non-obese women. On the other hand, obese women at baseline had a moderately increased risk of depression at the follow-up period compared with baseline normal-weight women also for new onset of depression (Pan et al. 2012).

Few studies examined participants in weight reduction trials to determine whether moderate weight loss would be associated with incident symptoms of depression and suicidal ideation, and whether symptoms of depression at baseline would limit weight loss. An accurate research seems to demonstrate that mild (or greater) symptoms of depression at baseline do not prevent overweight/obese individuals with type 2 diabetes from achieving significant weight loss (Faulconbridge et al. 2012).

Several studies have demonstrated a relationship between chronic stress, depression, and ingestion of caloric dense *comfort foods*. This association likely contributes to the higher risk of cardiovascular disease in anxiety and depressive disorders.

For better prevention and treatment it would be useful to clarify which biological systems and/or lifestyle factors underlie these links. Many recent biological studies highlight the role of inflammatory factors and of the gut-brain peptide ghrelin. For example, Arianne Van Reedt Dortland and coworkers found that chronic low-grade inflammation (biological stress factor) and smoking (lifestyle factor) partly explain increased dyslipidemia and abdominal obesity risk in patients with severe anxiety disorders and depression (van Reedt Dortland et al. 2012). In addition, we have to remark that tricyclic antidepressants may increase metabolic risk through enhanced sympathetic and decreased parasympathetic autonomic nervous system activity.

As regards ghrelin, according to the conclusions of a recent review article, the ghrelinergic system seems to be a critical factor at the interface of homeostatic control of appetite and reward circuitries, modulating the hedonic aspects of food intake (Schellekens et al. 2012). Craving for and addictions to alcohol and sweet foods were both found to be associated with increased levels of ghrelin. Animal and human studies suggest that ghrelin plays an important role in addictive behaviors, particularly in alcohol and food seeking behaviors. Consumption of alcohol and sweet food is associated with the activation of ghrelin receptors in the hippocampus and probably triggers chemical processes which are experienced as rewarding by the individual. (Leggio et al. 2011, 2012). Perhaps future anti-obesity pharmacotherapies should focus also on the ghrelinergic system to modulate the rewarding properties of food.

10.5 Obesity and Negative Body Image

The first systematic studies on body image in subjects with obesity date back to the 1960s (Stunkard and Burt 1967; Stunkard and Mendelson 1967) and their clinical relevance has continued to increase over the years.

Such factors as gender, age, degree of obesity, onset age of overweight, ethnicity, social class, history of childhood teasing and parental criticism about weight, history of weight cycling, and presence of binge eating all show important modulating effects on body uneasiness (Sarwer and Thompson 2002; Schwartz and Brownell 2002; Sarwer et al. 2005; Dalle Grave et al. 2007; Marsh et al. 2007; Franko et al. 2012). In the last two decades, several studies have remarked that individuals with BED-obesity present greater dissatisfaction and distress about their body appearance than people with non-BED-obesity (Wilfley et al. 2000a, b; Cuzzolaro et al. 2008).

Patients with BED may present weight and shape concerns comparable to bulimia nervosa patients and higher than anorexia nervosa patients (Wilfley et al. 2000a, b). A negative body image could drive obese binge eaters to seek bariatric surgery more often than non-BED obese subjects, other conditions being equal. Therefore, assessment and treatment of persons with obesity and BED should consider cognitive as well as behavioral dysfunctional attitudes and outcome studies should be focused not only on eating behavior but on body image as well.

10.6 Obesity and Cognitive Function

In spite of some relevant discrepancies in findings between studies, possibly due to participant sampling and methodological differences, obesity, in particular central obesity, seems to be a risk factor for faster cognitive decline and a number of epidemiological studies suggest that being obese in midlife is a risk factor for dementia in later life (Xu et al. 2011; Chang et al. 2012; Luchsinger et al. 2012). Remarkably, subjects with anorexia nervosa and obesity present similar dysfunctional executive profiles: decision making, response inhibition, and cognitive flexibility (Fagundo et al. 2012). Obesity adversely affects frontal lobe brain structure and function. Individuals suffering from obesity have significantly smaller total brain volumes and specifically, significantly reduced total gray matter volume (Brooks et al. 2012).

Midlife obesity may be an important modifier of brain atrophy in individuals who are developing cognitive impairment and dementia, while it seems to have little effect on structural brain integrity in non-demented older adults (Driscoll et al. 2012).

Is obesity without metabolic abnormalities linked to normal brain decline?

Archana Singh-Manoux recently analyzed with other British researchers cognitive function in 6,401 adults using a battery of tests, three times over 10 years, to study the effects of obesity and metabolic health on cognitive decline. The authors found that the obese subjects (BMI $\geq$ 30) tended to lose their mental powers faster than the thinner subjects. The fastest decline in memory and thinking skills was observed in those with both obesity and metabolic abnormalities (high blood pressure, high LDL cholesterol, etc.) who experienced a 22.5 % faster decline on their cognitive test scores over the decade than those who were healthy. However, among the obese, decline over 10 years was similar in the metabolically normal and abnormal groups (Singh-Manoux 2012).

10.7 Psychiatric Disorders Subsequent to Weight Loss Surgery

The current and lifetime rates of psychopathology are high in bariatric surgery candidates, in particular the lifetime rates of mood and alcohol use disorders; the BED prevalence is about 10 % (Mitchell et al. 2012) and BED is associated with an increased prevalence of axis I psychopathology (Jones-Corneille et al. 2012).

Weight loss surgery candidates may have a greater lifetime risk of alcohol use disorders. After surgery many patients continue to consume alcohol regardless of greater sensitivity to the intoxicating effects of alcohol (Heinberg et al. 2012) and not rarely they increase quantities. Some patients seem to experience *addiction transfer*. New compulsive behaviors and new addictions may blossom to replace overeating: alcohol, smoking, drugs, gambling, shopping, etc. However, adequate studies of this phenomenon are not yet available. In the last few decades several case studies and case series have investigated the emergence of eating disorders after bariatric surgery. In most cases diagnostic criteria for anorexia nervosa or bulimia nervosa full syndrome were not satisfied and Segal and coworkers proposed a new category, Post-Surgical Eating Avoidance Disorder, PSEAD (Powers et al. 1999; Segal et al. 2004). Emotional eating and compulsive grazing are eating patterns that can persist or appear after bariatric surgery and negatively affect outcome (Saunders 2004; Chesler 2012). The post-operative development of eating disorders in weight-loss surgery patients seems a quite rare outcome but it is likely that such symptoms are underreported (Marino et al. 2012). Bariatric surgery patients with comorbid psychiatric disorders (e.g., severe personality disorder, intense body image disturbance) may show greater difficulties in adapting to the postoperative way of life and psychological and/or psychiatric treatment may be necessary (Kinzl 2010).

10.8 Obesity, Serious Mental Illnesses, and Psychotropic Drugs

The prevalence of obesity in people diagnosed as having a mental illness is very high. Persons with serious mental illness (mood and anxiety disorders, schizophrenia, post-traumatic stress disorder, personality disorders, etc.) are 1.5 to more than 3 times as likely to be obese as the general population (Simon et al. 2006; Petry et al. 2008; Pagoto et al. 2012).

The prevalence of obesity (BMI > 30) in a sample of inpatients suffering from schizophrenia was 42.08 %, 3.5 times that of the Canadian average (12 %) and over three-quarters of women were overweight or obese (Coodin 2001).

A recent Australian study examined all male and female inpatients in a public mental health service center and found that the percentage of obese persons (30.3 %) was much higher than that of the general population (21.4 %). Females showed a higher propensity toward obesity than males. Most diagnostic categories had a mean BMI in the overweight range, whereas personality disorders had a mean BMI in the obese range with a percentage of individuals with obesity higher than that observed in patients suffering from psychosis (Stanley et al. 2012).

Overweight/obesity and mental disorders (depression, psychotic disorders, attention deficit hyperactivity disorder, anxiety disorders, etc.) appear frequently together also in children and adolescents (Haw and Bailey 2012; Kalarchian and Marcus 2012).

Psychiatrists who disregard the physical health of their patients and focus only on treating the patient's psychiatric condition are not only neglectful but are also contributing to the epidemic of obesity among people with serious mental illness (Radke et al. 2010).

Obesity and metabolic syndrome are now well-documented side effects of second-generation antipsychotic use (Nasrallah 2003; Shin et al. 2008; Ellingrod et al. 2012). However, the association between schizophrenia and the metabolic syndrome may be observed independently from the risk of induction of this syndrome by antipsychotic drugs (Bou Khalil 2012) and iatrogenic weight gain is only a piece of the obesity-mental illness dyad (Taylor et al. 2012a, b).

The interaction of genetic factors, socioeconomic status, lifestyle, and medications likely accounts for the high risk of overweight, metabolic syndrome, and premature mortality in people with serious mental illness.

Undoubtedly most psychotropic drugs—not only antipsychotics but antidepressants and mood stabilizers as well—are associated with the potential to induce weight-gain and obesity-related disorders such as diabetes type 2, hypercholesterolemia, and hypertension. In addition, weight gain contributes to patient noncompliance with pharmacotherapy. Genetic and gender differences may be significant factors: women are more susceptible to weight gain (Dent et al. 2012; Nihalani et al. 2012). On the other hand, very few psychotropics are associated with weight loss. For example, zonisamide, a sulfonamide anticonvulsant and mood stabilizer, has been shown to cause weight loss instead of weight gain (Wellmer et al. 2009).

Clinicians should take into account potential for weight gain when they prescribe any psychotropic. They should monitor metabolic parameters in their patients, and educate them and caregivers about the risks and how to prevent them (Ganguli and Strassnig 2011; Copeland et al. 2012). Early intervention is the key to preventing significant drug-related weight gain. Patients should be told about weight gain as a potential adverse effect before they begin treatment. Their weight and waist circumference should be monitored as a standard of care and routine monitoring for hyperglycemia, hyperlipidemia, and hypertension should be performed (Lin et al. 2012; Panagiotopoulos et al. 2012). Ideally, a diet and exercise plan should be initiated to prevent weight gain before medically significant overweight occurs. Diet and physical exercise require commitment and motivation on the part of the patient. This is a difficult task but not impossible (Umbricht et al. 2001; Ganguli and Strassnig 2011; Taylor et al. 2012a, b).

10.9 Obesity, Psychopathology, and Disability

Obesity and mental disorder each accounts for considerable burden of disease, social prejudice and discrimination, and internalized stigma. It is well known that also after controlling for physical disease severity, psychopathology is consistently associated with increased multidimensional disability in everyday living skills, structured daily activities and functional domains with a dose–response relationship between severity of mental illness and level of disability. Patient-reported and interviewer-rated disability, global burden of disease, number of disability days, disability-adjusted life years (DALYs) are the main outcome measures. Three domains of real-world functioning are usually explored: community and household activities, work skills, and interpersonal relationships. Disability is prominent among patients suffering from schizophrenia and other psychotic disorders, bipolar disorder, major depression and other mood disorders, panic disorder and other anxiety disorders, post-traumatic stress disorder (Ormel et al. 1994; Andrews et al. 1998; Sareen et al. 2007; Patel 2008; Bowie et al. 2010; Miller et al. 2012). As the prevalence of obesity in people diagnosed as having a mental illness is very high, obesity and obesity-related diseases contribute to the total disability level. In addition, a double social and internalized stigma impairs the quality of life of individuals who are mentally ill and overweight (Brownell et al. 2005; Carels et al. 2009; Paterson et al. 2012). Persons with intellectual disabilities present a higher prevalence of obesity than the general population, and a trend to an increase in the prevalence of excess weight. Furthermore, mental disorders frequently coexist and give a relevant contribution to the total disability level (Einfeld et al. 2011; Gazizova et al. 2012; Taua et al. 2012).

In conclusion, a multifaceted relationship links psychopathology, obesity, social discrimination, and disability with circular connections. For example, it has been observed that depressed mood seen in Class III obese surgery-seeking individuals may be most related to weight-related stigma rather than binge eating status, or weight-related physical disability (Chen et al. 2007). Implications for rehabilitation research and intervention require further investigation and a holistic approach should be more and more implemented.

References

Acosta MC, Manubay J et al (2008) Pediatric obesity: parallels with addiction and treatment recommendations. Harv Rev Psychiatry 16(2):80–96

Allison DB, Downey M et al (2008) Obesity as a disease: a white paper on evidence and arguments commissioned by the council of the obesity society. Obesity (Silver Spring) 16(6):1161–1177

American Association of Clinical Endocrinologists (AACE) (2011) (Published on AACE Online Newsroom (http://media.aace.com) on 7/29/11 9:23 am). American Association of Clinical Endocrinologists (AACE) Declares Obesity a Disease State

American Psychiatric Association (2000) Diagnostic and statistical manual of mental disorders, DSM IVTR, Text Revised, 4th edn. American Psychiatric Association, Washington

Andrews G, Sanderson K et al (1998) Burden of disease. Methods of calculating disability from mental disorder. Br J Psychiatry 173:123–131

Ashton K, Drerup M et al (2009) Brief, four-session group CBT reduces binge eating behaviors among bariatric surgery candidates. Surg Obes Relat Dis 5(2):257–262

Ashton K, Heinberg L et al (2011) Positive response to binge eating intervention enhances postoperative weight loss. Surg Obes Relat Dis 7(3):315–320

Avena NM (2010) The study of food addiction using animal models of binge eating. Appetite 55:734–737

Bou Khalil R (2012) Metabolic syndrome in patients with schizophrenia independently from atypical antipsychotics intake. Presse Med 41(5):e238–e243

Bowie CR, Depp C et al (2010) Prediction of real-world functional disability in chronic mental disorders: a comparison of schizophrenia and bipolar disorder. Am J Psychiatry 167(9):1116–1124

Brooks SJ, Benedict C et al (2012) Late-life obesity is associated with smaller global and regional gray matter volumes: a voxel-based morphometric study. Int J Obes (Lond) Advance online publication, 31 Jan 2012; doi: 10.1038/ijo.2012.13

Brownell K, Puhl R et al (eds) (2005) Weight bias. Nature, consequences and remedies. Guilford, New York

Carels RA, Young KM et al (2009) Internalized weight stigma and its ideological correlates among weight loss treatment seeking adults. Eat Weight Disord 14(2–3):e92–e97

Castelnuovo G, Manzoni GM et al (2011) Brief strategic therapy vs cognitive behavioral therapy for the inpatient and telephone-based outpatient treatment of binge eating disorder: the STRATOB randomized controlled clinical trial. Clin Pract Epidemiol Ment Health 7:29–37

Cebolla A, Perpina C et al (2012) Prevalence of binge eating disorder among a clinical sample of obese children. An Pediatr (Barc) 77(2):98–102

Chang WS, Won KH et al (2012) The relationship between obesity and the high probability of dementia based on the body mass index and waist circumference. Korean J Fam Med 33(1):17–24

Chen EY, Bocchieri-Ricciardi LE et al (2007) Depressed mood in class III obesity predicted by weight-related stigma. Obes Surg 17(5):669–671

Chesler BE (2012) Emotional eating: a virtually untreated risk factor for outcome following bariatric surgery. Sci World J 2012:365961

Colles SL, Dixon JB et al (2008) Loss of control is central to psychological disturbance associated with binge eating disorder. Obesity (Silver Spring) 16(3):608–614

Coodin S (2001) Body mass index in persons with schizophrenia. Can J Psychiatry 46(6):549–555

Copeland LA, Pugh MJ et al (2012) Use of obesity-related care by psychiatric patients. Psychiatr Serv 63(3):230–236

Cuzzolaro M, Bellini M et al (2008) Binge eating disorder and body uneasiness. Psychol Top 17(2):287–312

Cuzzolaro M, Vetrone G (2009) Overview of evidence on the underpinnings of binge eating disorder and obesity. Evidence based treatments for eating disorders: children, adolescents and adults. In: Dancyger I, Fornari V (eds) Nova Science Publishers, New York, pp 53–70

d'Autume C, Musher-Eizenman D et al (2012) Eating behaviors and emotional symptoms in childhood obesity: a cross-sectional exploratory study using self-report questionnaires in 63 children and adolescents. Arch Pediatr 19(8):803–810

Dalle Grave R, Cuzzolaro M et al (2007) The effect of obesity management on body image in patients seeking treatment at medical centers. Obesity (Silver Spring) 15(9):2320–2327

Dent R, Blackmore A et al (2012) Changes in body weight and psychotropic drugs: a systematic synthesis of the literature. PLoS ONE 7(6):e36889

Devlin MJ (2007) Is there a place for obesity in DSM-V? Int J Eat Disord 40:S83–S88

Donini LM, Cuzzolaro M et al (2010) Obesity and eating disorders. Indications for the different levels of care. An Italian expert consensus document. Eat Weight Disord 15(1–2 Suppl):1–31

Driscoll I, Beydoun MA et al (2012) Midlife obesity and trajectories of brain volume changes in older adults. Hum Brain Mapp 33(9):2204–2210

Einfeld SL, Ellis LA et al (2011) Comorbidity of intellectual disability and mental disorder in children and adolescents: a systematic review. J Intellect Dev Disabil 36(2):137–143

Ellingrod VL, Taylor SF et al (2012) Risk factors associated with metabolic syndrome in bipolar and schizophrenia subjects treated with antipsychotics: the role of folate pharmacogenetics. J Clin Psychopharmacol 32(2):261–265

Elliott CA, Tanofsky-Kraff M et al (2010) An examination of the interpersonal model of loss of control eating in children and adolescents. Behav Res Ther 48(5):424–428

Fagundo AB, de la Torre R et al (2012) Executive functions profile in extreme eating/weight conditions: from anorexia nervosa to obesity. PLoS ONE 7(8):e43382

Faulconbridge LF, Wadden TA et al (2012) One-year changes in symptoms of depression and weight in overweight/obese individuals with type 2 diabetes in the Look AHEAD study. Obesity (Silver Spring) 20(4):783–793

Franko DL, Coen EJ et al (2012) Considering J. Lo and "Ugly Betty": a qualitative examination of risk factors and prevention targets for body dissatisfaction, eating disorders, and obesity in young Latina women. Body Image 9(3):381–387

Gallant AR, Lundgren J et al (2012) The night-eating syndrome and obesity. Obes Rev 13(6):528–536

Ganguli R, Strassnig M (2011) Prevention of metabolic syndrome in serious mental illness. Psychiatr Clin North Am 34(1):109–125

Gazizova D, Puri BK et al (2012) The overweight: obesity and plasma lipids in adults with intellectual disability and mental illness. J Intellect Disabil Res 56(9):895–901

Gearhardt AN, Yokum S et al (2011) Neural correlates of food addiction. Arch Gen Psychiatry 68(8):808–816

Goldbacher EM, Grunwald HE et al (2012) Factor structure of the emotional eating scale in overweight and obese adults seeking treatment. Appetite 59(2):610–615

Guerdjikova AI, McElroy SL et al (2007) Comparison of obese men and women with binge eating disorder seeking weight management. Eat Weight Disord 12(1):e19–e23

Hartmann AS, Czaja J et al (2012) Psychosocial risk factors of loss of control eating in primary school children: a retrospective case-control study. Int J Eat Disord 45(6):751–758

Haw C, Bailey S (2012) Body mass index and obesity in adolescents in a psychiatric medium secure service. J Hum Nutr Diet 25(2):167–171

Heinberg LJ, Ashton K et al (2012) Alcohol and bariatric surgery: review and suggested recommendations for assessment and management. Surg Obes Relat Dis 8(3):357–363

Herman CP, Polivy J (1990) From dietary restraint to binge eating: attaching causes to effects. Appetite 14(2):123–125; discussion 142–123

Herpertz S, Petrak F et al (2006) Eating disorders and diabetes mellitus. Ther Umsch 63(8):515–519

Heshka S, Allison DB (2001) Is obesity a disease? Int J Obes Relat Metab Disord 25(10):1401–1404

Hudson JI, Hiripi E et al (2007) The prevalence and correlates of eating disorders in the national comorbidity survey replication. Biol Psychiatry 61(3):348–358

Jones-Corneille LR, Wadden TA et al (2012) Axis I psychopathology in bariatric surgery candidates with and without binge eating disorder: results of structured clinical interviews. Obes Surg 22(3):389–397

Kalarchian MA, Marcus MD (2012) Psychiatric comorbidity of childhood obesity. Int Rev Psychiatry 24(3):241–246

Karasu SR (2012) Of mind and matter: psychological dimensions in obesity. Am J Psychother 66(2):111–128

Kinzl JF (2010) Morbid obesity: significance of psychological treatment after bariatric surgery. Eat Weight Disord 15(4):e275–e280

Koball AM, Meers MR et al (2012) Eating when bored: revision of the emotional eating scale with a focus on boredom. Health Psychol 31(4):521–524

Leggio L, Addolorato G et al (2011) Role of feeding-related pathways in alcohol dependence: a focus on sweet preference, NPY, and ghrelin. Alcohol Clin Exp Res 35(2):194–202

Leggio L, Ferrulli A et al (2012) Ghrelin system in alcohol-dependent subjects: role of plasma ghrelin levels in alcohol drinking and craving. Addict Biol 17(2):452–464

Lin CC, Yu SC et al (2012) Measurement of waist circumference at different sites affects the detection of abdominal obesity and metabolic syndrome among psychiatric patients. Psychiatry Res 197(3):322–326

Luchsinger JA, Cheng D et al (2012) Central obesity in the elderly is related to late-onset Alzheimer disease. Alzheimer Dis Assoc Disord 26(2):101–105

Marcus MD, Kalarchian MA (2003) Binge eating in children and adolescents. Int J Eat Disord 34(Suppl):S47–S57

Marino JM, Ertelt TW et al (2012) The emergence of eating pathology after bariatric surgery: a rare outcome with important clinical implications. Int J Eat Disord 45(2):179–184

Marsh HW, Hau KT et al (2007) Childhood obesity, gender, actual-ideal body image discrepancies, and physical self-concept in Hong Kong children: cultural differences in the value of moderation. Dev Psychol 43(3):647–662

Michels N, Sioen I et al (2012) Stress, emotional eating behaviour and dietary patterns in children. Appetite 59(3):762–769

Milano W, De Rosa M et al (2012) Night eating syndrome: an overview. J Pharm Pharmacol 64(1):2–10

Miller CJ, Abraham KM et al (2012) Quality of life among patients with bipolar disorder in primary care versus community mental health settings. J Affect Disord Advance online publication, 12 Sept 2012; doi: 10.1016/j.jad.2012.08.045

Mitchell JE, Selzer F et al (2012) Psychopathology before surgery in the longitudinal assessment of bariatric surgery-3 (LABS-3) psychosocial study. Surg Obes Relat Dis 8:533–541

Molinari E, Baruffi M et al (2005) Binge eating disorder in obesity: comparison of different therapeutic strategies. Eat Weight Disord 10(3):154–161

Nasrallah H (2003) A review of the effect of atypical antipsychotics on weight. Psychoneuroendocrinology 28(Suppl 1):83–96

Nihalani N, Schwartz TL et al (2012) Obesity and psychotropics. CNS Neurosci Ther 18(1):57–63

Obesity Society Council (2008) Obesity as a disease: the obesity society council resolution. Obesity (Silver Spring) 16(6):1151

Orford J (2001) Excessive appetites: a psychological view of addictions. Wiley, New York

Ormel J, VonKorff M et al (1994) Common mental disorders and disability across cultures. Results from the WHO collaborative study on psychological problems in general health care. JAMA 272(22):1741–1748

Pagoto SL, Schneider KL et al (2012) Association of post-traumatic stress disorder and obesity in a nationally representative sample. Obesity (Silver Spring) 20(1):200–205

Pan A, Sun Q et al (2012) Bidirectional association between depression and obesity in middle-aged and older women. Int J Obes (Lond) 36(4):595–602

Panagiotopoulos C, Ronsley R et al (2012) Waist circumference is a sensitive screening tool for assessment of metabolic syndrome risk in children treated with second-generation antipsychotics. Can J Psychiatry 57(1):34–44

Patel D (2008) Depression is the most common mental disorder in community settings, and is a major cause of disability. Occup Med (Lond) 58(7):453

Paterson L, McKenzie K et al (2012) Stigma, social comparison and self-esteem in adults with an intellectual disability. J Appl Res Intellect Disabil 25(2):166–176

Petry NM, Barry D et al (2008) Overweight and obesity are associated with psychiatric disorders: results from the national epidemiologic survey on alcohol and related conditions. Psychosom Med 70(3):288–297

Powers PS, Perez A et al (1999) Eating pathology before and after bariatric surgery: a prospective study. Int J Eat Disord 25(3):293–300

Radke AQ, Parks J et al (2010) A call for improved prevention and reduction of obesity among persons with serious mental illness. Psychiatr Serv 61(6):617–619

Ramacciotti C, Coli E et al (2013) Therapeutic options for binge eating disorder: a review. Eat Weight Disord

Ricca V, Castellini G et al (2009) Zonisamide combined with cognitive behavioral therapy in binge eating disorder: a one-year follow-up study. Psychiatry (Edgmont) 6(11):23–28

Riva G, Bacchetta M et al (2006) Is severe obesity a form of addiction? Rationale, clinical approach, and controlled clinical trial. Cyberpsychol Behav 9(4):457–479

Sanchez-Villegas A, Field AE et al (2012) Perceived and actual obesity in childhood and adolescence and risk of adult depression. J Epidemiol Community Health Advance online publication, 5 July 2012; doi: 10.1136/jech-2012-201435

Sareen J, Cox BJ et al (2007) Physical and mental comorbidity, disability, and suicidal behavior associated with posttraumatic stress disorder in a large community sample. Psychosom Med 69(3):242–248
Sarwer D, Thompson J (2002) Obesity and body image disturbance. In: Wadden T, Stunkard A (eds) Handbook of obesity treatment. Guilford, New York, pp 447–464
Sarwer DB, Thompson JK et al (2005) Body image and obesity in adulthood. Psychiatr Clin North Am 28(1):69–87, viii
Saunders R (2004) Post-surgery group therapy for gastric bypass patients. Obes Surg 14(8):1128–1131
Schellekens H, Finger BC et al (2012) Ghrelin signalling and obesity: at the interface of stress, mood and food reward. Pharmacol Ther 135(3):316–326
Schwartz D, Brownell K (2002) Obesity and body image. In: Body image: a handbook of theory, research and clinical practice. Guilford, New York, pp 200–209
Segal A, Kinoshita Kussunoki D et al (2004) Post-surgical refusal to eat: anorexia nervosa, bulimia nervosa or a new eating disorder? A case series. Obes Surg 14(3):353–360
Shin L, Bregman H et al (2008) An overview of obesity in children with psychiatric disorders taking atypical antipsychotics. Harv Rev Psychiatry 16(2):69–79
Simon GE, Von Korff M et al (2006) Association between obesity and psychiatric disorders in the US adult population. Arch Gen Psychiatry 63(7):824–830
Singh-Manoux A, Czernichow S et al (2012) Obesity phenotypes in midlife and cognition in early old age: the Whitehall II cohort study. The Whitehall II cohort study. Neurology 79:755–762
Stanley SH, Laugharne JD et al (2012) Assessing overweight and obesity across mental disorders: personality disorders at high risk. Soc Psychiatry Psychiatr Epidemiol Advance online publication, 4 July 2012; doi: 10.1007/s00127-012-0546-1
Stunkard A, Allison K (2003) Binge eating disorder: disorder or marker? Int J Eat Disord 34(Suppl):S107–S116
Stunkard A, Allison K et al (2008) Issues for DSM-V: night eating syndrome. Am J Psychiatry 165(4):424
Stunkard A, Burt V (1967) Obesity and the body image: II. Age of onset of disturbances in the body image. Am J Psychiatry 123:1443–1447
Stunkard A, Grace W et al (1955) The night-eating syndrome: a pattern of food intake among certain obese patients. Am J Med 19:78–86
Stunkard A, Lu XY (2010) Rapid changes in night eating: considering mechanisms. Eat Weight Disord 15(1–2):e2–e8
Stunkard A, Mendelson M (1967) Obesity and the body image. I. Characteristics of disturbances in the body image of some obese persons. Am J Psychiatry 123(10):1296–1300
Tanofsky-Kraff M, Marcus MD et al (2008) Loss of control eating disorder in children age 12 years and younger: proposed research criteria. Eat Behav 9(3):360–365
Tasca GA, Balfour L et al (2012) Outcomes of specific interpersonal problems for binge eating disorder: comparing group psychodynamic interpersonal psychotherapy and group cognitive behavioral therapy. Int J Group Psychother 62(2):197–218
Taua C, Hepworth J et al (2012) Nurses' role in caring for people with a comorbidity of mental illness and intellectual disability: a literature review. Int J Ment Health Nurs 21(2):163–174
Taylor VH, McIntyre RS et al (2012a) Beyond pharmacotherapy: understanding the links between obesity and chronic mental illness. Can J Psychiatry 57(1):5–12
Taylor VH, Stonehocker B et al (2012b) An overview of treatments for obesity in a population with mental illness. Can J Psychiatry 57(1):13–20
Umbricht D, Flury H et al (2001) Cognitive behavior therapy for weight gain. Am J Psychiatry 158(6):971
van Reedt Dortland AK, Vreeburg SA et al (2012) The impact of stress systems and lifestyle on dyslipidemia and obesity in anxiety and depression. Psychoneuroendocrinology Advance online publication, 18 June 2012; doi: 10.1016/j.psyneuen.2012.05.017

Vanderlinden J, Adriaensen A et al (2012) A Cognitive-behavioral therapeutic program for patients with obesity and binge eating disorder: short- and long-term follow-up data of a prospective study. Behav Modif Advance online publication, 27 April 2012; doi: 10.1177/0145445512439313

Vannucci A, Tanofsky-Kraff M et al (2012) Construct validity of the emotional eating scale adapted for children and adolescents. Int J Obes (Lond) 36(7):938–943

Wellmer J, Wellmer S et al (2009) The impact of zonisamide on weight. A clinical study in 103 patients with epilepsy. Acta Neurol Scand 119(4):233–238

Wilfley DE, Friedman MA et al (2000a) Comorbid psychopathology in binge eating disorder: relation to eating disorder severity at baseline and following treatment. J Consult Clin Psychol 68(4):641–649

Wilfley DE, Schwartz MB et al (2000b) Using the eating disorder examination to identify the specific psychopathology of binge eating disorder. Int J Eat Disord 27(3):259–269

World Health Organization (1992) The ICD-10 Classification of mental and behavioural disorders. Clinical descriptions and diagnostic guidelines. World Health Organization, Geneva

Xu WL, Atti AR et al (2011) Midlife overweight and obesity increase late-life dementia risk: a population-based twin study. Neurology 76(18):1568–1574

Chapter 11
Principles and Protocols in Nutritional Rehabilitation

Lorenzo M. Donini and Maria Letizia Petroni

Abstract Obesity is a clinical condition characterized by fat mass reserves increased to such an extent as to represent a health risk. The clinical diagnosis of obesity requires a careful nutritional status valuation to get an early diagnosis of malnutrition (undernutrition, overnutrition, or mixed malnutrition); to establish nutrition requirements; to identify those patients who require a nutritional intervention for global or selective nutrient deficiencies; and to check the effectiveness of such an intervention. The main components which characterize nutritional status are energy balance (nutritional habits and physical activity), body composition, body function, and inflammatory status. Nutrition intervention will focus on achieving a weight loss of at least 10 % from baseline body weight with a significant reduction in body fat and preservation of lean mass; reconstructing long-term healthy eating habits (quality, quantity, rate) based on the canons of the Mediterranean Diet; obtaining a patient's compliance adequate to achieve the established objectives. The nutritional intervention should be included into a process of therapeutic education aimed at the implementation of the knowledge about the disease and its management, and to change behaviors related to it for better management. In addition, therapeutic education allows to understand and to manage the psychological aspects related with the disease itself. An increasing number of new inpatient admissions in Rehabilitation settings consist of either failed or complicated post-surgical cases characterized by malnutrition, functional impairment and psychological uneasiness. Nutritional rehabilitation may be effective in reducing short-term and long-term complications related to bariatric surgery in improving the results of bariatric surgery and the recovery from psychological and functional impairments.

L. M. Donini (✉)
Experimental Medicine Department, Medical Pathophysiology, Food Science and Endocrinology Section, Food Science and Human Nutrition Research Unit, Sapienza University of Rome, Rome, Italy
e-mail: lorenzomaria.donini@uniroma1.it

M. L. Petroni
Clinical Nutrition Laboratory, Istituto Auxologico Italiano IRCCS, Verbania, Italy
e-mail: marialetizia.petroni@gmail.com

M. L. Petroni
Department of Biomedical Sciences, University of San Marino, San Marino, Republic of San Marino

P. Capodaglio et al. (eds.), *Disabling Obesity*, DOI: 10.1007/978-3-642-35972-9_11,

11.1 Evaluation of Nutritional Status in the Obese Subject

Obesity is a clinical condition characterized by fat mass (FM) reserves increased to such an extent as to represent a health risk.

Despite the lack of a universally accepted consensus, obesity may be defined as FM exceeding 25 % in men or 35 % in middle-aged women (WHO 1995; Deurenberg et al. 1998, 1999). The overall increase in body weight is therefore just a sign—not always present—of an increased fat mass. Body mass index (BMI) represents a useful indicator of overweight from an epidemiological point of view, but it is not always sufficient for clinical diagnosis of obesity, which requires a careful nutritional status evaluation (NSE) in order to be accurate. The NSE allows, among others, to get an early diagnosis of malnutrition (undernutrition, overnutrition, or mixed malnutrition); to establish nutrition requirements; to identify those patients who require a nutritional intervention for global or selective nutrient deficiencies; and to check the effectiveness of such an intervention. Regardless of the chosen modality (e.g., oral supplements, diet therapy, artificial nutrition), for both clinical and ethical considerations the nutritional intervention must always be preceded by a diagnosis supported by an NSE. Indeed, the routine use of NSE allows to achieve better and more effective nutritional treatment, to verify the efficacy and, when needed, to modify the type of intervention according to results so far obtained and to experiment newer treatment modalities. Those groups that have more intensively studied NSE in the past years have defined the main components which, while with different characteristics and meanings, characterize nutritional *status: energy balance (nutritional habits and physical activity), body composition, body function*, and *inflammatory status* (Cheryl 2001; Bedogni et al. 1996; Dwyer 1999; Hill 1992; Soeters et al. 2009).

Of course, the NSE must be framed within a wider clinical evaluation, and signs and symptoms of a possible state of over- or undernutrition be interpreted in the light of the overall clinical picture of the patient.

11.1.1 Obesity

11.1.1.1 Nutrient and Energy Balance

In clinical practice, energy requirements must be evaluated through basal energy expenditure (measured by indirect calorimetry (Weir 1949) or estimated through predictive equations generally taking into account gender, age, weight, and height (SINU 2012; Harris et al. 1919), corrected for physical activity levels (SINU 2012) and for presence or absence of concurrent diseases (Long et al. 1979). As for requirements of single nutrients, in some cases this can be assessed quite accurately (e.g. nitrogen balance via urinary and fecal nitrogen determination), while in the other cases reference must be made to the national recommended dietary allowance (RDA).

Requirements must then be compared with the intake, evaluated by food records (e.g., 3 days dietary record with or without food weighting, 24 h dietary recall) and/or evaluation of habitual dietary behavior (food frequency questionnaire). Through food composition databases it is then possible to assess subject's energy and nutrients intake (INRAN 2003). Generally speaking, in the obese subjects, mostly so during the phase of the FM increase, there is a largely positive energy balance characterized by excess absolute or relative energy intake and a reduced energy expenditure due to sedentary behavior. Noteworthy, an energy intake higher than 10 % of the requirement, in a subject requiring 2000 kcal (8374 kJ) a day will end up increasing his/her body weight by almost 1 kg body weight (made for about 80 % of fat) per month.

- In a subject with a reference weight of 70 kg this would lead to accumulation of 12 kg of body weight, that is 10 kg of FM, within a year.
- Total body fat mass of the subject would rise (assuming as "normal" baseline fat mass: 15–17 % in a man, 24–27 % in a woman) to 28 % if male, 34 % if female.

11.1.1.2 Body Composition

As mentioned above, although the widespread use of BMI for obesity classification, this is best defined, especially in clinical practice, according to relative fat mass either measured or estimated by double X-ray densitometry (DEXA), bioelectrical impedance (BIA, after evaluation of the state of hydration of the patient), and anthropometry. Each procedure has limitations and strengths (Hill 1992). Table 11.1 shows the reference values for the FM from various sources. The differences are justified by the type of population and age group considered (Kyle et al. 2003; Coin et al. 2008). Some authors suggest the use of fat mass index (FMI) corresponding to a BMI = 30 kg/m^2 or 75 % percentile of the distribution of a reference population (Coin et al. 2008; Schutz et al. 2002) (Table 11.2).

The risk associated with visceral adiposity can be estimated through waist circumference measurement. This value should ideally measure less than 94 cm in men and less than 80 cm in women. Values between 94 and 102 in men and between 80 and 88 for women should draw the attention of the physician and patient to prevent a further increase in fat mass and suggest the correction of lifestyle errors. Values over 102 in man and 88 women urge nutritional intervention because of their correlation with significant increase in visceral fat mass and with the risk of occurrence of cardiovascular or metabolic diseases (Lean et al. 1995).

Another measure that can estimate the risk related to the visceral fat is the sagittal abdominal diameter (measured in the supine position), measure validated by computed tomography which was shown to correlate significantly with visceral fat compartment (r = 0.8 in men and 0.83 in women) (Zamboni et al. 1998). Also, the thickness of subcutaneous folds can be used as an index of distribution of body fat as an alternative to the waist circumference. Among the four major skinfolds

Table 11.1 Cut-off values for Fat Mass defining obesity

Reference	Sample	Males	Females
WHO–Physical Status-(1995)	BMI equal to 30 kg/m^2 in a sample of Danish subjects	30 % at 20 years age 40 % at 60 years age	40 % at 20 years age 50 % at 60 years age
De Lorenzo et al. (2003)	596 women and 294 men from Italy apparently healthy aged 18–83 years	Normal <25 % increased >25 % high >30 %	Normal <35 % increased >35 % high >40 %
Movsesyan et al. (2003)	404 women from Denmark apparently healthy aged 18–75 years		The cut-off value between "normal" and overweight comprised between 35 and 43 %; that between overweight and obesity between 40 and 50 %
Gallagher et al. (2000)	380 women and 291 men; 254 Afroamericans and 417 Caucasians	BMI $\geq$ 30 kg/m^2 20–39 years: 25 % 40–59 years: 28 % 60–79 years: 30 %	BMI $\geq$30 kg/m^2 20–39 years 39 % 40–59 years 40 % 60–79 years 42 %
Heitmann et al. (1913)	735 men aged 60 years; follow-up of 22 years	BMI 30.2 $\pm$ 2.2 kg/m^2 $\Rightarrow$ FM kg 33.1 $\pm$ 7.9 (35.8 %)	
Deurenberg et al. (1989)	35 men and 37 women apparently healthy aged 60–83 years	BMI 25 $\pm$ 2 kg/m^2 $\Rightarrow$ FM 31 %	BMI 25.9 $\pm$ 3 kg/m^2 $\Rightarrow$ FM 44 %
Zhu et al. (2003)	4684 Caucasians (2238 men e 2446 women)	Metabolic syndrome risk BMI 30 kg/m^2 $\Rightarrow$ FM >28 % BMI 35 kg/m^2 $\Rightarrow$ FM >35 %	Metabolic syndrome risk BMI 30 kg/m^2 $\Rightarrow$ FM >37 % BMI 35 kg/m^2 $\Rightarrow$ FM >43 %
Kyle et al. (2003)	2986 men and 2649 women apparently healthy aged 15–98 years	BMI 30 kg/m^2 $\Rightarrow$ FM 28.8 %	BMI 30 kg/m^2 $\Rightarrow$ FM 40 %
Davison et al. (2002)	1526 men and 1391 women >70 years, apparently healthy NHANES III	FM >18.9 kg/m^2 (2 top quintiles)	FM >26.4 kg/m^2 (2 top quintiles)
Baumgartner et al. (2004)	229 apparently healthy subjects aged 18–40 years	FM >28 % (value above 60° percentile of the distribution of reference population)	FM >40 % (value above 60° percentile of the distribution of reference population)
Coin et al. (2008)	1866 Italian subjects aged 20–80 years, not severely diseased	75° percentile: 26.5 %	75° percentile: 38.4 %

Legend FM: fat mass, BMI: body mass index

Table 11.2 Fat mass index (FMI) reference values

		FMI	
		Males	Females
Kyle UG et al. (2005)	5635 apparently healthy subjects	8.3 kg/m^2 corresponds to BMI = 30 kg/m^2	11.8 kg/m^2 corresponds to BMI = 30 kg/m^2
Coin A et al. (2008)	1866 Italian subjects aged 20–80 years, not severely	75° percentile: 7 kg/m^2	75° percentile: 10.6 kg/m^2
Schutz Y et al. (2002)	5635 Caucasian subjects aged 18–98 years, not severely diseased	75° percentile: 5.9 kg/m^2	75° percentile: 7.8 kg/m^2

(triceps-TSF, biceps-BSF, subscapular-SSSF, and suprailiac-SISF) the SSSF seems to best estimate visceral adiposity and the risk of developing hypertension. The SSSF/TSF ratio and the ratio between the sum of skinfolds of the trunk (SSSF and SISF) and that of the limbs (TSF and BSF) are indices of central distribution of adipose tissue and useful prognostic indicators of cardiovascular risk associated with overweight (Bedogni et al. 2001; Seidell et al. 1990, 1992).

11.1.1.3 Body Function

As recalled earlier, a proper NSE cannot be carried out without assessment of body function. A carefully collected medical history with appropriate questions, a clinical examination, and a few laboratory and instrumental tests are often sufficient to give a full clinical picture. Diagnosis (established and to be confirmed) and body function parameters should be accurately recorded. The nutritionist should have full access to the patients' records and also have the opportunity to discuss the case with the physicians in charge of the patients and with the other members of the multidisciplinary team.

The main indicators of altered body function in a subject with excess fat mass are the following:

- Metabolic parameters (i.e., blood lipids, glycemic control);
- Motor function (e.g., muscular strength, walking test);
- Organic complications.

In the case of the obese patient, often there will be some of the following:

(1) Metabolic complications

(a) Altered glycemic control

- impaired glucose tolerance (IGT) glucose 140–200 mg/dl (7.8–11 mmol/L), at 2 h after an oral glucose load,
- diabetes mellitus: fasting glucose ≥126 mg/dl (≥6.9 mmol/L) or ≥200 mg/dl (≥11 mol/L) at 2 h after an oral glucose load.

(b) Altered blood lipids
  - total cholesterol >200 mg/dl (>5.17 mmol/L),
  - HDL <40 mg/dl (<1.03 mmol/L) in men, <50 mg/dl (<1.29 mmol/L) in women,
  - triglycerides >150 mg/dl (>1,69 mmol/L),
  - LDL >130 mg/dl (>3.3 mmol/L).

(c) Hyperuricemia: uric acid >6 mg/dL (>360 μmol/L) in women, >6.8 mg/dL (>400 μmol/L) in men; there is a high risk of developing gout for values exceeding 9 mg/dL (535 μmol/l).

(2) Impaired motor function, as described in Chaps. 2, 3, and 4.
(3) Cardiovascular and/or respiratory complications represented by:

(a) Atherosclerotic disease (e.g. ischemic heart disease, stroke),
(b) Hypertension (SBP >130 mmHg or DBP >85 mmHg or on antihypertensive medication) (see Chap. 7),
(c) Obstructive sleep apnea syndrome (OSAS), restrictive respiratory failure (obesity-hypoventilation syndrome), asthma (see Chap. 8).

(4) Musculoskeletal complications (ankle sprains, flat feet, *tibia vara*, slipped capital femoral epiphysis, increased risk and burden of degenerative joint disease).
(5) Miscellaneous complications: gastrointestinal system (gallstones, non-alcoholic liver disease, gastro-esophageal reflux), endocrine function (infertility and sub-fertility), renal (glomerulosclerosis, proteinuria), venous and lymphatic circles (lymphedema, venous insufficiency, phlebitis) dermatological symptoms (intertrigo, achanthosis nigricans, pressure ulcers, skin tags, hirsutism, *striae distensae*), neurological diseases (*pseudotumor cerebri*).

11.1.1.4 Inflammatory Status

Obesity is characterized by a low-grade inflammatory state that is associated with an increased risk of atherosclerosis. High-sensitivity CRP determination may give useful indications to assess the risk of atherosclerosis in adults, normal concentration in healthy human serum being usually lower than 10 mg/L. Also, high CRP concentration (>40 mg/dl) can help in the detection of infection or inflammation.

11.1.2 Sarcopenic Obesity

More and more frequently an increase in FM is associated with a reduction of fat-free mass (FFM), a scenario termed sarcopenic obesity (SO), which may carry the cumulative risk derived from each of the two individual body composition phenotypes (Prado et al. 2012; Roubenoff et al. 2004). This is especially true for elderly subjects or in younger obese subjects with severe disabilities, following bariatric surgery without nutritional supervision and in response to long-lasting incongruous dietary regimen. In this group of subjects, requiring more specific nutritional rehabilitation protocols, in addition to the standard NSE there will be need to check specific nutritional parameters.

11.1.2.1 Body Composition

In sarcopenic obesity, along with the increase in FM there is a depletion of FFM. The reduction in lean body mass (sarcopenia) can be evaluated in clinical practice through indices which relate the amount of lean body mass to body size or body weight [fat-free mass index (FFMI); skeletal mass index (SMI), appendicular lean mass index (ALMI): values lower than the mean −2 SD of the value found in a control sample consisting of subjects aged <40 years are suggestive for sarcopenia (Table 11.3) (Kyle et al. 2003a; Coin et al. 2008; Schutz et al. 2002; Janssen et al. 2002, 2004; Tichet et al. 2008; Kyle et al. 2003b; Zamboni et al. 2008).

In our view, a definition of sarcopenia based on the reduction of FFM in "absolute" terms is limited in the case of obesity. In this condition there is in fact, at least in younger adults, a physiological increase of FFM (Passmore et al. 1955; Norgan et al. 1980; Gallagher et al. 2011; Elia et al. 1999) which enables to maintain, for a relatively long time, a good metabolic control and a satisfactory physical efficiency. Nevertheless, signs of sarcopenia appear at an early stage in the obese subjects; as a matter of fact, a relative, and not an absolute reduction of FFM is sufficient to determine a reduced functional performance. It is therefore proposed, as an index of sarcopenia in obese subjects, a reduction in the ratio actual FFF/ideal FFM less than 0.9. Studies to verify this threshold are underway.

Note Ideal FFM: 85 % of the reference weight in men and 75 % in women + 25 % of the weight further accumulated during fattening (example: man with reference weight of 70 kg who weighs 100 kg: ideal FFM = (0.85 * 70) + (0.25 * 30) = 67 kg).

11.1.2.2 Strength

The deterioration of lean body mass (with possible fatty infiltration) is evaluated through a progressive worsening of functional parameters such as aerobic capacity, muscle strength, walking speed, ability to maintain balance (Short Physical Performance Battery—SPPB: chair stands, gait speed standing balance test) (see also Chap. 6).

11.1.2.3 Body Function

Broadly, in line with what has already been described above for the non-sarcopenic obese subject (metabolic parameters, motor function, and organic complications), the subjects with sarcopenic obesity will be evaluated through indicators of an altered body function in relation to an altered nutritional status, consequently to the combination of an excess FM and a reduced FFM:

- metabolic parameters (plasma protein),
- immunocompetence (lymphocyte count),
- organic complications.

Table 11.3 Cut-off values for estimated/measured fat-free mass (FFM) for diagnosis of sarcopenia

Reference	Sample	Definition of sarcopenia
Janssen et al. (2002)	4504 subjects aged >60 years	Class I sarcopenia: SMI 31–37 % men, 22–28 % women Class II sarcopenia: SMI <31 % men, <22 % women
Janssen (2004)	4499 elderly subjects (>60 years) NHANES III	Men: ALMI normal >10.76 kg/m^2 Mild sarcopenia: 8.51–10.75 kg/m^2 Severe sarcopenia ≤8.5 kg/m^2 Women: ALMI normal >6.76 kg/m^2 Mild sarcopenia: 5.76–6.75 kg/m^2 Severe sarcopenia ≤5.75 kg/m^2
Tichet et al. (2008)	780 French subjects <40 years, 888 subjects 40–59 years, 218 subjects 60–78 aa	ALMI <6.2 kg/m^2 women, 8.6 kg/m^2 men SMI <26.6 % women 34.4 % men (mean –2 SD of values in control sample of subjects <40 years)
Kyle UG et al. (2003)	3549 men and 3184 women, Caucasians, apparently healthy aged 18–98 years	Correspondence between BMI and FFMI men BMI 18.5 ⇒ FFMI 16.7 kg/m^2 women BMI 18.5 ⇒ FFMI 14.6 kg/m^2
Schutz et al. (2002)	5635 Caucasians apparently healthy aged 18–98 years	BMI 25° percentile men ⇒ FFMI 18.2 kg/m^2 women ⇒ FFMI 15 kg/m^2
Coin A et al. (2008)	1866 Italians aged 20–80 years, not severely diseased	25° percentile FFMI men 18.7 kg/m^2 women 14.9 kg/m^2
Kyle et al. (2005)	5635 apparently healthy subjects	FFMI: men 17.4 kg/m^2 women 15 kg/m^2 Corresponds to a BMI = 20 kg/m^2
Baumgartner et al. (2004)	229 apparently healthy subjects 18–40 years	ALMI <5.45 kg/m^2 women, 7.26 kg/m^2 men (mean – 2 SD of values in control sample)
Delmonico et al. (2007)	1433 men, 1543 women enrolled in the Health, Aging and Body Composition Study (Health ABC)	ALMI <7.25 kg/m2 men, 5.67 kg/m2 women (mean – 2 SD of values in control sample)

Legend BMI: body mass index; FFMI: fat-free mass index; SMI: skeletal mass index: ALMI: appendicular lean mass index

Table 11.4 Metabolic and immunocompetence parameters linked to protein–calorie malnutrition Gallagher et al. (2011)

	Half-life	Mild malnutrition	Moderate malnutrition	Severe malnutrition
Metabolic parameters				
Prealbumin (mg/dl)	48 h	18–22	10–17	<10
Transferrin (mg/dl)	8–9 days	200–150	100–149	<100
Albumin (g/l)	29 days	30–35	25–29	<25
	Normal immunocompetence	Mild malnutrition	Moderate malnutrition, hypoergia	Severe malnutrition, anergia
Immunocompetence parameters				
Lymphocyte count (cells/ml)	>1500	1500–1200	1200–800	<800

A subject with sarcopenic obesity could often present with altered plasma proteins and immunocompetence parameters as for protein–calorie malnutrition (SINPE 2002) (Table 11.4) and organic complications. In addition to the already mentioned complications of obesity, in the case of concurrent sarcopenia, there can occur complications related to immobilization syndrome (pressure ulcers, venous thromboembolism) and/or reduced immunocompetence (respiratory sepsis).

11.2 Nutritional Intervention

Nutrition intervention will focus on:

(a) achieving a weight loss of at least 10 % from baseline body weight with a significant reduction in body fat and preservation of lean mass;
(b) reconstructing long-term healthy eating habits (quality, quantity, rate) based on the canons of the Mediterranean Diet (or reference national healthy diet: for Italy see http://www.piramidealimentare.it) also in an educational logic;
(c) obtaining a patient's compliance adequate to achieve the established objectives (Donini et al. 2010).

11.2.1 Nutritional Intervention and Therapeutic Education

It is important that nutritional intervention may contribute to correct previous unhealthy eating habits on a permanent basis. For this purpose there is need for frequent professional contact with the patient aimed at assessing the level of understanding of the given dietary recommendations, the eradication of bad eating behaviors related to so called uncorrect "dietary beliefs", the verification of the satisfaction level of the dietary plan (meal organization to suits the patient's needs; as far as reasonably possible, respect for customary dietary habits and tastes while encouraging at the same time novel healthier choices) (Breland et al. 2012).

As a matter of fact, the nutritional intervention should be included into a process of therapeutic education. This is an essential phase in the management of chronic patients: the aim is to implement the knowledge about the disease and its management, and to change behaviors related to it for better management. In addition, therapeutic education allows to understand and to manage the psychological aspects related with the disease itself. Besides the role of providing the patient with information about practical aspects of disease management, therapeutic education also aims to help improve his/her quality of life (Standard italiani per la cura dell'obesità 2012).

The nutritional intervention, inserted within a framework of therapeutic education, is therefore designed to:

- inform about healthy nutrition and physical activity, with emphasis on regaining the physiological stimuli of hunger and satiety, recognizing the biological significance of food, rediscovering a sense of comfort with physical activity,
- train for managing and control nutrition and physical activity, even while under stress or in anxiety,
- train in the management and control of simple clinical parameters (e.g., blood glucose, blood pressure),
- increase the sense of responsibility in illness and care (illness behavior),
- promote, strengthen, and maintain sufficient motivation to change, and
- contribute to the therapeutic compliance.

Therapeutic education for obesity must be guaranteed within the team, by professionals (doctor, nurse, dietician, community health educator, psychiatrist, psychologist, exercise physiologists) specifically qualified and trained for educational activity ((Standard italiani per la cura dell'obesità 2012).

Motivation is essential for obtaining adhesion to treatment and a stable weight loss. For this purpose, therapeutic education bases most of its intervention through techniques acting on conscious mental processes, some of which are derived from cognitive-behavioral therapy, such as therapeutic alliance, problem solving, and empowerment (Standard italiani per la cura dell'obesità 2012).

11.2.2 Procedures for Nutritional Intervention

The following guidelines for hypocaloric diet composition have recently been published in the Italian Standards for the Treatment of Obesity (Standard italiani per la cura dell'obesità 2012):

- Hypocaloric diets should secure a caloric intake equal to ± 10 % of basal metabolism.
- Carbohydrates, mainly of complex type (fiber-rich or containing slow absorption starch) should provide 65–70 % of non-protein calorie intake ($\geq$150 g/

day), simple sugars should not exceed 10 % of total caloric intake (favoring the consumption of foods that contain natural sugars such as fruit and limiting the consumption of added sucrose). Note that at the moment there is no evidence to support the use of low-carbohydrate diets (≤120–130 g/day) in patients with obesity.

- Foods with low glycemic index are preferable, especially for weight maintenance after a hypocaloric diet.
- The remainder energy should be covered from fat, with an optimal 10 % monounsaturated, 10 % polyunsaturated, 10 % saturated fatty acids.
- The intake of polyunsaturated and in particular of n3 fatty acids (for a beneficial effect on the prevention of cardiovascular risk) is ensured through the introduction of at least two servings per week of fish.
- The daily intake of cholesterol should not exceed 300 mg/day in adults.
- The use of trans fatty acids must be drastically reduced (≤2.5 g/day) because it is associated with increased body weight, waist circumference, and BMI in population-based studies.
- The recommended protein intake is 0.8–1 g/kg ideal body weight (which corresponds to a BMI of 22.5 kg/m^2).
- Proteins must be of good biological value and must come equally from both animal and vegetable protein sources.
- Fiber intake is ideally around 30 g/day in order to obtain both functional-type effects (i.e. intestinal function) and metabolic-type effects (glucose and lipid metabolism).
- Vitamin needs are generally covered by the consumption of fruits and vegetables generally prescribed in the diet (provided that calorie intake is at least 1000–1200 kcal/day).
- Maximum sodium intake should approximately be 3 g/day (equivalent to 7.5 g of NaCl), this should further be reduced in patients with hypertension or a family history of hypertension.
- Calcium requirements are not generally covered by a low-calorie diet, unless two servings/day of dairy products plus if possible integration with calcium-rich waters is provided.
- Alcohol is not recommended during weight loss because it provides readily available energy (7 kcal/g), without satiating power nor significant advantages with regard to the provision of other nutrients.

Not surprisingly, in the Italian guidelines the choice of foods to be prescribed in hypocaloric diets is based on the Mediterranean dietary pattern: grains, legumes, vegetables, fruits, olive oil, lean meats and fish, semi-skimmed milk and dairy products derived from it. This is because in diet therapy of obesity the Mediterranean model seems ultimately to be more efficient both in terms of results (body weight, metabolic parameters) and in setting the conditions for the long-term maintenance of a healthy lifestyle. Diets with unbalanced intake of macronutrients (i.e., very low carbohydrate or very low fat diets, high-protein diets) did not give significantly better results than the Mediterranean hypocaloric diet. Moreover, these dietary regimens could potentially expose the patient to side effects (e.g., metabolic acidosis, loss of lean body mass) and do not provide appropriate guidance to weight loss maintenance. Some guidelines warn against unbalanced regimes and/or lacking in certain nutrients, as well as against uncontrolled use of meal replacements in the absence of medical supervision. About the distribution of meals during the day it seems appropriate to propose a pattern characterized by a relatively abundant breakfast (semi-skimmed milk, cereals, fruit, yogurt) and two main meals; some patients might require, at least at the beginning of treatment, two snacks (mid-morning and mid-afternoon), in order to reduce the feeling of hunger between meals and to prevent hypoglycemia if diabetics on glucose-lowering drugs.

In selected subjects, in particular in whom there are no signs of undernutrition and within hospital (intensive residential rehabilitation) settings, it is also possible to use dietary low calorie diets (LCD) capable of ensuring a more rapid loss of body fat with a favorable risk–benefit profile. The LCD diet can be made of natural foods (milk, vegetables), dehydrated foods, and supplements. The LCD cycles usually last 15 days alternating with 15 days of classical moderate-calorie diet. The most recent reviews of the literature do not acknowledge a higher effectiveness of LCD versus standard diets if not in the short term. Nevertheless, LCD should be considered as the first choice in morbid obese patients with life-threatening conditions, or before major surgery. They can also play a role in the preparation of the patient for bariatric surgery to reduce anesthesiological risk, once the patient has completed its therapeutic educational training in healthy lifestyle (Clinical guidelines—Expert panel 1998; Clinical guidelines NIH 1998; Practical guide—NIH 2000; Canadian clinical practice guidelines 2007).

11.2.3 Rehabilitation for Bariatric Surgery

Bariatric surgery may consist in either restrictive (e.g., gastric banding, vertical-banded gastroplasty) or restrictive-malabsorpitive (gastric bypass) or malabsorpitive (e.g. biliopancreatic diversion) interventions. The degree of weight loss is generally higher with procedures having a malabsorptive component, at the expense of a higher surgical mortality and short-term and long-term complications (Kendrick et al. 2006). Metabolic complications (diabetes, lipid abnormalities) usually abate at an early stage (about −10 % weight loss) (Narbro et al. 2002) while for improvement of pulmonary complications such as obstructive sleep apnoea syndrome or obesity-hypoventilation syndrome a greater (−15 and 25 %, respectively) degree of weight loss may be required (Busetto et al. 2005; Sugerman et al. 1992). Improvement in orthopedic complications is greatly variable, depending on previous severity of arthritis, age, and residual muscle mass (Karason et al. 2005; Murr et al. 1995). Restrictive procedures may be ineffective in patients with deranged eating behavior (e.g. sweet eaters) or may induce vomit and electrolyte abnormalities in patients with eating disorders or unable to cope with drastically changed eating habits (Fujioka et al. 2005).

Outcomes of gastric bypass (one of the most popular procedures in the US) in terms of length of hospital stay, 30-day readmission, morbidity, and costs were found to be more favorable in hospitals treating more than 100 cases per year, than in hospitals treating fewer patients, therefore suggesting that expertise, likely both on clinical and organizational grounds, is a key factor to improve treatment outcome (Nguyen et al. 2004). However, although bariatric surgery has been proven to improve mortality, morbidity, and quality of life compared to standard medical treatment, yet a substantial number of patients experience both short-term and long-term complications related to surgery (Virji et al. 2006), have little or no benefit from treatment (Suter et al. 2006), or gain weight back over time (Christou et al. 2006). Furthermore, even in those with substantial weight loss (more than 20 %) and improvement of some complications of obesity, disability may not or may only partially subside, due to residual physical or psychological impairment (Velcu et al. 2005). There is a surprising lack of studies in the literature on ways to improve outcome of surgical treatment. This is likely to reflect a cultural gap on considering Morbid Obesity as a disabling condition requiring multidimensional rehabilitation and the lack of experience on non-surgical rehabilitation protocols for morbid obesity (Fujioka et al. 2005; Pain et al. 2006; Resnik et al. 2005). Due to the widespread diffusion of surgical techniques worldwide, it is no surprise that nowadays at obesity rehabilitation services an increasing number of new inpatient admissions consist of either failed or complicated post-surgical cases referred from other hospitals. Rehabilitation in these patients is often very difficult; malnutrition due to malabsorption or vomiting may require semi-intensive care and/or prolonged artificial nutrition in addition to oral nutritional rehabilitation (i.e., meals with specific rheological characteristics administered under active supervision and support of dieticians), physical therapy (to treat functional impairment due to carential neuropathies, postural unbalances or joint pain due to excessively fast weight loss, and consequent muscle wasting), psychological and psychiatric support therapy for quite a long time. The aim is to try and improve patient functioning and compliance to dietary treatment and/or to reduce surgical risk and improve outcome of reintervention (either reconversion of a malabsorptive intervention or shift from a restrictive to a malabsorptive intervention); nevertheless, this goal is seldom fully reached.

A completely different scenario is that represented by those patients who had been referred to surgical treatment and in whom a pre-and post-surgical, rehabilitation cycle had been carried out.

In our experience, pre-surgery reduction of body weight (and therefore reduction of anesthesiological and surgical risk) but mostly training the patient to improve his/her nutritional skills, to cope with changes in feeding habits, to increase self-efficacy, physical conditioning, and treating psychological comorbidity and eating disorders are associated with a lower (compared to the intervention-specific Italian Obesity Surgery Registry data) early and late-onset morbidity and with a higher rate of patients achieving and maintaining a −20 % weight loss (unpublished data).

In their paper on "Best Practice Updates for Multidisciplinary Care in Weight Loss Surgery" Apovian and coworkers have set the set the standard evidence-based best practices in multidisciplinary care with emphasis on preoperative medical evaluation, patient education, appropriate perioperative care, and long-term follow-up (Apovian et al. 2009). The above quoted "best practices" also include detailed recommendations on pre-surgery evaluation and clinical management of complications like hypoglycemia, ischemic heart disease, metabolic liver disease, chronic renal insufficiency, prophylaxis of deep venous insufficiency and pulmonary embolism, and also about plastic (body contouring) surgery following stabilized weight loss.

Additional psychosocial support, such as cognitive behavioral therapy or targeted psychoeducation, has been suggested to improve patient completion of the pre-surgery assessment phase (Sockalingam et al. 2012). There is encouraging preliminary evidence from randomized trials on the effectiveness of nutrition intervention in patients following weight loss surgery. Nijamkin and coworkers (Nijamkin et al. 2012) have shown how intensive nutrition and lifestyle educational intervention result in higher weight loss for gastric bypass surgery patients as compared with those receiving standard care.

Exercise and physical activity also play a fundamental role for management of bariatric surgery patients. Patients with low exercise tolerance and low fitness are at higher risk of developing postoperative complications than patients with better physical conditioning (McCullough et al. 2006). Exercise can prevent the loss of lean body mass due to the surgical procedure (which is greater following malabsorptive procedures) (Metcalf et al. 2005). The 6-min walking test has been validated in laparoscopic adjustable gastric band patients as a submaximal index of functional capacity (Maniscalco et al. 2006). Unfortunately, preoperative exercise counseling does not result in postoperative treatment adherence in bariatric surgery patients (Lier et al. 2012), thus prompting the need for post-surgical interventions. The risk of nutrient deficiency after bariatric surgery is well documented (Shah et al. 2006, Shikora et al. 2007). In Table 11.5 are summarised the main nutrients at risk of depletion following bariatric surgery. Aside from well-known complications like anaemia and osteoporosis, nutrient deficiency can lead to threatening and insidious conditions difficult to diagnose e.g. Wernicke encephalopathy (Shikora et al. 2007) and neural tube defects in offspring (Pelizzo et al.

Table 11.5 Nutrients at risk of deficiency following bariatric surgery

Surgical intervention	Nutrients at risk of deficiency
Gastric banding Vertical gastroplasty Sleeve gastrectomy	Iron, calcium, zinc, magnesium, vitamin B12, thiamine
Gastric by-pass	Proteins, calcium, iron, zinc, vitamin A, thiamine, folic acid, vitamin B12, vitamin D
Bilio-pancreatic diversion	Proteins, essential fatty acids, calcium, iron, zinc, vitamin A, thiamine, vitamin B12, vitamin D, vitamin E, vitamin K, selenium

2012). Multivitamin multimineral supplementation is mandatory for life, and compliance should be regularly checked, especially in women of fertile age who represent particularly vulnerable subjects. For the prevention and treatment of nutritional deficiencies in post-surgical patients there are algorithms and guidelines specifically dedicated to this topic (Schweitzer & Posthuma 2008, Mechanic et al. 2009, Heber et al. 2010).

11.3 Protocols for Multidisciplinary Rehabilitation in Bariatric Surgery Patients

In the above quoted paper by Apovian and coworkers, among pre-surgery recommendations, those more strictly linked to rehabilitation and to clinical nutrition are the following:

- preoperative weight loss of 5–10 % of initial body weight, especially for patients with a BMI ≥ 50,
- more prospective randomized controlled studies to determine optimal preoperative, weight loss and improved supervision of preoperative weight reduction and weight loss surgery education programs,
- strong encouragement of smokers to stop smoking prior to weight loss surgery,
- availability of smoking cessation advice and treatment at the institution or through the weight loss surgery program,
- pre- and postoperative monitoring for deficiencies in vitamin D, thiamin, calcium (including parathyroid hormone), iron, vitamin B12, and folic acid, with repletion as indicated,
- encouragement of bariatric surgery patients to increase pre- and postoperative physical activity and low-to-moderate intensity exercise,
- guidance and periodic monitoring to help bariatric surgery patients remain physically active.

On the basis of our experience, we would recommend the following three phases for bariatric surgery assessment and rehabilitation:

Phase 1: Assessment, eligibility, and planning

This can be carried out in either an out-patient, day-care, or in-patient setting (the latter is especially recommended for patients with severe complications or with psychiatric comorbidity).

All the following steps should be taken into account and documented in patients' clinical notes:

(a) Is the patient a true "non responder" to medical treatment? There should be documented previous attempts to weight loss in appropriate medically supervised settings (of whom at least one is in a multidisciplinary obesity rehabilitation service);
(b) Individual evaluation of risk–benefit ratio (risk factors for surgical mortality and morbidity as compared to the potential benefit from surgery);

(c) Group and individual information about bariatric surgery given by the multidisciplinary team (internist, dietician, psychologist, surgeon, nurse);
(d) Evaluation of patient motivation and of her/his willingness to adhere to long-term follow-up with emphasis on:
    (i) evaluation of family and social support,
    (ii) making sure that the patient has fully understood the benefits, consequences, and risks of the surgical option and the need for a lifetime follow-up,
    (iii) making sure that the patient has fully understood the limits of surgery (avoid expectancies on non-realistic outcomes),
    (iv) making sure that the patient has fully understood the dietary modifications which will be requested following surgery,
    (v) making sure that the patient has fully understood the need for vitamin and mineral assumption following surgery.
(e) Counseling in women in fertile age: bariatric surgery is contraindicated in pregnant women or in those who seek getting pregnant within 18 months from surgery (36 months following malabsorptive procedures);
(f) Psychological and psychiatric assessment: screening tests for eating disorders, body image disturbances, other psychiatric conditions followed by interview by a clinical psychologist (also by a psychiatrist for confirmation of diagnosis and prescriptions);
(g) Evaluation of nutritional status (see Paragraph 11.1);
(h) Metabolic assessment: attention to previous episodes of hypoglycemia and to hypothyroidism which represent contraindication to some types of surgical interventions; to identify diabetic patients who might face drastic changes in insulin requirements following surgery;
(i) Pneumological, cardiovascular, anesthesiological assessment (including instrumental exams and 6 min walking test);
(j) Physiatric assessment for exercise prescription in all patients and for evaluation of patients with limited mobility;
(k) Surgical assessment (including liver size).

Phase 2: Risk reduction and preparation for surgery

To achieve these goals we recommend day-care multidisciplinary rehabilitation (two weekly sessions for 2–3 months) and/or a 3 week inpatient intensive rehabilitation according to individual needs:

(a) Pre-surgical reduction and stabilization of metabolic, cardiovascular, and respiratory risk factors: weight loss, gradual physical reconditioning, smoking cessation, respiratory exercises, etc.;
(b) Cessation or at least reduction of binge eating (Ashton et al. 2011) or other dysfunctional eating patterns through a combined intervention by psychologist and dietician;
(c) Adaptation to future dietary changes following surgery: dietary training (smaller meal size, proper chewing, slower eating, etc.) by supervised meals of the dietician;

(d) Brief psychotherapy and psychoeducational interventions in individual cases;
(e) Nurse-led patient education in preparation to surgery;
(f) Pre-surgery weight loss of at least 5–10 % of initial body weight by a very-low calorie diet (especially for patients with BMI $\geq$ 50 kg/m^2 or who need a significant reduction of liver size).

Both group and individual interventions can be scheduled according to each patient's needs. Examples of schedule for a day-care rehabilitation session can be found in Table 11.6.

Phase 3: Rehabilitation following surgery

If no major complication has occurred and if the patient has undergone appropriate pre-surgical rehabilitation and training, this can often be carried out on an outpatients' basis following discharge.

Visits with the surgeon, nutritionist, and internist, blood tests and examinations are scheduled according to current guidelines (Mechanick et al. 2009; Fried et al. 2008) and to individual patient's needs. In the first few weeks, particular attention is

Table 11.6 Examples of schedule for a day-care rehabilitation session

Time	Activity	Personnel	Location
Day 1			
8:00–8.45	Aerobic and isometric reconditioning (group) (heart rate, blood pressure, and Sat O2 monitoring during exercise)	Physiotherapist Nurse	Gym
8.45–9.15	Respiratory exercises (group)	Physiotherapist	Gym
9:30–10.15	Patient education group	Nurse	Meeting room
10.15–10.30	Snack (low or very-low calorie)	Nurse	Common room
10.30–11.15	Individual psychotherapy or Smoking cessation session	Psychotherapist Psychologist Nurse	Room
11.30–12.30	Supervised meal (resembling post-surgical meal) with dietary training	Dietician	Dining room
Day 2			
8:00–8.45	Aerobic and isometric reconditioning (group) (heart rate, blood pressure, and Sat O2 monitoring during exercise)	Physiotherapist Nurse	Gym
8.45–9.15	Respiratory exercises (group)	Physiotherapist	Gym
9:30–10.15	Individual session with nutritionist	Dietician	Room
10.15–10.30	Low-calorie snack	Nurse	Common room
10.30–11.30	Group psychotherapy	Psychotherapist	Room
11.15–11.30	Carbon monoxide monitoring (smokers only) Weigh-ins	Nurse	Nurses' room
11.30–12.30	Supervised very-low or low calorie meal	Dietician	Dining room

paid to early detection of surgical complications, to weaning from slurry to semisolid foods, to adaptation of dosage medications (especially insulin). After a few weeks physical exercise is gradually restarted and psychological well-being is monitored especially in patients with known eating disorders or other psychiatric conditions (by psychometric tests and/or psychological interview and/or psychiatric assessment).

Between 6 and 15 months following bariatric surgery the patient usually sets its long-term lifestyle changes, and it is especially important to have her/him engaged in regular physical activity (possibly supervised by an exercise physiologist or by a physiotherapist) and trained to have an overall active lifestyle. Attention should be paid to maximize weight loss (since in most patients the nadir weight is reached between 12 and 15 months from bariatric surgery). On the other hand it is mandatory to prevent malnutrition by checking compliance to supplementations and by monitoring blood levels of vitamins and minerals (Pain et al. 2006; Lier et al. 2012; Fugioka et al. 2005). The patient should also be monitored for early signs of maladaptive eating patterns—like Post-Surgical Eating Avoidance Disorder (Segal 2004). Cognitive-behavioral techniques (Van Dorsten 2011) can be used to support and motivate the patient through this phase.

Yet, if the patient is malnourished, or has difficulty in weaning from liquid to solid food, or reports postcibal vomiting in the presence of a normal post-surgical transit of the contrast media through the alimentary canal at X-rays, she/he should be admitted to in-patient or day-care for further assessment and nutritional and psychological rehabilitation.

From 15 months onwards usually no further weight loss is detected and there is a risk for weight gain over time. The patient should be followed up at regular time intervals and instructed to check weight regularly (weekly) to prevent weight recurrence (Butryn 2007).

References

Apovian CM, Cummings S, Anderson W, Borud L, Boyer K, Day K, Hatchigian E, Hodges B, Patti ME, Pettus M, Perna F, Rooks D, Saltzman E, Skoropowski J, Tantillo MB, Thomason P (2009) Best Practice Updates for Multidisciplinary Care in Weight Loss Surgery. Obesity (Silver Spring) 17:871–879

Ashton K, Heinberg L, Windover A, Merrell J (2011) Positive response to binge eating intervention enhances postoperative weight loss. Surg Obes Relat Dis 315–20

Baumgartner RN, Wayne SJ, Waters DL, Janssen I, Gallagher D, Morley JE (2004) Sarcopenic obesity predicts instrumental activities of daily living disability in the elderly. Obes Res 12:1995–2004

Bedogni G, Battistini N, Severi S, Borghi A (1996) The physiological bases of the assessment of nutritional status. Clin Diet 23:141–146

Bedogni G, Borghi A, Battistini NC (2001) Manuale di valutazione antropometrica dello stato nutrizionale. EDRA. pp 72–73

Breland JY, Fox AM, Horowitz CR, Leventhall H (2012) Applying a common-sense approach to fighting obesity. J Obesity, Article ID 710427, 8 pages, doi:10.1155/2012/710427

Busetto L, Enzi G, Inelmen EM, Costa G, Negrin V, Sergi G, Vianello A (2005) Obstructive sleep apnea syndrome in morbid obesity: effects of intragastric balloon. Chest 128:618–623

Butryn ML, Phelan S, Hill JO, Wing RR (2007) Consistent self-monitoring of weight: a key component of successful weight loss maintenance. Obesity 15:3091–3096

Canadian clinical practice guidelines on the management and prevention of obesity in adults and children CMAJ (2007) 176 (8): S1–13

Cheryl CHC (2001) A concept analysis of malnutrition in the elderly. J Adv Nurs 36(1):131–142

Christou NV, Look D, Maclean LD (2006) Weight gain after short- and long-limb gastric bypass in patients followed for longer than 10 years. Ann Surg 244:734–740

Clinical guidelines on the identification (1998) evaluation, and treatment of overweight and obesity in adults: executive summary. Expert Panel on the Identification, Evaluation, and Treatment of Overweight in Adults. Am J Clin Nutr 68(4):899–917

Clinical guidelines on the identification, evaluation, and treatment of Overweight and obesity in Adults - The Evidence Report, National Institutes of Health (1998) Obes Res 6 Suppl 2: 51S-209S

Coin A, Sergi G, Minicucci N, Giannini S, Barbiero E, Manzato E, Pedrazzoni M, Minisola S, Rossini M, Del Puente A, Zamboni M (2008) Fat-free mass and fat-mass reference values by DEXA in a 20–80 year-old Italian population. Clin Nutr 27:87–94

Davison KK, Ford ES, Cogswell ME, Dietz WH (2002) Percentage of body fat and body mass index are associated with mobility limitations in people aged 70 and older from NHANES III. JAGS 50:1802–2002

De Lorenzo A, Deurenberg P, Pietrantuono M, Di Daniele N, Cervelli V, Andreoli A (2003) How fat is obese? Acta Diabetol 40:S254–S257

Delmonico MJ, Harris TB, Lee JS, Visser M, Nevitt M, Kritchevsky SB, Tylavsky FA, Newman AB; Health, Aging and Body Composition Study (2007) Alternative definitions of sarcopenia, lower extremity performance, and functional impairment with aging in older men and women.JAGS 55:769–774

Deurenberg P, van der Kooy K, Hulshof T, Evers P (1989) Body mass index as a measure of body fatness in the elderly. Eur J Clin Nutr 43:231–236

Deurenberg P (1999) Deurenberg Yap M, Wang J, Lin FP, Schmidt G.: The impact of body build on the relationship between body mass index and percent body fat. Int J Obes Relat Metab Disord 23(5):537–542

Deurenberg P, Yap M, van Staveren WA (1998) Body mass index and percent body fat: a meta analysis among different ethnic groups. Int J Obes Relat Metab Disord 22(12):1164–1171

Donini LM, Cuzzolaro M, Spera G, Badiali M, Basso N, Bollea MR, Bosello O, Brunani A, Busetto L, Cairella G, Cannella C, Capodaglio P, Carbonelli MG, Castellaneta E, Castra R, Clini E, Contaldo F, Dalla Ragione L, Dalle Grave R, D'Andrea F, Del Balzo V, De Cristofaro P, Di Flaviano E, Fassino S, Ferro AM, Forestieri P, Franzoni E, Gentile MG, Giustini A, Jacoangeli F, Lubrano C, Lucchin L, Manara F, Marangi G, Marcelli M, Marchesini G, Marri G, Marrocco W, Melchionda N, Mezzani B, Migliaccio P, Muratori F, Nizzoli U, Ostuzzi R, Panzolato G, Pasanisi F, Persichetti P, Petroni ML, Pontieri V, Prosperi E, Renna C, Rovera G, Santini F, Saraceni V, Savina C, Scuderi N, Silecchia G, Strollo F, Todisco P, Tubili C, Ugolini G, Zamboni M (2010) Obesità e disturbi dell'Alimentazione. Indicazioni per i diversi livelli di trattamento. Documento di Consensus. Eat Weight Disord. 15(1–2):1–31

Dwyer JT (1999): Nutritional Status. Encyclopaedia of Human Nutrition. Sadler MJ, Strain JJ, Caballero B (eds) Acedemic Press, San Diegop 1347–1389

Elia M, Stubbs RJ, Henry CJK (1999) Differences in fat, carbohydrate, and protein metabolism between lean and obese subjects undergoing total starvation. Obesity Res. 7:597–604

Fried M, Hainer V, Basdevant A, Buchwald H, Deitel M, Finer N, Greve JW, Horber F, Mathus-Vliegen E, Scopinaro N, Steffen R, Tsigos C, Weiner R, Widhalm K (2008) Interdisciplinary European guidelines on surgery of severe obesity. Obes Facts. 1:52–59

Fujioka K (2005) Follow-up of nutritional and metabolic problems after bariatric surgery. Diabetes Care

Fugioka K (2005) Follow-up of nutritional and metabolic problems after bariatric surgery. Diabetes Care 28:481–484

Gallagher D, DeLegge M (2011) Body Composition (Sarcopenia) in Obese Patients: implications for care in the intensive care unit. JPEN 35(Supplement 1):21S–28S

Gallagher D, Heymsfield SB, Heo M, Jebb SA, Murgatroyd PR, Sakamoto Y (2000) Healthy percentage body fat ranges: an approach for developing guidelines based on body mass index. Am J Clin Nutr 72:694–701

Harris, JA; Benedict, FG (1919) A biometric study of basal metabolism in man. Carnegie Institute of Washington, Washington DC, Publ 279

Heber D, Greenway FL, Kaplan LM, Livingston E, Salvador J and Still C (2010) Endocrine and nutritional management of the post-bariatric surgery patient: an endocrine society clinical practice guideline. J Clin Endocrinol Metab 95:4823–4843

Heitmann BL, Erikson H, Ellsinger BM, Mikkelsen KL, Larsson B.: Mortality associated with body fat, fat-free mass and body mass index among 60-year-old swedish men-a 22-year follow-up. The study of men born in 1913 (2000) Int J. Obesity 24:33–37

Hill GL (1992) Body composition research: implications for the practice of clinical nutrition. JPEN 16:197–218

INRAN Istituto Nazionale di Ricerca per gli Alimenti e la Nutrizione (2003) Linee Guida per una sana alimentazione italiana, http://www.inran.it

Janssen I, Baumgartner RN, Ross R, Rosenberg IH, Roubenoff R (2004) Skeletal muscle cutpoints associated with elevated physical disability risk in older men and women. Am J Epidemiol 159:413–421

Janssen I, Heymsfield SB, Ross R (2002) Low relative skeletal muscle mass in older persons is associated with functional impairment and physical disability. JAGS 50:889–896

Karason K, Peltonen M, Lindroos AK, Sjostrom L, Lonn L, Torgerson JS (2005) Effort-related calf pain in the obese and long-term changes after surgical obesity treatment. Obes Res 13:137–145

Kendrick ML, Dakin GF (2006) Surgical approaches to obesity. Mayo Clin Proc 81(10 Suppl):S18–S24

Kyle UG, Morabia A, Schutz Y, Richard C (2003a) Sedentarism affects body fat mass index and fat-free mass index in adults aged 18–98 years. Nutrition 20:255–260

Kyle UG, Pirlich M, Lochs H, Schuetz T, Pichard C (2005) Increased length of hospital stay in underweight and overweight patients at hospital admission: a controlled population study. Clin Nutr 24:133–142

Kyle UG, Schutz Y, Dupertuis YM, Pichard C (2003b) Body composition interpretation: contributions of the fat-free mass index and the body fat mass index. Nutrition 19:597–604

Lean MEJ, Han TS, Morrison CE (1995) Waist circumference as a measure for indicating need for weight management. BMJ 311:158–161

Lier HØ, Biringer E, Stubhaug B, Tangen T (2012) The impact of preoperative counseling on postoperative treatment adherence in bariatric surgery patients: a randomized controlled trial. Patient Educ Couns 201(87):336–42

Livelli di Assunzione Raccomandati di Energia e Nutrienti per la Popolazione Italiana - LARN - Revisione 2012 - Società Italiana di Nutrizione Umana

Long CL, Schaffel N, Geiger JW, Schiller WR, Blakemore WS (1979) Metabolic response to injury and illness: estimation of energy and protein needs from indirect calorimetry and nitrogen balance. JPEN J Parenter Enteral Nutr, 3(6):452–456

Maniscalco M, Zedda A, Giardiello C et al (2006) Effect of bariatric surgery on the six-minute walk test in severe uncomplicated obesity. Obes Surg 16:836–841

McCullough PA, Gallagher MJ, Dejong AT et al (2006) Cardiorespiratory fitness and short-term complications after bariatric surgery. Chest 130:517–525

Mechanick JI, Kushner RF, Sugerman HJ, Gonzalez-Campoy JM, Collazo-Clavell ML, Spitz AF, Apovian CM, Livingston EH, Brolin R, Sarwer DB, Anderson WA, Dixon J, Guven S (2009) American Association of Clinical Endocrinologists. The Obesity Society, and American Society for Metabolic & Bariatric Surgery medical guidelines for clinical practice for the perioperative nutritional, metabolic, and nonsurgical support of the bariatric surgery patient. Obesity 17(Suppl 1):S1–70

Metcalf B, Rabkin RA, Rabkin JM, Metcalf LJ, Lehman-Becker LB (2005) Weight loss composition: the effects of exercise following obesity surgery as measured by bioelectrical impedance analysis. Obes Surg 15:183–186

Movsesyan L, Tankó LB, Larsen PJ, Christiansen C, Svendsen OL (2003) Variations in percentage of body fat within different BMI groups in young, middle-aged and old women. Clin Physiol Funct Imaging 23:130–133

Murr MM, Siadati MR, Sarr MG (1995) Results of bariatric surgery for morbid obesity in patients older than 56 years. Obes Surg 5:399–402

Narbro K, Agren G, Jonsson E, Naslund I, Sjostrom L (2002) Swedish Obese Subjects Intervention Study. Pharmaceutical costs in obese individuals: comparison with a randomly selected population sample and long-term changes after conventional and surgical treatment: the SOS intervention study. Arch Intern Med 162:2061–2069

National Heart Lung and Blood Institute (NHLBI) (2000) North American Association for the Study of Obesity (NAASO). Practical Guide to the identification, evaluation and treatment of overweight and obesity in adults. National Institutes of Health, Bethesda

Nguyen NT, Paya M, Stevens CM, Mavandadi S, Zainabadi K, Wilson SE (2004) The relationship between hospital volume and outcome in bariatric surgery at academic medical centers. Ann Surg 240:586–593

Nijamkin MP, Campa A, Sosa J, Baum M, Himburg S, Johnson P (2012) J Comprehensive nutrition and lifestyle education improves weight loss and physical activity in Hispanic Americans following gastric bypass surgery: a randomized controlled trial. Acad Nutr Diet 112:382–390

Norgan NG, Durnin JVGA (1980) The effect of 6 weeks of overfeeding on the body weight, body composition, and energy metabolism of young men. Am J Clin Nutr 33:978–988

Pain H, Wiles R (2006) The experience of being disabled and obese. Disabil Rehabil 28:1211–1220

Passmore R, Meiklejohn AP, Dewar AD, Thow RK (1955) An analysis of the gain in weight of overfed thin young men. Br J Nutr 9(1):27–37

Pelizzo G, Nakib G, Alfei A, Iasci A, Cena H, Locatelli D, Mosconi M, Zappoli F, Calcaterra V (2012) Fetal neural tube defects in pregnant women previously submitted to bariatric surgery: more attention to a new emerging entità. Prenat Diag 32:1–2

Prado CM, Wells JC, Smith SR, Stephan BC, Siervo M (2012) Sarcopenic obesity: a Critical appraisal of the current evidence. Clin Nutr. 31(5):583–601

Physical Status: the use and interpretation of anthropometry (1995) Report of a WHO Expert Committee. WHO Technical report Series–854–Geneva

Resnik L, Lapane KL, Allen SM (2005) Obesity and receipt of personal care assistance for people with mobility impairments. Obes Res 13:1307–1310

Roubenoff R (2004) Sarcopenic obesity: the confluence of two epidemics. Obes Res 12(6):887e8

Schutz Y, Kyle UG, Pichard C (2002) Fat-free mass index and fat mass index percentiles in Caucasians aged 19–98 y. Int J Obes 26:953–960

Schweitzer DH, Posthuma EF (2008) Prevention of vitamin and mineral deficiencies after bariatric surgery: evidence and algorithms. Obes Surg 18:1485–1488

Segal A (2004) Kinoshita Kussunoki D, Larino MA: Post-surgical refusal to eat: anorexia nervosa, bulimia nervosa or a new eating disorder? A case series Obes Surg 14:353–360

Seidell JC, Cigolini M et al (1992) Fat distribution in European men: a comparison of anthropometric measurements in relation to cardiovascular risk factors. Int J Obes 16:17–22

Seidell JC, Cigolini M et al (1990) Fat distribution in European women: a comparison of anthropometric measurements in relation to cardiovascular risk factors. Int J Epidemiol 19:303–308

Shah M, Simha V, Garg A (2006) Long-term impact of bariatric surgery on body weight, comorbidities and nutritional status. J Clin Endocrinol Metab 91:4223–4231

Shikora SA, Kim JJ, Tarnoff ME (2007) Nutrition and gastrointestinal complications of bariatric surgery. Nutr Clin Pract 22:29–40

SINU Società Italiana di Nutrizione Umana (2012) Livelli di Assunzione di Riferimento di Nutrienti ed energia per la popolazione, http://www.sinu.it

SINPE Società Italiana Nutrizione Parenterale ed Enterale (2002) Linee Guida SINPE per la Nutrizione Artificiale Ospedaliera (Guidelines for Hospital Nutrition Support Therapy)

Sockalingam S, Cassin S, Crawford SA, Pitzul K, Khan A, Hawa R, Jackson T, Okrainec A (2012). Psychiatric predictors of surgery non-completion following suitability assessment for bariatric surgery. Obes Surg Sep 9

Soeters PB, Rejiven PLM, van Bokhorst-de van der SChuren MAE, Schols JMGA, Halfens RJG, Meijers JMM, van Gemert WG (2008) A rational approach to nutritional assessment. Clin Nutr 27(5):706-16

"Standard Italiani per la Cura dell'Obesità" a cura SIO-Soc It dell'Obesità e ADI – Ass It Dietetica e Nutrizione Clinica – 2012-2013. P Sbracca, R Vettor eds, cap 3.c

Sugerman HJ, Fairman RP, Sood RK, Engle K, Wolfe L, Kellum JM (1992) Long-term effects of gastric surgery for treating respiratory insufficiency of obesity. Am J Clin Nutr 55(2 Suppl):597S–601S

Suter M, Calmes JM, Paroz A (2006) Giusti V.A 10-year experience with laparoscopic gastric banding for morbid obesity: high long-term complication and failure rates. Obes Surg 16:829–835

Tabelle di Composizione degli Alimenti. Istituto Nazionale della Nutrizione. 2000; 106–108

Tichet J, Vol S, Goxe D, Salle A, Berrut G, Ritz P (2008) Prevalence of sarcopenia in the French senior population. JNHA 12:202–206

Van Dorsten B, Lindley EM (2011) Cognitive and behavioral approaches in the treatment of obesity. Med Clin North Am 95:971–988

Velcu LM, Adolphine R, Mourelo R, Cottam DR, Angus LD (2005) Weight loss, quality of life and employment status after Roux-en-Y gastric bypass: 5-year analysis. Surg Obes Relat Dis 1:413–416

Virji A, Murr MM (2006) Caring for patients after bariatric surgery. Am Fam Physician 73:1403–1408

Weir JB (1949) New methods for calculating metabolic rate with special reference to protein metabolism. J Physiol 109:1–9

Zamboni M, Mazzali G, Fantin F, Rossi A, Di Francesco V (2008) Sarcopenic obesity: a new category of obesity in the elderly. NMCD 18:388–395

Zamboni M, Turcato E, Armellini F, Kahn HS, Zivelonghi A, Santana H, Bergano-Andreis IA, Bosello O (1998) Sagittal abdominal diameter as a practical predictor of visceral fat. Int J Obes Relat Metab Disord 22(7):655–660

Zhu S, Wang Z, Shen W, Heymsfield SB, Heshka S (2003) Percentage body fat ranges associated with metabolic syndrome risk: results based on the third National Health and Nutrition Examination Survey (1988–1994). Am J Clin Nutr 78:228–235

Chapter 12
Bariatric Surgery: Current Techniques and Results

Joel Faintuch, Shirley A. F. Souza, Sonia M. Fabris and Ivan Cecconello

Abstract Half a century ago obesity was not a public health problem, however, gastroduodenal ulcers were ubiquitous. Many gastrectomies were conducted at that time and patients eventually lost weight. That is how bariatric surgery commenced, naturally expanding to a variety of techniques and accesses. Success has been both bigger and smaller than expected. Yes, bariatric operations became so popular that they already represent one of the five most performed major operations in some countries. They are followed by strong secondary benefits particularly concerning diabetes remission, to the point that a new subspecialty has arisen, metabolic surgery for diabetes. No, they are not carefree and they have not solved the problem of severe obesity, at least from the epidemiological point of view. They were actually not designed to be a mass treatment. The pursuit of new therapeutic avenues and the reinforcement of old ones is still mandatory, if the worldwide obesity epidemic is to be curtailed.

12.1 Introduction

Bariatric operations were devised to create carefully dimensioned restrictive and/or disabsorptive mechanisms, aiming to eliminate at least 50 % of excess body weight/EBW (difference between ideal and current weight), but not much more

J. Faintuch (✉) · S. M. Fabris · I. Cecconello
Department of Gastroenterology, Hospital das Clinicas, São Paulo University Medical School, São Paulo, Brazil
e-mail: blfaintuch@hotmail.com

S. M. Fabris
e-mail: sofabris@hotmail.com

I. Cecconello
e-mail: departamentogastro@gmail.com

S. A. F. Souza
School of Physical Therapy, State University of Londrina, Parana, Brazil
e-mail: shirleysouza@uel.br

P. Capodaglio et al. (eds.), *Disabling Obesity*, DOI: 10.1007/978-3-642-35972-9_12,

Table 12.1 Requirements of a *successful bariatric intervention*

Feature	Description
Safety	Mortality under 2 % (currently ≤0.5 %)
Efficacy	Elimination of >50 % excess body weight (EBW)
Durability	Results should remain stable along the years, with rare recurrencies
Tolerability	Minor side effects, compatible with good quality of life
Age range	Initially 18–65 years, adolescents and the aged still questioned
BMI range	Classically >35 kg/m^2 with comorbidities, <35 kg/m^2 debated ([a])

([a]) In Asia BMI < 35 kg/m^2 acceptable, and <30 kg/m^2 under investigation

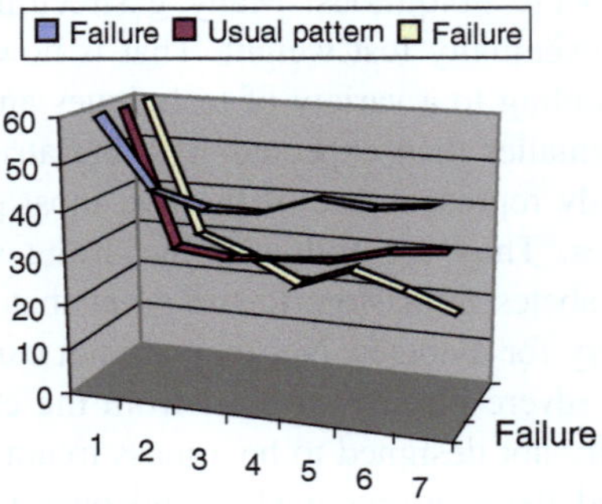

Fig. 12.1 Theoretical patterns of postoperative course after bariatric surgery (Body mass index versus years of follow up). Usual course (successful operation): >50 % loss of excess body weight, modest late catch-up (>5 years). Surgical failure (too little reduction): Patient does not sufficiently respond, effective loss <50 % of excess body weight, BMI still in the severe obesity range. Clinical failure (malnutrition): Patient loses too much weight and becomes undernourished

than 80 %. These results are expected within 2–3 years and should be long-lasting, thus only a minority of the subjects achieve ideal weight, most remaining 20–25 % above it.

This is quite acceptable, as patients are able to lead productive and rewarding lives, at least in comparison with their former status of morbid obesity (Table 12.1).

A minority progresses to underweight (1–4 %) (Faintuch et al. 2004), whereas as many as 15–40 % will recover some weight after 5–10 years, although still remaining below their preoperative status (Fig. 12.1).

12.2 Surgical Techniques

The most commonly used procedures are indicated in Fig. 12.2. Gastric banding and sleeve gastrectomy are purely *restrictive interventions* because just the gastric reservoir is diminished, without any further changes. Roux-en-Y gastric bypass is considered a mixed modality because it includes a malabsorptive component

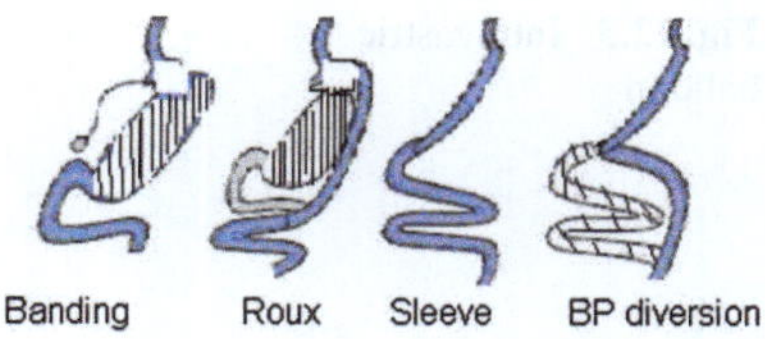

Fig. 12.2 Major bariatric procedures

(duodenal exclusion). In the biliopancreatic diversion remodeling of intestinal architecture predominates, therefore this is classified as a *malabsorptive treatment.*

Adjustable *gastric banding* was the most popular modality in Europe during nearly two decades, however, it is becoming less used. It consists in an external device that is introduced around the upper part (fundus) of the stomach, compressing this segment and thus creating an hourglass stomach. Roux-en-Y *gastric bypass* is considered the gold standard in most of the world. A small gastric chamber is directly connected to the jejunum, whereas 90–95 % of the stomach along with the entire duodenum is excluded and does not receive food.

Sleeve gastrectomy is a 50–60 % vertical gastrectomy. *Biliopancreatic diversion* includes a sleeve gastrectomy, plus a gastro-ileal anastomosis. Most of the jejunum and ileum is excluded, anastomosis occurring only one meter before the ileo-cecal valve. Some techniques respect gastrointestinal integrity, whereas during sleeve gastrectomy and biliopancreatic diversion, part of the stomach is permanently removed.

All are accepted by professional societies and most national reimbursement systems, however, technical features and long-term results vary. Most surgeons adopt just one of them as a routine, reserving the others for special circumstances.

12.3 Endoluminal or Minimally Invasive Interventions

Endoscopists are much interested in obesity treatment and have described a number of techniques. Two representative ones are shown in Figs. 12.3 and 12.4.

The *balloon* is a temporary device, usually kept in place for 6 months. An empty prosthesis is endoscopically introduced inside the stomach and filled with 500 ml of saline. It is abandoned there and freely floats in the cavity, occupying space and thus eliciting early satiety. Weight loss is substantial, however, less than after surgical interventions, and weight regain is a strong possibility, unless the balloon intervention is accompanied by lifestyle changes such as exercise and dieting. The balloon may be an option for patients who do not accept operation or present very high risk such as extreme obesity or organ failures.

The *endoprosthesis* is a waterproof plastic tube which is endoscopically anchored at the pylorus by means of tiny hooks, and introduced inside the duodenum and proximal jejunum, thus lining the respective mucosa. Food is consequently prevented from mixing with upper gastrointestinal enzymes (biliopancreatic secretions), therefore digestion and absorption are delayed and reduced.

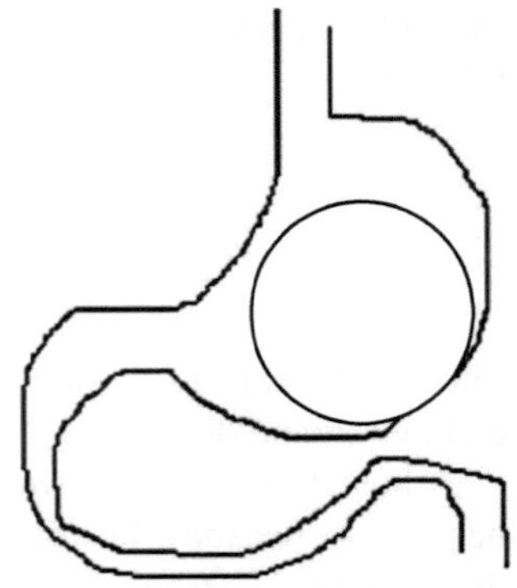

Fig. 12.3 Intragastric balloon

Fig. 12.4 Duodeno-jejunal endoprosthesis

Similar to the gastric balloon this is a temporary method, the device being usually removed after 6–12 months (Escalona et al. 2012).

12.4 Endoluminal Operations

More ambitious approaches target endoscopic vertical gastroplasty, transoral gastroplasty, and transluminal sleeve construction (Kethu et al. 2012). Specific devices being tested for *endoscopic gastroplasty* include the EndoCinch Suturing System (C. R. Bard, Murray Hill, NJ, USA), the Transoral Gastroplasty System (TOGA) (Satiety, Inc, Palo Alto, CA, USA), and the Trans-oral Endoscopic Restrictive Implant System (TERIS; BaroSense, Redwood City, CA, USA). Also announced are a gastric sleeve (ValenTx bypass sleeve; ValenTx, Inc, Carpinteria, CA, USA) and a radiofrequency machine for endoscopic gastroplasty. Some of them create mechanical barriers such as a small pouch at the upper part of the stomach, similar to surgical gastric banding, or an artificial sleeve gastrectomy. Others somehow restrict gastric capacity or interfere with gastric physiology, in order to inhibit food intake and elicit early satiety.

12.5 Botulinum Toxin Injection

Dermatologists and plastic surgeons have long realized that *botulinum toxin* (Botox, Allergan, USA) promotes temporary denervation in the area where it is infiltrated. This may be useful in the skin for wrinkle elimination, however, visceral effects have also been investigated. Cholinergic denervation of the gastric mucosa, particularly in the fundus along with the prepyloric antrum, is followed by weight loss in obese subjects. Relatively frequent endoscopies with multiple injections are necessary for lasting results, however, the technique is simple, well tolerated, and could be used even in small centers, especially with high risk patients who are not candidates for more aggressive options (Kethu et al. 2012).

12.6 Gastric Pacemaker

One original proposal is the electronic stimulation of the stomach, in order to interfere with local physiology and diminish appetite. Though still not available in daily practice, these apparatuses, which to some extent mimic heart pacemakers employed for serious arrhythmia, might represent an interesting combination of minor surgical trauma with long-term weight loss (Fig. 12.5).

http://www.freepatentsonline.com/6564101.html

Subserosal electrodes are positioned on the stomach and an external battery-operated pacemaker, subcutaneously implanted, modulates gastric physiology and hormone production. Preliminary results suggest up to 40 % weight reduction, and

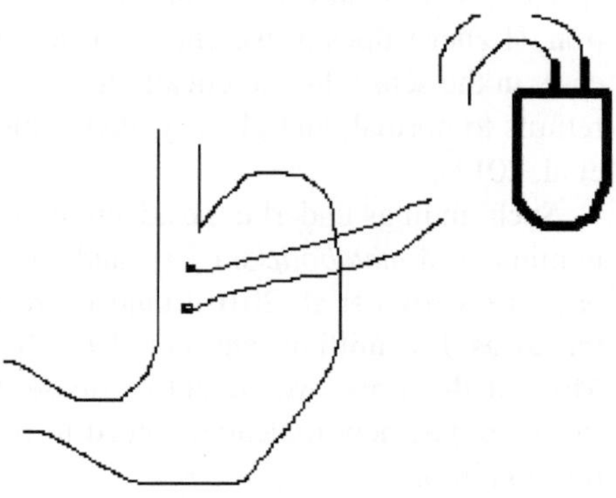

Fig. 12.5 Gastric electronic stimulation

improvement of comorbidities such as diabetes, however, not all patients respond and long-term results are not available (Gallas et al. 2011).

12.7 Deep Brain Stimulation

Cerebral electrodes and deep brain electrical stimulation are nowadays familiar to neurologists and neurosurgeons, as they have been used for Parkinson's disease, cerebral palsy, heroin addiction, and other conditions. Devices are marketed in many countries and implantation is well standardized, with good tolerance and encouraging long-term therapeutic results, at least for the most common indications.

Experience with obesity is limited and one may only speculate whether it will eventually be successful and replace other options. The scientific rationale is nevertheless stronger than for operations targeting the stomach or intestine. Obesity is more a behavioral disorder than an ailment of the digestive viscera, therefore it makes more sense to modulate the hypothalamus than to perform gastric bypass or intestinal diversion (Halpern et al. 2011).

12.8 Metabolic Surgery: The Diabesity Patient

Diabetes in not the first association in morbidly obese subjects. That role belongs to arterial hypertension, usually present in twice as many bariatric candidates (20–40 % vs. 60–80 %). If one includes prediabetes, insulin resistance, and other minor evidence of dysfunctional glucose homeostasis the numbers will substantially increase, rivaling hypertension.

Moving to another angle, obesity certainly is the most important comorbidity of type 2 diabetes, affecting as many as 80 % of such population. Even more important, in both the bariatric and diabetic setting, weight loss of as little as 10 % will improve management and strongly reduce glucose-lowering drug consumption. Higher values of weight loss often induce partial or full remission of the disease, in the sense that practically no prescription is required, fasting blood glucose returns to normal, and glycosylated hemoglobin HbA1c remains <6 % (Pournaras et al. 2012).

Such findings underlie the advent of a new subspecialty within bariatric surgery, nominally of *metabolic surgery*, and the creation of a new syndrome, the diabesity patient (Arroyo et al. 2010). Indeed, the number of diabetics in the world is estimated as 170 million, and could reach 300 million within one or two decades. Most of them are overweight or obese, but with a body mass index <35 kg/m^2. In this sense, new indications need to be established for the surgical approach to this condition.

At this moment antidiabetes operations in patients with BMI < 35 kg/m^2 are still considered experimental, but there are robust reasons to believe that they will soon convert to mainstream, because of mounting positive results with comparatively few complications, in the same range as conventional bariatric procedures.

Indeed the majority of such patients are handled by techniques quite similar to those adopted for the severely obese, namely adjustable gastric band, Roux-en-Y gastric bypass, minigastric bypass, biliopancreatic diversion, and sleeve gastrectomy. Specific operations geared toward diabetes therapy have been devised in the last decade, such as ileal interposition and duodenal-jejunal bypass (Arroyo et al. 2010; Reis et al. 2012).

Postulated mechanisms for the positive metabolic achievements include calorie restriction, duodenal bypass (foregut theory), early arrival of the food bolus to the ileum (hindgut theory), and as a consequence of such anatomic changes, hormonal changes particularly involving glucagons-like peptide 1(GLP-1) and peptide YY (PYY).

A recent review of 29 articles dealing with a variety of metabolic operations, both standard and purpose-designed, surveyed a total population of 675. BMI decreased from 30.0 ± 0.5 to 24.8 ± 0.4 kg/m^2, fasting blood glucose from 208 ± 9 to 114 ± 5 mg/dL, and HbA1c from 8.9 ± 0.2 to 6.4 ± 0.2 %. Diabetes resolution (HbA1c < 7 % without drugs) occurred in 84.0 % (Reis et al. 2012).

Perspectives might not remain as optimistic if the operation becomes popular and is indicated with less strict care and follow-up, however, some predict it to be quite routinely employed in the near future. In an unpublished yet study, our group has observed that even after conventional gastrectomy for cancer, previously diabetic cases exhibited a reduction of fasting blood glucose from 149 ± 67 to 107 ± 32 mg/dL, although patients were not obese and weight reduction was modest (25.2 ± 3.8 to 22.9 ± 4.0 kg/m^2).

Many diabesity subjects and also lean diabetics are fed-up with therapeutic regimens and anxious for more promising avenues. Poor drug compliance and increased risk of complications are old problems in chronic diseases and specifically in diabetes mellitus, and successful surgery might alleviate these shortcomings.

12.9 Costs of Obesity and of Bariatric Surgery

Obesity and its comorbidities is an extremely onerous problem, in some countries representing the first or second largest national public-health expenditure. Bariatric operations are not inexpensive either, demanding major investment in hospital facilities, specialized equipment, outpatient care, and a highly trained team (Table 12.2).

It is estimated that organizing a surgical weight-loss center will initially cost at least 50,000 US dollars per bed, with additional investment for staff training, inventory stocking, storage place, and periodic updates (Lautz et al. 2009). Patient or medical

Table 12.2 Personnel and equipment for an anti-obesity medical and surgical center (Modified from Lautz et al. 2009)

Resource	Description
Health care team	Bariatric surgeon, bariatric anesthetist, bariatric intensive care unit, endocrinologist, internal medicine consultants (cardiology, pneumology, gastroenterology), dietitian, nurse, physical therapist, psychologist, social worker
Operating rooms	Purpose-built operating tables, lifts and transferring equipment, venous compression devices, appropriate surgical material (laparoscopes, retractors, instrument sets for the obese)
Hospital wards, outpatient facilities	Wide doors, rooms and corridors, weight-resistant beds and furniture, reinforced toilets and lavatories, large elevators, wheelchairs, stretchers, walkers, obesity-dimensioned scales, oversize gowns and robes
Imaging methods	Radiology, ultrasonography, tomography and magnetic resonance machines with 200 kg capacity
Central pharmacy	Familiarity and availability of anticoagulant, antiarrhythmic and vasoactive prescriptions for the obese, appropriate dosing guidelines, especial coding for high-dose medications (barocode)
Miscellaneous	Wide blood pressure cuffs, extra-long airway equipment, biphasic defibrillators, automated medication administration systems, continuous positive airway pressure therapy
Staff safety	Lifts and transferring equipment widely available, well-trained personnel, courses on how to avoid injury when dealing with morbidly obese patients, possible dedicated lift team

insurance disbursement for the operation is proportional, ranging from 4–6,000 up to 15–20,000 US dollars, depending on the country and the procedure.

All these problems notwithstanding, anti-obesity operations have been admitted as cost-effective, however, long-term studies are scarce. Hayashi et al. investigated a cohort during 5 years, namely the preoperative year along with 4 years of follow-up. Outpatient visits and laboratory tests diminished by approximately half with the antiobesity treatment, but not total hospitalizations or emergency admissions (Hayashi et al. 2011). In another Brazilian series expenditures with drugs, professional care, and laboratory tests also progressively decreased after the operation, the change becoming more remarkable in cases suffering from more comorbidities (Sussenbach et al. 2012).

12.10 Limitations of Bariatric Surgery

It is estimated that <5 % of the morbidly obese population currently undergo bariatric operations, and in most countries <1 %. In the USA, among 7–10 million eligible candidates, about 220,000 are actually submitted to such therapy, or 2–3 % (Buchwald et al. 2009). These numbers are expected to grow but not very

Table 12.3 Patient-oriented criteria of success in a bariatric program (Modified from Nguyen et al. 2012)

Variable	Defining parties
Weight loss	Patient + medical team
Amelioration of comorbidities	Patient + medical team
Increased quality of life	Patient
Better social, psychological and employment status	Patient
Acceptable morbidity and mortality (short and long-term)	Patient + medical team
Widely available, low cost program	Patient + medical team

dramatically, because of barriers related to costs, risks, specialized facilities, and patient acceptance. In other words, the vast majority of seriously obese patients will still be waiting for other modalities of treatment.

Moreover, the role of bariatric surgery is debated in certain circumstances such as body mass index/BMI < 35 kg/m^2 and age <18 or >65 years. Also controversial are candidates to nonbariatric major surgery especially cardiovascular, orthopedic, or organ transplantation, as well as general medical and obstetric/gynecologic patients at risk of life-endangering complications because of abnormal body weight.

Nguyen et al. have suggested a cluster of new parameters for assessing *success of bariatric management*, conflicting with the standard reduction of 50 % of excess body weight (Table 12.3). Weight decrease is definitely one of them, but not the overarching criterium. In their list they emphasize outcomes that are a priority for the patient, not just for the medical team, such as quality of life and employment status (Nguyen et al. 2012). Yet they fail to mention a low relapse rate, however, this variable should not be overlooked. Most patients come to the obesity specialist sick and tired, sometimes literally, from dieting, exercising and following other therapies, only to again become frustrated if within a short period the weight disorder recurs.

It is true that even classic bariatric procedures may not be ideal after 5–10 years (Yamaguchi et al. 2012; Dalcanale et al. 2010), however, currently they unquestionably exceed other modalities with regard to degree and durability of response.

12.11 Synthesis

Weight loss operations are an old idea that took decades to mature but was eventually embraced, in some contexts with overwhelming enthusiasm, by most of the world. It is not a comfortable embrace despite the huge medical and social success, both the receiving and the offering parties resenting the multiple and occasionally costly and painful strings attached to such option.

Welcome or not, this is a true paradigm shift that may stay with us for a long time, until a *magic bullet* is crafted to remove this scourge simply and effortlessly. In 1998 Hirsch wrote: "Who would not rejoice to find a magic bullet that we could fire into obese people to make them permanently slim and healthy?" Realistically admitting that such a development was remote and might never materialize, as few drugs and genetic treatments could be envisaged with the necessary qualities, the article still encouraged people to pursue that road, in order to shed more light on the pathophysiology of deranged energy balance (Hirsch 1998).

Major centers are indeed engaged in serious research aiming to reduce morbidity and mortality of bariatric modalities, likewise striving to increase availability and popularization of the new techniques. Nevertheless, a gap remains to be bridged for millions of patients who need effective treatment, however, they are not ready for surgery (or surgery is not ready for them), as well as for national health care systems unable to support such a complex and expensive endeavor.

Until Hirsch's bullet is manufactured, new and creative pathways encompassing not only conventional surgery but particularly *minimally invasive* and endoscopic maneuvers, along with even more conservative routes such as transcutaneous gastric electrical stimulation (Wang et al. 2010) will be required, in order to enlarge the spectrum and enhance the results of antiobesity management.

References

Arroyo K, Kini SU, Harvey JE, Herron DM (2010) Surgical therapy for diabesity. Mt Sinai J Med 77:418–430. doi:10.1002/msj.20209

Buchwald H, Oien DM (2009) Metabolic/bariatric surgery worldwide 2008. Obes Surg 19:1605–1611. doi:10.1007/s11695-009-0014-5

Dalcanale L, Oliveira CP, Faintuch J, Nogueira MA, Rondó P, Lima VM, Mendonça S, Pajecki D, Mancini M, Carrilho FJ (2010) Long-term nutritional outcome after gastric bypass. Obes Surg 20:181–187. doi:10.1007/s11695-009-9916-5

Escalona A, Pimentel F, Sharp A, Becerra P, Slako M, Turiel D, Muñoz R, Bambs C, Guzmán S, Ibáñez L, Gersin K (2012) Weight loss and metabolic improvement in morbidly obese subjects implanted for 1 year with an endoscopic duodeno-jejunal bypass liner. Ann Surg 255:1080–1085. doi:10.1097/SLA.0b013e31825498c4

Faintuch J, Matsuda M, Cruz ME, Silva MM, Teivelis MP, Garrido AB Jr, Gama-Rodrigues JJ (2004) Severe protein-calorie malnutrition after bariatric procedures. Obes Surg 14:175–181. doi:10.1381/096089204322857528

Gallas S, Fetissov SO (2011) Ghrelin, appetite and gastric electrical stimulation. Peptides 32:2283–2289 doi:org/10.1016/j.peptides.2011.05.027

Halpern CH, Torres N, Hurtig HI, Wolf JA, Stephen J, Oh MY, Williams NN, Dichter MA, Jaggi JL, Caplan AL, Kampman KM, Wadden TA, Whiting DM, Baltuch GH (2011) Expanding applications of deep brain stimulation: a potential therapeutic role in obesity and addiction management. Acta Neurochir (Wien) 153:2293–2306. doi:10.1007/s00701-011-1166-3

Hayashi SY, Faintuch J, França JI, Cecconello I (2011) Four-year hospital resource utilization after bariatric surgery: comparison with clinical and surgical controls. Obes Surg 21:1355–1361. doi:10.1007/s11695-011-0404-3

Hirsch J (1998) Magic bullet for obesity. BMJ 317:1136–1138 PMCID: PMC1114112

Kethu SR, Banerjee S, Barth BA, Desilets DJ, Kaul V, Pedrosa MC, Pfau PR, Pleskow DK, Tokar JL, Wang A, Wong Kee Song LM, Rodriguez SA. Endoluminal bariatric techniques. Gastrointest Endosc 12 May 2012 doiorg/10.1016/j.gie.2012.02.020

Lautz DB, Jiser ME, Kelly JJ, Shikora SA, Partridge SK, Romanelli JR, Cella RJ, Ryan JP (2009) An update on best practice guidelines for specialized facilities and resources necessary for weight loss surgical programs. Obesity (Silver Spring) 17:911–917. doi:10.1038/oby.2008.581

Nguyen N, Champion JK, Ponce J, Quebbemann B, Patterson E, Pham B, Raum W, Buchwald JN, Segato G, Favretti F (2012) A review of unmet needs in obesity management. Obes Surg 22:956–966. doi:10.1007/s11695-012-0634-z

Pournaras DJ, Aasheim ET, Søvik TT, Andrews R, Mahon D, Welbourn R, Olbers T, le Roux CW (2012) Effect of the definition of type II diabetes remission in the evaluation of bariatric surgery for metabolic disorders. Br J Surg 99:100–103. doi:10.1002/bjs.7704

Reis CE, Alvarez-Leite JL, Bressan J, Alfenas RC (2012) Role of bariatric-metabolic surgery in the treatment of obese type 2 diabetes with body mass index <35 kg/m^2: a literature review. Diabetes Technol Ther 14:365–372. doi:10.1089/dia.2011.0127

Sussenbach SP, Padoin AV, Silva EN, Benzano D, Pufal MA, Barhouch AS, Chatkin R, Ramos RJ, Balestro A, Mottin CC (2012) Economic benefits of bariatric surgery. Obes Surg 22:266–270. doi:10.1007/s11695-011-0558-z

Wang J, Song J, Hou X, Liu J, Chen JD (2010) Effects of cutaneous gastric electrical stimulation on gastric emptying and postprandial satiety and fullness in lean and obese subjects. J Clin Gastroenterol 44:335–339. doi:10.1097/MCG.0b013e3181d34572

Yamaguchi CM, Faintuch J, Hayashi SY, Faintuch JJ, Cecconello I (2012) Refractory and new-onset diabetes more than 5 years after gastric bypass for morbid obesity. Surg Endosc. doi:10.1007/s00464-012-2256-x

Chapter 13
Disability in the Post-Obese Bariatric Patient: Old and New Problems

Joel Faintuch, Shirley A. F. Souza, Sonia M. Fabris, Alberto Rosenblatt and Ivan Cecconello

Abstract Sustained and lifelong weight loss for severely obese people is not anymore an impossible dream, and a Swedish journal has described bariatric surgery as the fairy tale about the ugly duckling (Olbers 2011). Of course this is a bittersweet remark, because although for millions such is an advantageous and even life-saving intervention, all of them have to cope with the post-obesity status. Obesity is a chronic incurable disease and the postobesity status is an attenuated albeit ongoing illness, not a mere sequela. Appropriate follow-up and secondary interventions, be they surgical, clinical, physiatric, dietary, or psycho-social, may be demanded. Obesity entails widespread disorders involving as far away organs, structures, needs and abilities as the teeth, the central nervous system, the gut microbiome, the susceptibility to cancer, the performance at the workplace, and the demand for health care resources. Subsequent weight gain and comorbidity relapse is a permanent possibility, as endogenous and environmental obesogenic stimuli are not suppressed, only weakened. Gastrointestinal restriction and bypass are highly successful maneuvers when correctly indicated and conducted, however, they do not signal the end of the battle. Patients have to be educated and followed for life. It is hoped that such experience along with general public-health initiatives will eventually trickle down to their families, their offspring, and society in general, so that the new generations might be born and nurtured with obesity prevention in mind.

J. Faintuch (✉) · S. M. Fabris · A. Rosenblatt · I. Cecconello
Department of Gastroenterology, Hospital das Clinicas, São Paulo University Medical School, São Paulo, SP, Brazil
e-mail: blfaintuch@hotmail.com

S. M. Fabris
e-mail: sofabris@hotmail.com

A. Rosenblatt
e-mail: albrose1@gmail.com

I. Cecconello
e-mail: departamentogastro@gmail.com

S. A. F. Souza
School of Physical Therapy, State University of Londrina, Parana, PR, Brazil

P. Capodaglio et al. (eds.), *Disabling Obesity*, DOI: 10.1007/978-3-642-35972-9_13,

13.1 Introduction

Morbidly obese young men may undergo a decrease of 13 years in their *life expectancy*, and women of 8 years (Schauer et al. 2010). These numbers do not generate effects during old age only, when somewhat longer or shorter life spans are mere demographic statistics. Quality of life diminishes much earlier, with multiple treatments, hospitalizations, impaired working ability, and assorted physical, social, economical, and psychological handicaps (Table 13.1).

Table 13.1 Health problems associated with advanced obesity (Kaul and Sharma 2011; Gunstad et al. 2011)

Diabetes mellitus
Arterial hypertension
Hyperlipidemia
Atherosclerosis
Myocardial infarction
Stroke
Obstructive sleep apnea
Asthma
Pulmonary hypoventilation
Varicose veins
Leg lymphedema
Venous thrombosis
Pulmonary embolism
Albuminuria
Renal failure
Gastroesophageal reflux disease
Nonalcoholic fatty liver disease
Liver cirrhosis
Gallbladder disease
Disorders of posture and gait
Knee, hip and back osteoarthritis
Gout
Psychiatric conditions (depression, anxiety)
Cognitive impairment
Intracranial hypertension
Erectile dysfunction
Gynecologic and obstetric complications
Cancer of multiple organs
Migraine
Social and economic troubles
Reduced life expectancy
Low quality of life

13.2 Clinical Amelioration After Bariatric Weight Loss

Most associated diseases and correlates of obesity namely arterial hypertension, diabetes mellitus, dyslipidemia, obstructive sleep apnea, systemic atherosclerosis, microinflammation, and possibly even cognitive impairment (Gunstad et al. 2011) are positively influenced by postoperative weight loss. Follow-up till 15 years is compatible with a definitely positive clinical and metabolic balance (Sjostrom et al. 2009, 2012). Nevertheless, there are reasons to believe that after the first 5 years, benefits will be partially offset by progressively increasing weight regain interfering with *comorbidities* and quality of life, although long-term alleviation of medical conditions still represents substantial profit (Yamaguchi et al. 2012; Dalcanale et al. 2010).

13.3 Neuropsychiatric Impact of Bariatric Intervention

Chronic *psychiatric conditions* of surgically treated morbidly obese patients such as anxiety and depression are sometimes refractory and do not respond to weight loss. There is an ongoing discussion whether such patients should be accepted in bariatric programs in the first place. Indeed, chances are that such patients could be disappointed by the operation, or will refrain from following dietetic instructions and returning to follow-up, thus exposing themselves to increased complications.

A curious phenomenon of addiction transfer/substitution is also occasionally detected after bariatric management. As psychological reward stemming from generous meals becomes anatomically impossible or is followed by shame, anguish, and weight regain, patients may seek gratification from alcohol and substance abuse (Ivezaj et al. 2012). Even irrational spending and credit card overdraft may rarely be precipitated as a form of vicarious and ineffective pleasure, in the attempt to replace overfeeding, as documented in our experience.

Neuroimaging studies employing positron emission tomography (PET), *functional magnetic resonance imaging* (fMRI) and magnetic resonance imaging (MRI) are now being employed to investigate brain regions implicated in reward, emotion, memory, appetite, cognitive control, and attention.

Ochner et al. revealed with the help of fMRI that after Roux-en-Y gastric bypass patients responded less to food cues 1 month after operation. Interestingly, such reduction was more remarkable to energy dense foods than to less caloric products, which were presented in the form of words or pictures. The authors suggest that gastrointestinal hormones such as PYY, GLP-1, and ghrelin could underlie such behavior, which is protective and should prevent patients from regaining weight (Ochner et al. 2011).

Other groups have unveiled changes in *dopamine type 2 receptors* in several areas of the brain, within the first weeks or months after the same gastric bypass (Steele et al. 2010; Dunn et al. 2010), however, with conflicting conclusions.

According to Dunn et al. receptors became about 10 % scarcer, and such decrease correlated with reduction of serum insulin and leptin as well as body weight. In the view of these authors increase in extracellular dopamine levels, followed by diminished receptors, would result in enhanced dopaminergic neurotransmission conducting to reduced hunger and improved satiety. Steele et al. however, documented increased dopamine-2 receptors within a similar context including improved satiety and weight loss.

13.4 General Consequences of the Operation

Social and economic outcome of the intervention may as well be a source of sadness, especially in young patients with excessive expectations regarding aesthetic harmony and personal success. *Suicide* in disappointed patients is not common, however, it is a recognized risk after antiobesity operations (Heneghan et al. 2012; Tindle et al. 2010).

It is true that obesity itself is a rather depressive condition accompanied by an excess of suicides, however, a clear trend in the bariatric population has been unveiled in large series (Tindle et al. 2010). Although most bariatric patients are female this is more typically a male complication, suggesting that either men are less adept at coping with frustration, or perhaps they are more prone to radical decisions. Nevertheless, women are not immune to this development either.

Venous stasis usually responds to weight loss, nevertheless varicose veins, lymphedema of the lower limbs, along with other dermatological blemishes may represent a continuing source of distress after operative weight loss, thus requiring further interventions or acceptance of undesirable sequelae.

The *metabolic profile* is definitely more privileged, provided body weight stabilizes at lower levels. Best therapeutic results correspond to diabetes, with highly favorable perspectives for arterial hypertension, obstructive sleep apnea, and hyperlipidemia as well (Table 13.2). Overall mortality including *cardiovascular diseases* diminishes and life expectancy increases, although not equaling the values of the regular population (Sjostrom et al. 2009, 2012; Schauer et al. 2010). Also aerobic capacity, general fitness, and ability to walk and perform daily activities should be counted among the major achievements of such operations, when compared to preoperative status (Souza et al. 2010; de Souza et al. 2009, 2010).

Interestingly the Swedish Obese Subjects Study (SOS), which is the largest (>4,000 subjects) and the longest longitudinal bariatric protocol in the world

Table 13.2 Remission of major *comorbidities* after Roux-en-Y gastric bypass (Ribeiro et al. 2009)

Abnormality	Preoperative (%)	One year later (%)
Diabetes	21.3	7.5
Hypertension	63.8	41.3
Hyperlipidemia	23.8	10.0
Use of antidiabetic drugs	15.0	6.3
Use of statins	13.8	6.3

(since 1987), documented reduction in *disability pension* incidence and duration in operated men, for up to 19 years. However, no change was detected in women, which represent about 80–85 % of the bariatric candidates (Gripeteg et al. 2012).

Observation: All changes were statistically significant.

13.5 Malignancies in the Obese Population

One theoretical concern within this context is cancer. As indicated in Table 13.3, risks are slightly or moderately increased for more than a dozen tumor types. No cancer epidemic has materialized in the postbariatric population, available evidence suggesting that to the contrary, risks steadily diminish (Sjostrom et al. 2009). Yet, given the long natural history of malignancies, even if body composition is corrected patients may still be at risk for the more common modalities such as colon, breast, lower esophagus, and endometrium. In this sense, long-term surveillance is advised for this population (Table 13.3).

13.6 Dental Changes

Is bariatric intervention good for *dental health*? In principle it should be, as patients eat globally less including a diminished amount of sweets, chocolate, and sugared soda drinks. On the other hand vegetables, especially uncooked or fiber rich, may only be consumed sparingly and with careful chewing, after operations that restrict the stomach such as gastric bypass, adjustable gastric banding, and sleeve gastrectomy.

That means that some bariatric patients avoid them altogether, in favor of soft food that may not be as healthy. At the same time it must be admitted that obese persons in general are not known as avid fans of vegetables, therefore the postoperative pattern will not be dramatically different from the preoperative profile.

Table 13.3 Increased risk for *cancer* development in obesity (Khandekar et al. 2011)

Modality	Additional risk (%)
Endometrium	59
Lower third of the esophagus	52
Kidney	24–34
Thyroid	14–33
Breast	12
Colon	9–24
Leukemia	8–17
Melanoma	−17
Myeloma	11
Pancreas	7–12
Non-Hodgkin's lymphoma	7
Rectum	2–9
Prostate	3 %

Table 13.4 Challenges for dental health after bariatric weight loss

Variable	Comments
Positive changes	
Less sugar and sweets	Diminished risk of cavities
Reduced fizzy drinks	Better protection for tooth enamel
More attention to health	More frequent office appointments, nursing surveillance
Remission of bulimia	Less vomiting, diminished exposure to gastric acid
Negative changes	
Reduced calcium intake	Deterioration of alveolar bone and teeth
Exchange of vegetables for more caloric food	Increased risk of periodontitis, cavities
Decreased salivation	Dry mouth, enhanced bacterial proliferation, periodontitis
Changes in mouth pH	Not well demonstrated
Use of ice popsicles	More common cracks

Fizzy drinks are also restricted by most bariatric teams as patients with small stomachs feel full and uncomfortable with large amounts of gas. The good news is that this might reduce tooth erosion, as many of such drinks have a low pH and attack the enamel.

It is known for a long time that bulimic patients suffer an excess of cavities and tooth decay, however, frequent vomiting with exposure to gastric acid is the obvious reason. Much less has been scrutinized in nonbulimic bariatric patients, which represent the majority, yet it is believed that the final outcome may not be entirely favorable (Table 13.4).

As alluded to, this is not a well-studied field and many questions remain. One recent report found that in adolescents, obesity itself is associated with more bacteria in the oral *subgingival biofilm*. The study was carefully controlled for all possible variables, such as presence of chronic disease, medication use, diagnosis of plaques or bleeding, salivary flow, and meal frequency (Zeigler et al. 2012).

In contrast calcium deficiency is uncommon in obesity, but, a recognized complication of gastric bypass and malabsorptive procedures. Dry mouth is also a comparatively frequent complaint, and the unpublished experience of our group indicates that quite a few patients lose teeth in the first 10 years after surgery because of cavities, fractures, and periodontitis. *Alveolar bone loss* has been specifically emphasized in this setting (de Moura Grec et al. 2012), even though patients are comparatively young.

13.7 Sexual Function

Sexual activity is highly relevant for health and quality of life in males and females, and has been associated with increased life span (Lindau and Gavrilova 2010). Obesity affects all levels of the hypothalamic–pituitary–gonadal axis, thus eliciting hormonal imbalance and infertility. In males, conspicuous androgen deficiency and erectile dysfunction are recognized.

In a study of our group, three populations were prospectively compared: males submitted to bariatric treatment more than 5 years earlier, nonoperated obese, and lean controls. *Low testosterone* values in the obese participants contrasted with a normalized androgenic profile in patients who lost weight after gastric bypass, reaching similar values as the lean population.

The same did not occur with erectile function. A couple of domains of the International Index of *Erectile Function*/IIEF questionnaire responded, however, the final score was not statistically different from the standard obese cases. The reasons for hormonal improvement without obvious recovery of sexual activity in these circumstances could be attributed to residual nutritional and metabolic problems.

Indeed, despite substantial weight loss many of the bariatric patients still displayed BMI above 30 kg/m^2, plasma lipids were not optimal, and diabetes, arterial hypertension, and obstructive sleep apnea exhibited suboptimal alleviation (Figs. 13.1, 13.2 and Table 13.5).

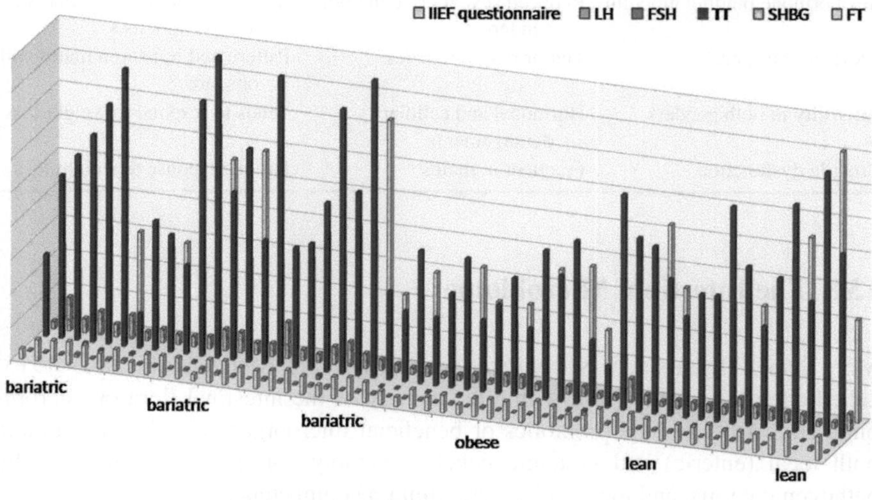

Fig. 13.1 Total and free testosterone in the three groups

Observation: Long columns correspond to total testosterone, and tips of the columns to free testosterone. Shorter columns behind them represent prolactin, estrogen, and other hormones. Only testosterone and sex hormone binding globulin were significantly improved. The first columns on the right are normal values (lean controls), those in the middle of the picture correspond to nonoperated obese males (depressed values), and the left third of the picture displays findings in the bariatrically treated patients. This last group was statistically comparable to the lean controls with regard to the mentioned improvements, and significantly higher than obese controls. (Alberto Fig. 1- HC)

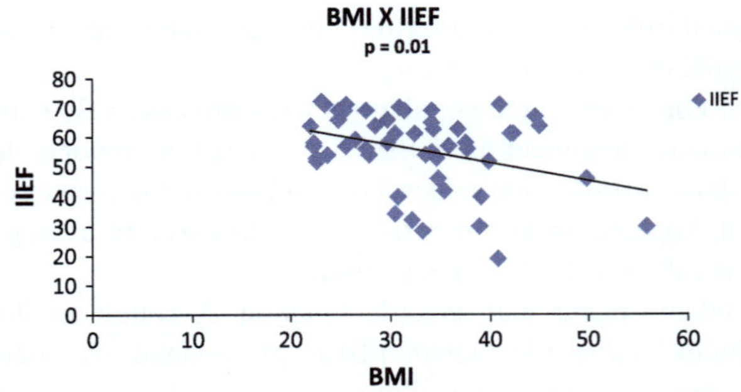

Fig. 13.2 Correlation between the international index of erectile function/IIEF and body mass index

Observation: The IIEF score negatively correlated with BMI, as expected. Nevertheless, bariatric patients with conspicuous weight loss did not exhibit significant improvement in the total score

Table 13.5 Sexual and *reproductive aberrations*

Variable	Disorder	Comments
Low total, free testosterone	Reduced sexual drive in both genders	Impotence in males—Improves after weight loss
Sex hormone binding globulin	Reduced sexual drive in both genders	Impotence in males—Improves after weight loss
Elevated estrogen	Tumor risk in females	Pattern and impact in males still obscure
Infertility in both genders	Hormonal and cellular derangements	Tends to respond to weight loss
Erectile dysfunction	Frequent in males	Partial response to weight loss

13.8 The Intestinal Microbiome

More than 100 years ago, Nobel Prize winner Ilya Ilich Metchnikoff (1845–1915) called attention to the world within us, namely of the intestinal flora or microbiome. His focus was the probiotics or beneficial microorganisms able to promote both local (enteric) and systemic benefits, notably concerning colonization by pathogenic germs, and local and general immune competence.

The idea remained more or less dormant until relatively recent years. It finally came of age in 2009, when the National Institutes of Health (USA) decided to invest in a huge international investigation called the Human *Microbiome Project* (Jumpstart Consortium 2012). No, the priority of the NIH was not the contribution of traditional fermented milks and yoghurts to longevity, as advocated by Metchnikoff but never proven. They were concerned with the association between selected flora patterns and chronic diseases such as obesity and diabetes, which are the scourge of our days.

How could two or three handfuls of innocent intestinal bacteria, with a total mass of a mere 700 g, lead to severe obesity with dozens of kilograms of excess body weight? *Germ-free mice* provide part of the answer. As known, these so-called gnotobiotic organisms maintained in especial laboratories are entirely devoid of microbes. Coincidentally, they never acquire obesity or metabolic syndrome.

If, however, these animals are conventionalized, which means exposed to conventional micro-organisms, such protection disappears and animals will gain a little weight. If germs from obese mice are inoculated, such gnotobiotic animals may become outright obese. Subsequent studies suggest that by means of fermentation of otherwise nonabsorbable fibers and intestinal mucus, along with other local metabolic changes, the microbiome may interfere with intestinal energy harvest and thus with energy balance. As much as 5–10 % of daily energy needs may stem from the by-products of such fermentation, notably short chain fatty acids *acetic, propionic, and butyric acid* (Krajmalnik-Brown et al. 2012). And it is known that even modest levels of positive calorie balance will lead to obesity, given sufficient time (Tilg and Kaser 2011; Borody and Khoruts 2011) (Fig. 13.3).

Experimental and clinical studies reveal that the microbiome has far-reaching effects, interfering with hormones, cytokines, inflammatory molecules, and other mediators. Appetite, satiety, and mood may be modulated by bacterial phyla with names like *Bacteroidetes* and *Proteobacteria* (Gram-negative), or *Actinobacteria* and *Firmicutes* (Gram-positive). Microbes may participate in the regulation of neuroendocrine systems such as the hypothalamic–pituitary–adrenal axis.

Multiple studies suggest that *Firmicutes* are obesity inducing and *Bacteroidetes* antagonize such tendency, however, there is certainly much more to it than the *Fir micutes/Bacteroidetes ratio. One focus of attention is the influence of bacterial genes on human gene expression. Just 700 g of intestinal microbioma contain 10 times more cells than a 70 kg man, respectively, 100 trillion versus 10 trillion, and 100 times more genes.*

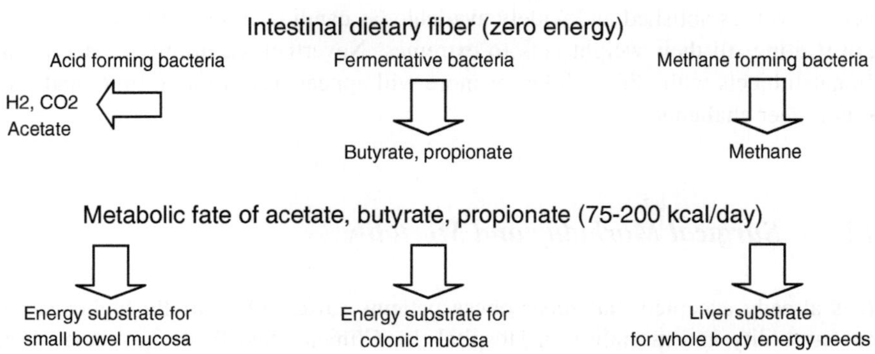

Fig. 13.3 Energy extraction from nondigestible food components

What happens after bariatric operation? Preliminary evidence with Roux-en-Y gastric bypass suggests that Bacteroidetes increase along with other shifts, a bonus able to enhance the antiobesity properties of the intervention. Nevertheless, other therapeutic interventions are being considered to modulate the microbiome both before and after bariatric treatment, particularly dietary changes and yes, the old suggestion of Metchnikoff's, namely probiotics. Diet may induce one or another phylum depending on the amount and composition of nutrients, and probiotics could be helpful in some contexts as well, by means of changes in flora ecology (Tilg and Kaser 2011; Borody and Khoruts 2011).

And yes, *fecal transplantation* is also being envisaged for obesity. No, this is not a typographical error. Healthy human stools may be transferred to another person via nasogastric tube, retention enema, or colonoscopy, after the indigenous flora was mechanically eliminated by whole bowel irrigation.

The current most common indication for this unusual intervention is recurrent or refractory infection by Clostridum difficile, however, the procedure could well mimic in obese patients the beneficial effects observed in laboratory animals. Removal of energy-harvesting, inflammation-triggering germs and replacement by benign strains is a tempting hypothesis. This is still not routine yet anywhere and requires careful planning. The gut may harbor highly pathogenic bacteria, fungi, parasites, and viruses, including human immunodeficiency (HIV) and hepatitis A, B, C, and D viruses. Donor selection and detailed laboratory tests are mandatory. Nevertheless, this could be an adjuvant maneuver for well-selected candidates, if somewhat scatologic, perhaps in association with *probiotics* (Weissman and Coyle 2012; Mattila et al. 2012; Borody and Khoruts 2011).

13.9 The Super Obese Bariatric Patient

Some groups advocate that patients with BMI > 50 kg/m^2 be classified as *super obese,* and those >60 kg/m^2 as super super obese. Most prefer to call super obese anybody with BMI > 50 kg/m^2.

These patients tend to become rare in countries and regions where bariatric surgery is well established and widely available, as candidates seek treatment earlier, not waiting till their weight gets to extremes. Nevertheless, in any service occasional subjects with 180, 200 kg, or more will appear from time to time, and present bigger challenges.

13.9.1 Surgical Morbidity and Mortality

It is already accepted that super obese patients suffer substantially higher surgical morbidity and mortality. In Hospital das Clinicas, Sao Paulo, a special ward in a long-term facility was created for such cases. They are admitted for up to

2–3 months and submitted to a low-calorie diet and specialized clinical supervision, till part of their excess body weight is eliminated and comorbidities such as arterial hypertension, heart insufficiency, respiratory troubles, and diabetes are well compensated. Only afterwards are they transferred to the main hospital for surgical intervention.

As alluded to, some of the endoluminal devices available in the market are partly geared toward such population. The advantage would be adequate preoperative compensation at home, avoiding costly and prolonged hospitalization.

13.9.2 Lower Limb Abnormalities

in a study of our group with bariatric candidates, stratified as morbidly obese (BMI 40–50 kg/m^2) and super obese (BMI > 50 kg/m^2), knee and foot radiologic assessment, baropodometric footprint measurement, and the questionnaires western ontario osteoarthritis index/womac and foot and ankle outcome score/faos were employed. The main outcome measures were imaging diagnosis of *knee osteoarthritis* and flatfoot, along with functional impact on activities of daily living estimated by the questionnaires knee osteoarthritis which was common (74.1 %), and the entire cohort suffered from *flatfoot* according to both footprint index and talar-first metatarsal radiographic angle, without difference between morbid obesity and super obesity. However, WOMAC and FAOS scores were markedly worse in the super obese, affecting joint pain, stiffness,and general mobility. Although amelioration is probable with weight loss, long-term orthopedic assistance might be essential for this extremely obese population.

13.10 The Obese and Postobese Worker

As often alluded to, obesity is a major deterrent for success in the workplace. Getting a job is more difficult, career advancement occurs after longer intervals or not at all, average salaries are lower, and general satisfaction suffers within this context.

Companies defend themselves with the excuse that obese employees cost more, perform less, are more often diseased and disabled, exhibit limited fitness for challenging jobs or frequent travel, and their general productivity is more disappointing. A few of these complaints have been scientifically documented, although a degree of prejudice cannot be entirely ruled out. Total medical costs tend to me more elevated as well (Table 13.6).

Bariatric weight loss should be an antidote against all those troubles, however, there are reasons to suspect that they may linger for many years, in a substantial proportion of operated subjects.

Table 13.6 Risks and handicaps of *obese employees* (Goetzel et al. 2010)

Variable	Remarks	Impact
Presenteism	Working despite sickness	Not measured however significant
Absenteeism	Missing working days	4.3 billion USD/year (USA)
Doctor visits	Office appointments	20 % higher
Emergency visits	Hospital Emergency Room	26 % increased
Excess medical costs (*)	Increased resource use	121 USD/worker/year
Productivity losses	Presenteism, absenteeism	524 USD/worker/year

(*) Only office and emergency visits; does not include drug costs, other health care needs

13.11 Health Care Resources

Within the bariatric context, Hayashi et al. investigated use of *hospital resources* after antiobesity surgery, during one year before operation in comparison with four years afterwards. Two separate control populations were selected, namely obese persons who did not undergo surgical treatment, as well as lean surgical patients who were submitted to colectomy because of colorectal cancer.

RYGB patients needed 45.8 % less *outpatient visits* and 53.0 % less laboratory tests than nonoperated moderately obese cases, even including aesthetic operations. Results were comparable to those observed after elective colorectal surgery and remained fairly stable during 4 years. No change in requirements for emergency or elective hospitalization were documented in the series, confirming that even after weight loss, these patients may still need more health care attention than the general population (Hayashi et al. 2011).

The cost/benefit and the risk/benefit ratio of bariatric surgery is unclear in certain patient populations, such as patients with lower body mass index (BMI, 30–35 kg/m^2), the high-risk super–super obese patients (BMI > 60 kg/m^2), the morbidly obese adolescent, and obese patients requiring weight reduction in preparation for other procedures, such as orthopedic, transplant, or vascular surgeries. In these circumstances, there might be a need for an effective but less complex treatment to bridge the gap between medical and surgical therapy.

13.12 Final Considerations

The recent discovery or near-discovery of the Higgs boson, a subatomic particle deemed to be responsible for the mass of all atoms, and therefore of all living and inanimate matter, did not escape the eye of neurologists, metabologists, and rehabilitation professionals.

Many well-humored scientists suggested the building of portable Great Hadron Colliders to smash body fat, until people emerged entirely lean. Anti-Higgs particles were suggested to take all mass away from obese people, Higgs boson diets have been proposed with weightless foods, and a Higgs McBoson sandwich was anticipated.

Jokes are a very primitive and unscientific tool, nevertheless, they represent a lighter and sometimes necessary approach to sad and depressing events such as the obesity epidemic. Major antiobesity public health policies have been adopted in several countries, however, a recent investigation in the USA encountered few reasons to rejoice, at least up to now (Trivedi et al. 2012).

It is no secret that physical conditioning and nutritional education programs, taxation of high-calorie (*"junk"*) *food*, and other large-scale interventions have been advocated almost worldwide during the past decade (Table 13.7).

In the USA more than 20 states provided funding for walking and bicycle trails, 23 states underwent Centers for Disease Control-sponsored initiatives to prevent health-risk behavior including sedentarism, legislation concerning antiobesity physical activity has been enacted in 28 states, and 48 states have standards for physical education. There is a dearth of official recommendations addressing *school cafeterias* and vending machines either. All these efforts notwithstanding, no correlation between such programs and obesity prevalence has been demonstrated. On the contrary, mean body weight continues to grow in the population (Trivedi et al. 2012).

"Light" and "zero" *soda drinks* are another sad example. Although commercialized and indeed popular in many countries, no reduction in the prevalence of obesity has been associated with their introduction. There is evidence that although similarly exciting taste buds, sugar and artificial sweeteners, are differently processed at the brain reward centers, energy regulation is not benefitted and may even be impaired (Green and Murphy 2012).

This could be just a matter of time, and the next decade might reserve agreeable surprises. Maybe not, because major cultural and social-economical shifts are involved. It is estimated that in just 10 years, between 1990 and 2000, Latin American supermarkets increased their participation in food sales from 15 to 60 %, and this is not an isolated phenomenon, occurring at different rates all over the world. In traditional local markets most foods are natural, nonindustrialized

Table 13.7 *Public health initiatives* to curtail the obesity epidemic

Policy	Aims
Physical fitness	Educate and train students from kindergarten till university
Bicycle trails, walking trails	Stimulate use of bicycles instead of automobiles, fight sedentarism
Nutrition seminars	Disseminate nutrition and fitness knowledge among students and teachers
Special food taxes	Limit consumption of sugared beverages, snacks, ice-cream, chocolate, and other high-calorie-density foods
School cafeteria	Removal of "junk" food, incentives for fruit and vegetables
Vending machines	Restrictions for non-nutritious foods
Weight-loss campaigns	Competitions in the workplace for 100 kg, 500 kg, or one ton collective weight shedding

produce, whereas in supermarket ready-to-use, frozen, canned, and otherwise processed products abound. These last dietary sources tend to be more caloric because fat is generously used, more tasty on account of added sugar, salt, spices, and other additives, and devoid of fibers, thus obesity inducing.

As women have access to better education and jobs, they will not toil in the kitchen all day as their grandmothers did, preferring for their families foods that are easy to prepare, are delivered to the door, or else opt for eating out. By the same token obese adults who are used since early childhood to delicious and *energy-rich snacks* and fast food available around the clock, will not easily adapt to comparatively unattractive and unsatisfying whole grains and fresh vegetables even if they become ubiquitous, and also beautifully packaged and advertised, which is hardly the case.

The new generations will have to be educated within the constraints of these new realities, learning that coping with obesity is a heavy burden and lifetime endeavor, perhaps requiring as *aggressive legislation* as for alcohol, tobacco, drugs, and other addictions. Success is distant in the horizon, therefore it is up to the health care profession to continue devising creative strategies and treatments, and to improve existing ones. In the Middle Ages people tended to think Plague, in the 1800s Tuberculosis, in the early 1900s Syphilis, and in more recent times HIV/AIDS. Currently, Obesity deserves a similar priority in everyone′s mindset.

References

Borody TJ, Khoruts A (2011) Fecal microbiota transplantation and emerging applications. Nat Rev Gastroenterol Hepatol 9:88–96. doi:10.1038/nrgastro.2011.244

Dalcanale L, Oliveira CP, Faintuch J, Nogueira MA, Rondó P, Lima VM, Mendonça S, Pajecki D, Mancini M, Carrilho FJ (2010) Long-term nutritional outcome after gastric bypass. Obes Surg 20:181–187. doi:10.1007/s11695-009-9916-5

de Moura-Grec PG, Marsicano JA, Rodrigues LM, de Carvalho Sales-Peres SH (2012) Alveolar bone loss and periodontal status ibn a bariatric patient: a brief review and case report. Eur J Gastroenterol Hepatol 24:84–89. doi:10.1097/MEG.0b013e32834bebb3

de Souza SA, Faintuch J, Fabris SM, Nampo FK, Luz C, Fabio TL, Sitta IS, de Batista Fonseca IC (2009) Six-minute walk test: functional capacity of severely obese before and after bariatric surgery. Surg Obes Relat Dis 5:540–543. doi:10.1016/j.soard.2009.05.003

de Souza SA, Faintuch J, Sant'anna AF (2010) Effect of weight loss on aerobic capacity in patients with severe obesity before and after bariatric surgery. Obes Surg 20:871–875. doi:10.1007/s11695-010-0109-z

Dunn JP, Cowan RL, Volkow ND, Feurer ID, Li R, Williams DB et al (2010) Decreased dopamine type 2 receptor availability after bariatric surgery: preliminary findings. Brain Res 1350:123–130. doi:10.1016/j.brainres.2010.03.064

Green E, Murphy C (2012) Altered processing of sweet taste in the brain of diet soda drinkers. Physiol Behav 11 May 2012 doi:10.1016/j.physbeh.2012.05.006

Goetzel RZ, Gibson TB, Short ME, Chu BC, Waddell J, Bowen J, Lemon SC, Fernandez ID, Ozminkowski RJ, Wilson MG, DeJoy DM (2010) A multi-worksite analysis of the relationships among body mass index, medical utilization, and worker productivity. J Occup Environ Med 52(Suppl 1):S52–S58. doi:10.1097/JOM.0b013e3181c95b84

Gripeteg L, Lindroos AK, Peltonen M, Sjöström L, Narbro K (2012) Effects of bariatric surgery on disability pension in Swedish obese subjects. Int J Obes (Lond) 36:356–362. doi:10.1038/ijo.2011.15

Gunstad J, Strain G, Devlin MJ, Wing R, Cohen RA, Paul RH, Crosby RD, Mitchell JE (2011) Improved memory function 12 weeks after bariatric surgery. Surg Obes Relat Dis 7:465–472. doi:10.1016/j.soard.2010.09.015

Hayashi SY, Faintuch J, França JI, Cecconello I (2011) Four-year hospital resource utilization after bariatric surgery: comparison with clinical and surgical controls. Obes Surg 21:1355–1361. doi:10.1007/s11695-011-0404-3

Heneghan HM, Heinberg L, Windover A, Rogula T, Schauer PR (2012) Weighing the evidence for an association between obesity and suicide risk. Surg Obes Relat Dis 8:98–107. doi:10.1016/j.soard.2011.10.007

Ivezaj V, Saules KK, Wiedemann AA (2012) "I Didn't See This Coming": why are Postbariatric patients in substance abuse treatment? Patients' perceptions of etiology and future recommendations. Obes Surg. doi:10.1007/s11695-012-0668-2

Jumpstart Consortium Human Microbiome Project Data Generation Working Group (2012) Evaluation of 16S rDNA-based community profiling for human microbiome research. PLoS One. 7:e39315 doi:10.1371/journal.pone.0039315

Kaul A, Sharma J (2011) Impact of bariatric surgery on comorbidities. Surg Clin North Am 91:1295–312. doi:10.1016/j.suc.2011.08.003

Khandekar MJ, Cohen P, Spiegelman RM (2011) Molecular mechanisms of cancer development in obesity. Nat Rev 11:886–895. doi:10.1038/nrc3174

Krajmalnik-Brown R, Ilhan ZE, Kang DW, DiBaise JK (2012) Effects of gut microbes on nutrient absorption and energy regulation. Nutr Clin Pract 27:201–214. doi:10.1177/0884533611436116

Lindau ST, Gavrilova N (2010) Sex, health, and years of sexually active life gained due to good health: evidence from two US population based cross sectional surveys of ageing. BMJ 340:c810. doi:10.1136/bmj.c810

Mattila E, Uusitalo-Seppala R, Wuorela M, Lehtola L, Nurmi H, Ristikankare M, Moilanen V, Salminen K, Seppala M, Mattila PS, Anttila VJ, Arkkila P (2012) Fecal transplantation, through colonoscopy, is effective therapy for recurrent Clostridium difficile infection. Gastroenterology 142:490–496. doi:10.1053/j.gastro.2011.11.037

Ochner CN, Kwok Y, Conceicao E, Pantazatos SP, Puma LM, Carnell S et al (2011) Selective reduction in neural responses to high calorie foods following gastric bypass surgery. Ann Surg 253:502–507. doi:10.1097/SLA.0b013e318203a289

Olbers T (2011) Bariatric surgery or the fairy tale about the ugly duckling. Lakartidningen 108:2570–2573

Ribeiro AG, Faintuch J, Dias MC, Cecconello I (2009) Euglycemia and normolipidemia after anti-obesity gastric bypass. Nutr Hosp 24:32–39. ISSN 0212-1611

Schauer DP, Arterburn DE, Livingston EH, Fischer D, Eckman MH (2010) Decision modeling to estimate the impact of gastric bypass surgery on life expectancy for the treatment of morbid obesity. Arch Surg 145:57–62. doi:10.1001/archsurg.2009.240

Sjöström L, Gummesson A, Sjöström CD, Narbro K, Peltonen M, Wedel H, Bengtsson C, Bouchard C, Carlsson B, Dahlgren S, Jacobson P, Karason K, Karlsson J, Larsson B, Lindroos AK, Lönroth H, Näslund I, Olbers T, Stenlöf K, Torgerson J, Carlsson LM (2009) Swedish Obese Subjects Study. Effects of bariatric surgery on cancer incidence in obese patients in Sweden (Swedish Obese Subjects Study): a prospective, controlled intervention trial. Lancet Oncol 10:653–662. doi:10.1016/S1470-2045(09)70159-7

Sjöström L, Peltonen M, Jacobson P, Sjöström CD, Karason K, Wedel H, Ahlin S, Anveden Å, Bengtsson C, Bergmark G, Bouchard C, Carlsson B, Dahlgren S, Karlsson J, Lindroos AK, Lönroth H, Narbro K, Näslund I, Olbers T, Svensson PA, Carlsson LM (2012) Bariatric surgery and long-term cardiovascular events. JAMA 307:56–65. doi:10.1001/jama.2011.1914

Souza SAF, Faintuch J, Cecconello I (2010) Spirometric function improves in the morbidly obese after 1-year post surgery. Obes Surg 20:1273–11. doi:10.1007/s11695-010-0142-y

Steele KE, Prokopowicz GP, Schweitzer MA, Magunsuon TH, Lidor AO, Kuwabawa H et al (2010) Alterations of central dopamine receptors before and after gastric bypass surgery. Obes Surg 20:369–374. doi:10.1007/s11695-009-0015-4

Tilg H, Kaser A (2011) Gut microbiome, obesity, and metabolic dysfunction. J Clin Invest 121:2126–2132. doi:10.1172/JCI58109

Tindle HA, Omalu B, Courcoulas A, Marcus M, Hammers J, Kuller LH (2010) Risk of suicide after long-term follow-up from bariatric surgery. Am J Med 123:1036–1042. doi:10.1016/j.amjmed.2010.06.016

Trivedi NJ, Fields J, Mechanick CH, Klein M, Mechanick JI (2012) Lack of correlation between anti-obesity policy and obesity growth rates: review and analysis. Endocr Pract 18:1–23. doi: 10.4158/EP11379.RA

Weissman JS, Coyle W (2012) Stool transplants: ready for prime time ? Curr Gastroenterol Rep 16 May 2012. doi:10.1007/s11894-012-0263-7

Yamaguchi CM, Faintuch J, Hayashi SY, Faintuch JJ, Cecconello I (2012) Refractory and new-onset diabetes more than 5 years after gastric bypass for morbid obesity. Surg Endosc. doi:10.1007/s00464-012-2256-x

Zeigler CC, Persson GR, Wondimu B, Marcus C, Sobko T, Modéer T (2012) Microbiota in the oral subgingival biofilm is associated with obesity in adolescence. Obesity (Silver Spring) 20:157–164. doi:10.1038/oby.2011.305

Chapter 14
Aesthetic Rehabilitation After Bariatric Surgery

Wilson Cintra Jr and Joel Faintuch

Abstract Sustained and lifelong weight loss for severely obese people is not anymore an impossible dream, and a Swedish article suggests, half jokingly, that this is the new fairy tale of the ugly duckling (Olbers, Lakartidningen 108:2570–2573, 2011). The example is appropriate because for health care professionals, resolution of obesity means reversal of diabetes, hyperlipidemia, and other cardiovascular risk factors. For the majority of the patients all these results are secondary. Priority is getting beautiful and achieving a rich and fulfilling emotional life. That is the reason why plastic interventions are the most frequent new operations after bariatric treatment, and as many as 85 % of this population desire subsequent body contouring maneuvers (Gusenoff et al., Plast Reconstr Surg 122:951–958, 2008; Highton et al. J Plast Reconstr Aesthet Surg 65:426–432, 2012). Fortunately this specialty has responded to the challenge, and both old and new techniques are being successfully adapted to the specificities of the patients who shed much weight and suffer from lax and redundant skinfolds and widespread lipodystrophy. Principal procedures such as abdominoplasty, mammaplasty, thigh lift, brachioplasty and others are described, emphasizing the experience of the Hospital das Clinicas Service during more than 15 years.

14.1 General Definitions

Body contouring procedures. Any operation aiming at a more shapely body profile or silhouette. Most plastic surgeries for weight loss after bariatric therapy are contouring operations (abdominoplasties, brachioplasties, mammaplasties). Face lifts and other localized interventions are not usually considered body contouring.

JrW. Cintra Jr (✉)
Plastic Surgery Service, Hospital das Clinicas, Sao Paulo University Medical School, Sao Paulo, Brazil
e-mail: wcintra@terra.com.br

J. Faintuch
Department of Gastroenterology, Hospital das Clinicas, São Paulo University Medical School, São Paulo, Brazil
e-mail: blfaintuch@hotmail.com

P. Capodaglio et al. (eds.), *Disabling Obesity*, DOI: 10.1007/978-3-642-35972-9_14,

Circumferential abdominoplasty. See *Lower body lift*.

Implant. The same as prosthesis. A foreign body (usually made of silicone) introduced to replace or expand missing or insufficient tissue, especially in the female breast.

Lift. The most ubiquitous problem after massive weight loss is sagging skin with collapse of the dermal architectural scaffold. Operations aim not only at removing excess tissue but also at lifting the pendulous anatomical segments (face, arm, thigh, breast). Pexy (mastopexy) has a rather similar connotation, namely suspension of low-lying structures, although conventionally using the patient's own tissues, without implants.

Lipodystrophy. Aberrant or exaggerated distribution of subcutaneous fat, creating an unalluring and dislikable morphology.

Liposuction. Any minimally invasive intervention that removes subcutaneous adiposity. May utilize syringes, mechanical suction cannulas, laser-assisted, radiofrequency-assisted, or ultrasound-assisted techniques.

Lower body lift. The same as circumferential abdominoplasty. A large operation in which a variable segment of tissue following the entire waistline is removed, eliminating redundancies and circumferentially raising the lax tissues of the lower half of the body.

Pannus. A large fold of dermis (skin and subcutaneous tissue) that appears in any body segment, causing aesthetic and functional impairment. May be part of generalized lipodystrophy.

Panniculitis. Inflammation or infection of the pannus.

Plasty. This is a nonspecific surgical expression. Any intervention on the thigh, abdomen, breast, or arm, aiming at thinning, reshaping, debulking, or suspending local dermis including or not myoaponeurotic tissues, may get the suffix plasty.

Rhytidectomy. See Rhytidoplasty.

Rhytidoplasty. Correction of face wrinkles and skin surpluses, including face lift.

Upper body lift. Association of mastoplasty and brachioplasty.

14.2 Historical Introduction

Beauty may be important to many creatures. Birds purposefully exhibit and ruffle their multicolored feathers during the prenuptial flight, and predators often use their vivid colors to charm prey. Nevertheless, use of make-up, fancy clothing or jewels, and concepts such as self-image, self-esteem, pride because of harmonious physical proportions, and shame because of an unappealing figure are intrinsically human, or at least have never been well documented in animals.

Plastic surgery seems contemporary but its roots are very old. The Edwin Smith Egyptian papyrus of 5,000 years ago mentions nose amputations as a terrible aesthetical blemish consequent to trauma or punishment, and teaches how to handle the injury with local dressings. Reconstruction was obviously primitive, with wooden splints and cloth. The same can be found in the writings of Hippocrates

(2,400 years ago); however, he attempted to reshape nasal bones in the case of fractures or deviation.

A major step forward is ascribed to Sushruta in India, actually one century before Hippocrates (2,500 years ago). This priest wrote an encyclopedia including information about how to pull and transfer skin grafts from the forehead and the cheek to reconstruct the nose, and devised surgical instruments to this aim. Flaps and grafts are obviously fundamental tools in most modalities of plastic surgery, including the bariatric context (Whitaker et al. 2007).

14.3 Early Plastic Interventions in Obesity

The principles of Sushruta migrated and were expanded in Rome. In De Medicina, the principal medical text of that era written by *Aulus Cornelius Celsus* (25 BCE–50 CE), ordinary and rotation flaps for variety of lesions were discussed. The Romans deserve mention because of another fact. Obesity became prominent for the first time in this empire, which reportedly enjoyed the most lavish and exuberant lifestyle of Antiquity. Of course it was restricted to royalty and other privileged classes, because ordinary people enjoyed few luxuries, ate sparingly, and had a life expectancy of 25 years only. *Pliny the Elder* (Gaius Plinius Secundus, 23–79 CE) wrote in his Naturalis Historia that the consul *Lucius Apronius* requested an abdominal operation for his son, because he was markedly obese and thus disabled for many activities. In the opinion of the consul such procedure might improve him both physically and psychologically, and thus enhance the ability to perform his professional and civic duties. Details of the intervention are not available, however, it was performed, probably dermolipectomy of the pendulous lower abdomen, and apparently successful. (Boriani et al. 2005). The Middle Ages saw relatively little progress, and indeed Pope Innocent III (1160–1216) explicitly prohibited all surgical operations. Nevertheless, Gaspare Tagliacozzi (1545–1599), professor of anatomy at the University of Bologna, is credited with the creation of the first distant pedicled skin flap, convenient for various modalities of reconstruction. Moreover, he wrote the first compendium of plastic surgery, De Curtorum Chirurgia per Insitionem. His boldness and daring did not come cheap. He was excommunicated, and his corpse was removed from the church graveyard to be interred somewhere else.

14.4 Obesity in the Last 100 Years

Two of the most famous and powerful heads of state in the beginning of the twentieth century, nominally President Howard Taft of the USA (1857–1930) and King Edward VII of Great Britain (1841–1920), were morbidly obese. Also, during World War I (1914–1918) it is reported that the German Imperial Army distributed scores of corsets to male officers with prominent waistlines. Actually no female soldiers existed at that time. Affluent societies were progressively affording five meals a day, including the higher classes of Argentina where three of those meals included meat. Still, for ordinary people this luxury (or disgrace, depending on the perspective) only became widespread half a century or more later, in the 1970s and the 1980s. Such a trend was accompanied by a cultural shift regarding ideal body image, especially for women. Paradoxically to such indulgement in calories, females were expected to become more and more slim, as summarized by Wallis Simpson, the Duchess of Windsor (1896–1986): "You can never be too rich or too thin." (Banja 2004). In Antiquity, as well as in the Renaissance and Post-Renaissance pictures of Titian, Rubens, and Rembrandt, full and exuberant feminine bodies were glorified. Not now, as female artists and notably models are required to exhibit such delicate profiles that they are reminiscent of anorexia nervosa. Of course many political, economical, and social circumstances underlay such changes. Motherly and fertile women with large families were being replaced by professional types with successful careers. Food and especially calories have never been so available and comparatively inexpensive, by virtue of agricultural revolutions and industrialization of food production. At the same time, the Internet and other media now reach the most remote planetary regions, disseminating the new patterns of universal beauty and elegance to virtually all nations and communities. In the face of so many powerful and conflicting influences, plastic surgery has become more and more necessary to reconciliate the extremes and to promote the metamorphosis that millions desire, from frog to prince or more specifically, from a fat, motherly figure to a sexy, young-looking lady.

14.5 Body Dysmorphism of the Postbariatric Patient

It is not a coincidence that the most frequent plastic procedure in this context is abdominoplasty, or repair of the sagging, pendulous abdomen by a variety of operations. Although obesity is classically divided into two major phenotypes, respectively, the *pear morphology* (feminine, with predominance of the buttocks) and the *apple-shaped* modality (male type, with a prominent abdomen), in both circumstances massive amounts of fat accumulate in the frontal half of the waistline. One should not overlook the fact that at the same time that body contouring improves body image, it may precipitate dissatisfaction with other anatomical regions. As patients become closer to their ideal this ideal may shift, thus justifying sequential procedures along the years (Modolin et al. 2003; Song et al. 2006; Cintra et al. 2008). As postoperative resorption of adiposity occurs, large and unaesthetical skinfolds appear everywhere, often interfering with dressing, washing, gait, posture, and personal hygiene; moreover, undergoing secondary erosion, intertriginous infection, or panniculitis due to friction, sweat, and bacterial accumulation. In patients suffering from severe lipodystrophy, repercussions on body image tend to loom high as well (Song et al. 2006; Kenkel 2006).

Degree of weight loss primarily influences severity of subsequent aesthetic abnormalities, nevertheless, texture and elasticity of the skin is important as well. Young patients often enjoy seamless recovery in contrast to elderly ones, and males with their thicker skin also suffer less skin surplus than females. All these variables notwithstanding, an intrinsic abnormality in the collagen fiber network of the abdomen was demonstrated in our service in postbariatric cases (Orpheu et al. 2010).

14.6 General Surgical Indications

Body perception is a highly personal construct, and a small amount of sagging skin may be considered a disgrace by some people, whereas others will display much more prominent aberrations without any visible discomfort. In spite of the subjective interpretations, an attempt to classify derangements and establish common surgical recommendations was recently published (Table 14.1). According to the authors (Iglesias et al. 2010), lack of objective parameters has led to many unnecessary interventions especially in public hospitals, with long waiting lines and a waste of insurance money and resources. In Hospital das Clinicas standard indications are not adopted, an individual evaluation being conducted by experienced professionals including a psychologist, followed by a case analysis. Emphasis is on functional rehabilitation, not exactly on achievement of physical beauty. Incidentally, these populations undergo as many as four or five different operations, and their total scars may be several meters long, a fact that is openly discussed with them and which will preclude picture-perfect bodies.

Table 14.1 Indications for body contour procedures after bariatric weight loss (modified from Iglesias et al. 2010)[a]

Body segment	Not advised	Optional	Recommended
Abdomen	Overhanging pannus above inguinal ligament	At the upper third of the thigh	Below the upper third
Arms, thighs	Loose flap up to 50 % of limb diameter	50–100 % of diameter	>100 %
Breast, lateral thorax	Breast or pannus in the upper third of the reference line[b]	In the middle third of that line[b]	Beyond the middle third[b]

a In the original protocol, the patient's BMI should be added to all measurements, however, this detail was omitted

[b] The distance between the sixth rib (or the xiphoid process, which roughly corresponds to it) and the upper edge of the iliac crest is divided into three. Each third corresponds to a surgical category

14.7 Specific Surgical Criteria

Bariatric weight loss is a dynamic phenomenon with a prolonged course. Typically, the most massive changes occur in the first 12 months, and the curve tapers toward stability in the next 12 months, or a little more in the super obese. Later on weight may stay the same, move down, or occasionally be followed by partial obesity recurrence (See Chap. 11). Sudden shifts in body size and shape may ruin the results of plastic intervention, therefore 12 months of stable body weight is the standard requirement for such procedures. Moreover, this time corresponds to maximum skin retraction, allowing better planning of the necessary resections.

14.8 Deflated and Nondeflated Patients

Depending on such variables as initial BMI, degree of weight loss and body fat distribution, dermal folds may exhibit a more deflated pattern, when adipose tissue is nearly entirely degraded and just redundant skin remains, or conversely a still thickened profile with important accumulation of subdermal fat. If this last alternative is the case, operation needs to be more carefully considered or even postponed, because danger of major bleeding as well as of postoperative seroma is considerable (Shermak et al. 2006).

14.9 Surgical Staging and Risks

Patients often list complaints in different anatomical regions, and would prefer one-stage correction of all of them. Nevertheless, extreme makeovers are best suited for television programs. In real life meticulous and lengthy techniques may be required, implying prolonged surgical time and general anesthesia, along with increased technical and metabolic pitfalls. In such context, the most annoying troubles are given priority, subsequent interventions being scheduled after full recovery.

Surgical complications are possible and should be clearly informed, such as venous thrombosis, pulmonary embolism, suture dehiscence or infection, and fluid collections (hematoma, lymphedema, seroma), some of them directly affecting aesthetic results. All incisions are precisely demarcated, nonetheless scars may migrate, become asymmetric, or widen. Even if weight regain is avoided skin may again become lax in the ensuing years, and redundant folds are apt to recur after some time. Future revisions are indispensable if serious troubles occur.

Last but not least, plastic treatment is expensive and not usually covered by private health organizations. In Hospital das Clinicas, Sao Paulo, which is partly financed by the Brazilian Public Health Insurance System, such assistance is delivered for free; however, this rule does not apply to all institutions. Aesthetic prostheses such as breast implants are rarely available because of high cost, and in any case prolonged waiting lines are the rule.

14.10 Psychological Interview

Obesity is associated with profound effects on body image, self-esteem, and psychosocial functioning. Plastic rehabilitation is part of the same mental process that leads patients to undergo bariatric therapy, namely restoration of everything that was lost, or presumably lost, on account of obesity. Unfortunately, many dreams that are simply beyond the reach or ability of the patients, be they familial, professional, or concerning social-economic status, are also common in the presence of excessive adiposity, and become part of the complaints and frustrations if not achieved after weight loss. Thus one should be prudent not to operate for the wrong reason, and not to engage in improbable and even dangerous commitments (Shermak et al. 2006; Kitzinger et al. 2012). Psychological evaluation is essential both to assess the impact of the deformity and lipodystrophy from the point of view of the patient, and to measure his or her desires and expectations regarding postoperative results (Kramer and Ward 2010). Anxious, manic, or depressive subjects, those with intolerance for less than full success, as well as persons who associate plastic maneuvers with specific personal goals such as getting a marriage, a new job, or a rewarding social life should be handled with attention, including possible contraindication of the operation. Expectations could materialize or not, and need to be explained to the candidate as a lucky bonus, not as a guaranteed outcome. Dramatic consequences including suicide have been reported in such circumstances, as obesity by itself is a risk factor for suicide, and surgical disappointment might trigger it (Heneghan et al. 2012).

14.11 Quality of Life

Certain interventions promptly enhance quality of life, such as augmentation mastoplasty which immediately improves body sillhouette. Within weeks or months, the new shape is incorporated into body image. Circumferential abdominoplasty or upper body lift is another operation that significantly impacts quality of life. The size of the incision and the multiple redundant tissues that are resected almost never fail to impress the patients, with repercussions concerning their quality of life.

Other interventions such as brachioplasty tend to be followed by more subtle responses, still most patients acknowledge the progress and report better quality of life as well. Especially, as regards body contouring techniques, many variables are involved: duration of obesity, degree of weight loss, extent of the operation, as well as emotional and interpersonal reactions (Cintra et al. 2008).

14.12 Anesthesia

Though many plastic interventions are office procedures conducted with local infiltration only, postbariatric interventions require general anesthesia. Operations are medium size or large, as a rule extensive undermining and resection is necessary. Duration can be quite long and not infrequently, position changes during the procedure will be essential to reach the lateral or dorsal areas of the chest, abdomen, or limbs, a maneuver that would be awkward and might lead to contamination in the awake person. Tracheal intubation is not obligatory when the intravenous route is selected; however, monitoring of the airway should not be overlooked. Blood pressure control is highly relevant, especially when the patient is moved from the prone to the supine position or vice versa, because postural hypotension may ensue.

14.13 Perioperative Controls

Obese patients suffer from systemic inflammation along with a procoagulatory status, as a result of expression of hormones, cytokines (adipokines), and other biomolecules originating from adipose tissue. Particularly active as a source of such mediators are fat-infiltrating lymphocytes of visceral fat; however, macrophages, adipocytes, and preadipocytes have also been incriminated in the production of chemotactic adipokines (Sell and Eckel 2010). By the time plastic corrections are conducted, a couple of years after the original bariatric intervention, most patients are not morbidly obese anymore, their body mass index being not much more than 30 kg/m^2, or even a little less. Such progress notwithstanding, anticoagulants are prescribed for 4 days, along with intermittent pneumatic compression of the lower limbs during the operation and immediate postoperative period, in order to inhibit the formation of venous clots and possible pulmonary embolism.

Body composition and water needs are difficult to estimate in the obese population, underlying the concern about adequate fluid replenishment. In this sense, a catheter in the urinary bladder is introduced to monitor fluid output and prevent dehydration. During prolonged operations a thermal mattress may be recommended, because of the danger of hypothermia.

14.14 Common Surgical Procedures

The principal modalities requested by the patients are abdominoplasty, mastoplasty, thigh lift, brachioplasty and rhytidoplasty (face lift, including management of periorbital wrinkles or crow's feet). Abdominoplasty is usually the first intervention, and as many as 90 % of the bariatric population demand such therapy, sooner or later in the postoperative period (Modolin et al. 2003; Cintra et al. 2008; Kitzinger et al. 2012).

14.15 Abdominoplasty Techniques

Multiple approaches are available for the reconstruction of the fallen abdomen and associated structures, depending on local anatomy as well as severity of local dysmorphism. The most frequent are the transverse infraumbilical resection, the transverse resection with a midline upper extension (anchor-shaped incision), along with the circumferential abdominoplasty (upper body lift) (Modolin et al. 2003). Figures 14.1, 14.2, 14.3, 14.4, 14.5, and 14.6 illustrate representative procedures.

In all of these contexts, fusiform segments of redundant skin and subcutaneous tissue are resected, in accordance with preoperative demarcation. Though not routinely employed, liposuction and myoaponeurotic layer plicature may also be appropriate in some circumstances, in order to diminish the waistline, improve lipodystrophy and shrink the lower abdomen. Navel detachment and reimplantation at the anatomical level is another detail that enhances abdominal wall reconstruction.

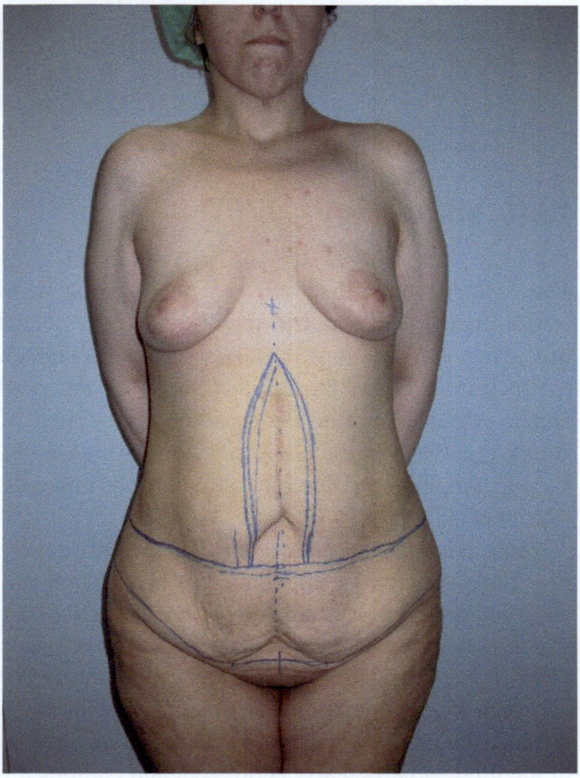

Fig. 14.1 Circumferential abdominoplasty with *anchor-shaped* extension

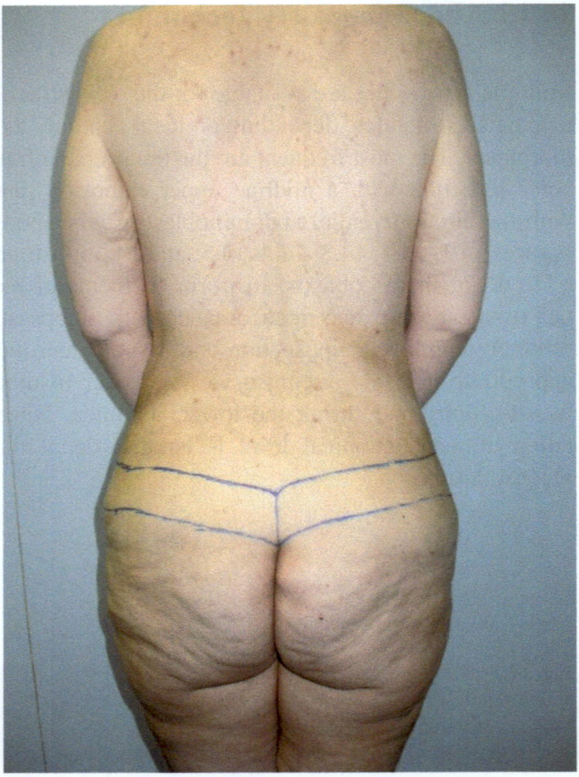

Fig. 14.2 Dorsal half of the abdominoplasty—Surgical demarcation (2 E)

14.16 Mastoplasty

The cosmetic importance of the female breast in Western society can hardly be overestimated, and bariatric surgery may have deleterious effects on the profile of this gland. Interventions usually target size (augmentation mastoplasty), position (correction of breast ptosis), and shape. In the postbariatric patient, this usually entails removal of skin excesses and remodeling of the remaining parenchyma. Breast prosthesis is the choice if mammary tissue is scarce, otherwise mastopexy is done, employing in the reconstruction the patient's own gland and dermal tissue (Rubin et al. 2009).

Most commonly, an inverted "T" shaped incision is practiced (Fig. 14.7 and Fig. 14.8). Men may also require cosmetic treatment of the breast, which is called adenomastectomy or correction of gynecomastia. Both dermal surplus and breast parenchyma may be removed, depending on the local anatomy (Stevens et al. 2007).

Thigh interventions aim at thinning, debulking, and suspending the proximal lower limbs, notably removing unsightly folds and redundancies. The internal face of the thigh is particularly addressed, and combined liposuction is often necessary.

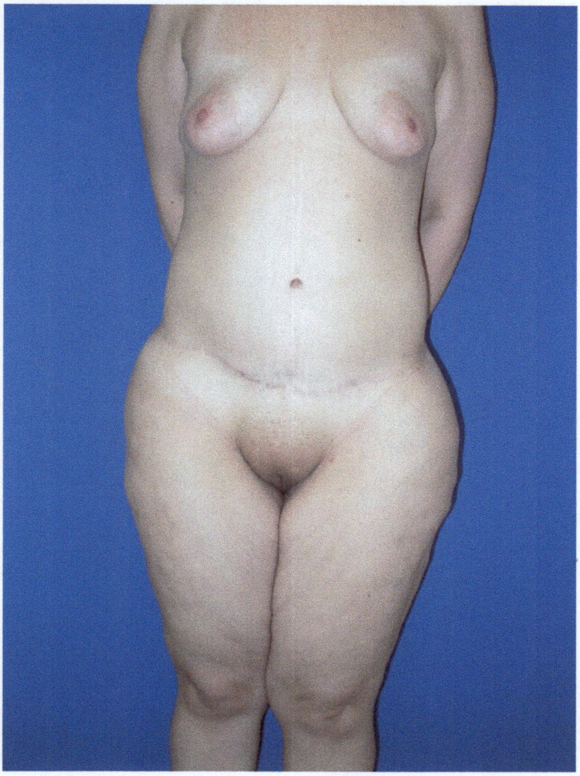

Fig. 14.3 Circumferential abdominoplasty—Postoperative result (abdomen) (2 F)

Three incisions can be considered: the transversal one, coinciding with the inguinal fold, the "T" shaped modality, when a vertical extension downward is associated, and the "S" procedure, when a sickle-shaped flap of the dermal excess in the internal face of the thigh is removed, leaving an "S"-like scar (Labardi et al. 2012). (Figs.14.9, 14.10, and 14.11).

The preference for all incisions is the medial face of the thigh, because scars will be least visible in this area. The length of the scar depends on the severity of thigh ptosis, occasionally extending from the inguinal fold till the knee.

14.17 Brachioplasty

General principles are rather analogous to those of the thigh. All incisions are directed toward the medial aspect of the arm where they become least prominent, and a fuse of dermis is removed accompanied or not by complementary liposuction.

A vertical incision is mostly practiced, starting at the anterior axillary line and proceeding downward, eventually till the elbow (medial epicondyle). The length of resection and amount of tissue removal are obviously dictated by the severity

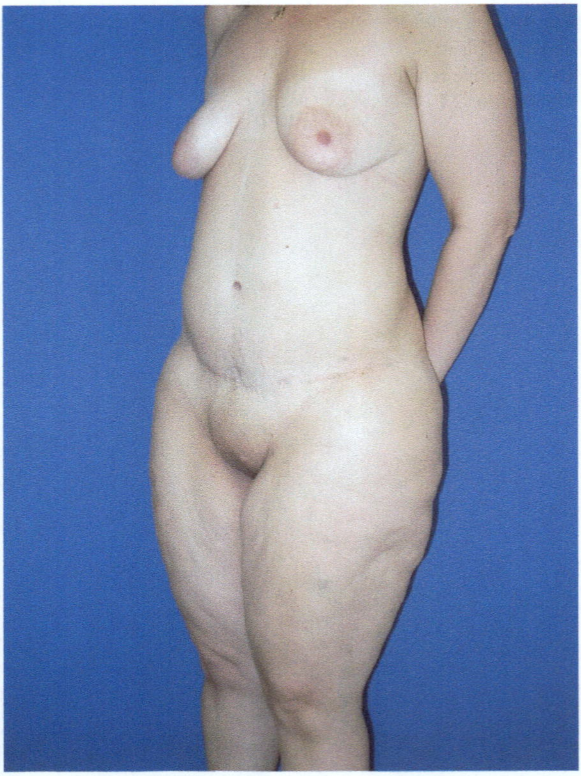

Fig. 14.4 Circumferential abdominoplasty—Postoperative result (*anterolateral view*) (2 G)

of brachial ptosis. Attention during closure of this procedure should be given to the formation of small skin redundancies ("dog ears") near the axilla. In order to correct such aberration a secondary horizontal fuse of skin is removed, coinciding with the anterior axillary line. In this sense, the final incision is actually converted into a "T", with very acceptable cosmetic outcome (Figs. 14.12 and 14.13) (Aly et al. 2006; Modolin et al. 2011).

The association of brachioplasty and mastoplasty has been called upper body lift.

14.18 Rhytidoplasty (Face Lift)

The face is the greeting card, the primary identifier of the person. It is also a window to the inner self, widely believed to convey personality traits and even general health status (Kramer and Ward 2010). An ugly, wrinkled, or unappealing likeness may elicit negative affective reactions or even outright rejection, which are deleterious to self-image and self-esteem. Whether bariatric operations are followed by unique face changes or not, they commonly accelerate facial ptosis with aged, tired, and even sickly features. Characteristic findings are loss of the

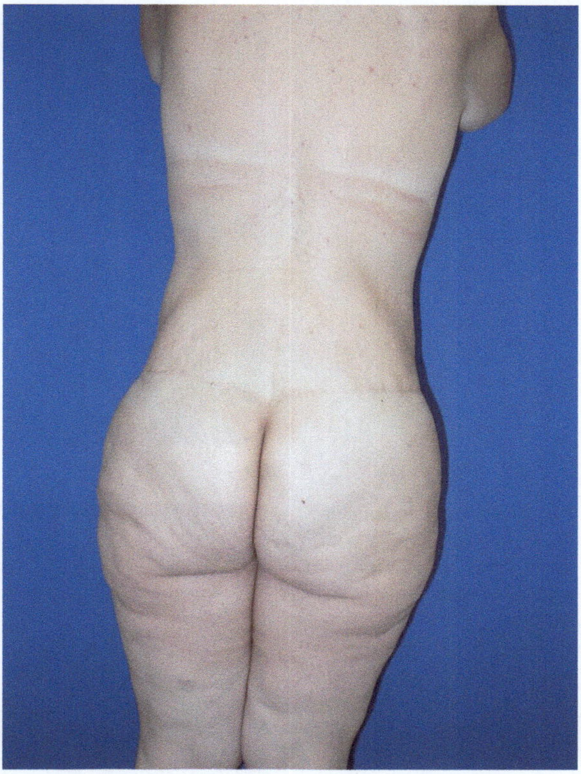

Fig. 14.5 Circumferential abdominoplasty—Postoperative result (*dorsal view*) (2 H)

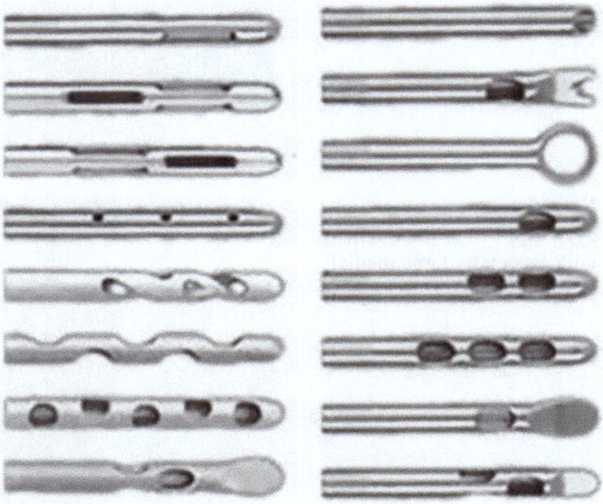

Fig. 14.6 Cannulas for complementary *liposuction*

cervicomental angle, reduction of platysmal tone, and jowling of the face and neck. Descent of the cheeks and deepening of the nasogenian sulcus also tend to be prominent. Rejuvenation and general lifting of lax soft tissues is therefore

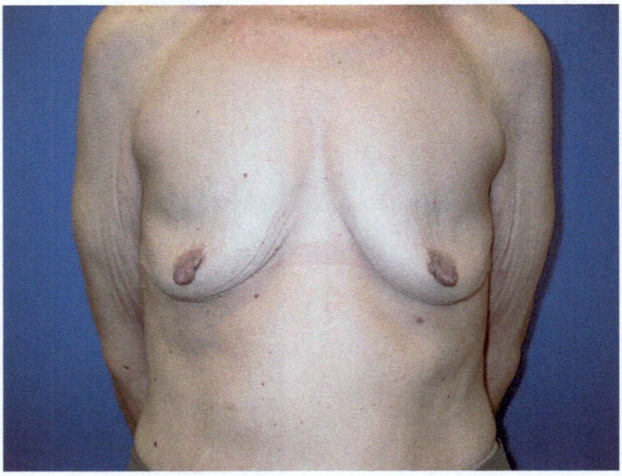

Fig. 14.7 Collapsed mammary glands in a bariatric patient. (Cintra breast 1)

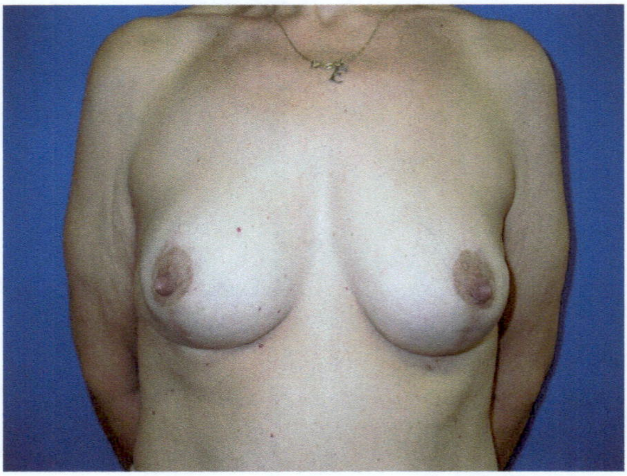

Fig. 14.8 Augmentation mastopexy (mastopexy with the inverted "T" incision associated with a prosthesis inclusion

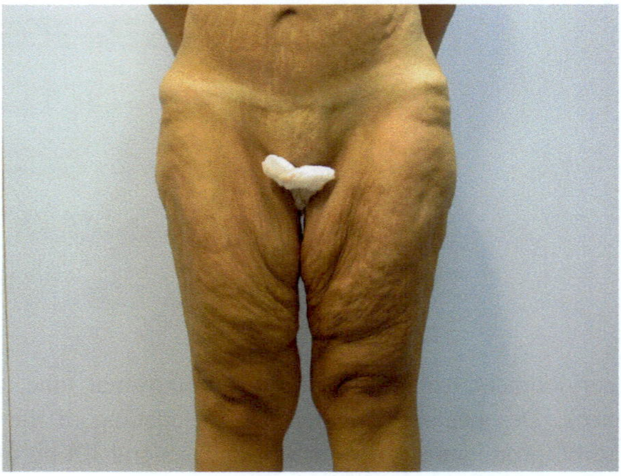

Fig. 14.9 Severe dysmorphism of the thigh, with especially large medial skinfolds

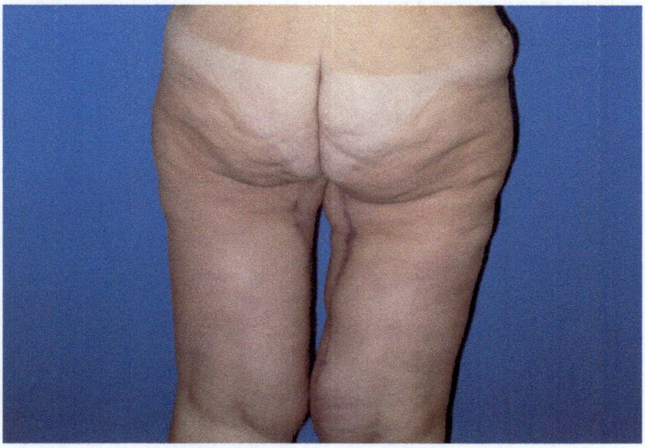

Fig. 14.10 In this dorsal postoperative view, an S-shaped flap was removed and the anatomy was reconstructed

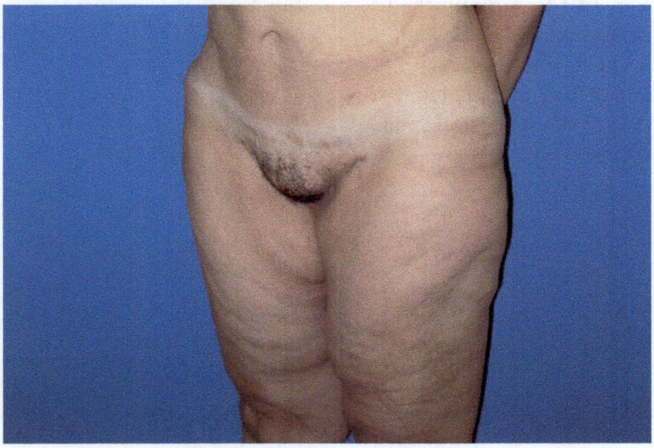

Fig. 14.11 Same patient, postoperative result, anterior view

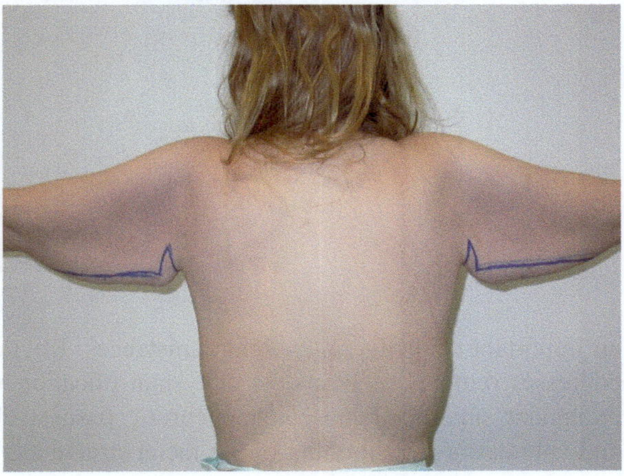

Fig. 14.12 Typical "bat wing" deformity of the arms, with demarcation of surgical incisions

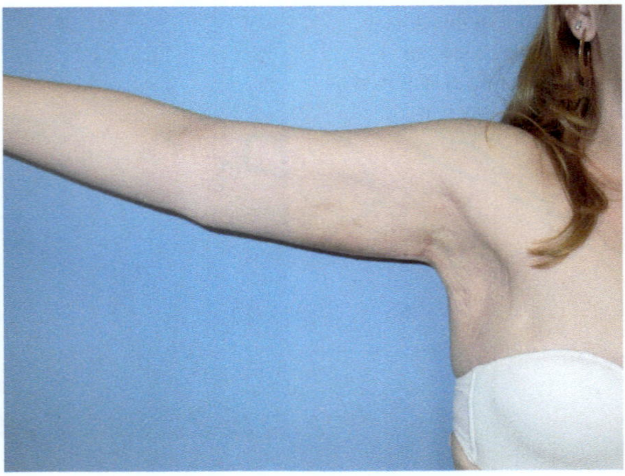

Fig. 14.13 After double-fuse dermolipectomy, arm anatomy is restored

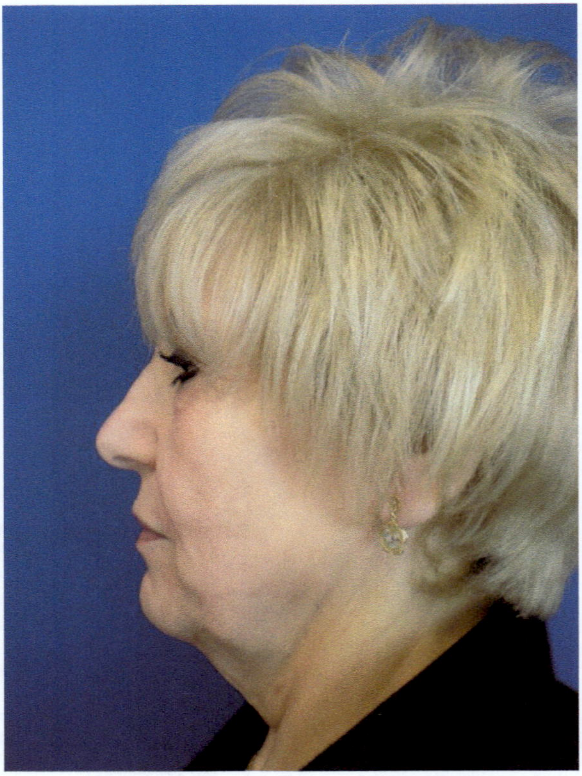

Fig. 14.14 A diminished cervicomental angle and descent of the cheeks conducting to a moderate degree of turkey neck, can be appreciated

an important challenge in such circumstances. Rhytidectomy (face lift) generally addresses removal of wrinkles, platysmal plication with or without resection of redundant superficial musculoaponeurotic tissues, elimination of skin excesses, and volumetric repositioning of the facial structures. Whenever the jawline is lost

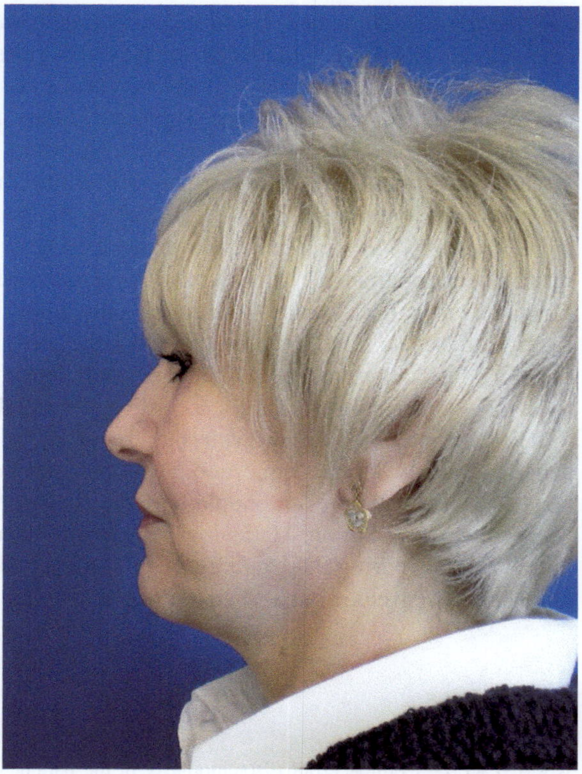

Fig. 14.15 The cervicomental angle was restored, lax musculoaponeurotic tissues were lifted, and a more attractive and younger-looking face was achieved

generating continuity between the chin and the neck, or is partially blunted by jowling (turkey neck), the cervicomental angle requires reconstruction (Sclafani 2005). The illustrations display a comparatively mild case of diminished cervicomental angle and jowling, with postoperative improvement of these traits (Figs. 14.14, 14.15).

14.19 Synthesis

A diminished body mass index and waist circumference, a more favorable cardiovascular risk score, and a relief for major comorbidities are absolutely valuable targets in bariatric treatment. Such achievements may properly be classified as a huge success. However, not necessarily for the patient, who may seek different priorities.

A brand new, masterly sculpted body or at least a markedly rejuvenated look could be the primary desire, and the medical team should carefully listen to the patient and exchange views and perspectives before undertaking any procedure. Bariatric intervention is not necessarily beauty-enhancing, and depending on the degree of weight loss, the intrinsic properties of skin and subdermal tissues, as well as the age, gender, ethnic origin, and degree of sedentarism, quite the opposite might result.

Plastic surgery is effective in many contexts and should be considered an integral component of the rehabilitation process. Once again, although aesthetic and functional gains almost always occur, ideal outcomes are not assured. Results need to be openly discussed before surgery, confronting the patient's expectations with the realities of operative techniques, including the possibility of complications and additional interventions. Specialized teams are highly recommended. Even though most of the operations are mere adaptations of standard, well-tested procedures, thus familiar to most professionals in the area, only a plastic surgeon with experience in the bariatric context will be able to predict the probable cosmetic result, to define the individualizations that many outcomes will depend on, and to establish the best timetable for plastic interventions, according to the deflated or nondeflated status.

There is nothing wrong with the desire of an alluring silhouette, and if achieved, such outcome unquestionably boosts self-confidence and quality of life; however, unrealistic dreams should be formally dispelled by objective and factual preoperative information. Meticulous attention to details will maximize results and inhibit failures during the handling of the patient's skin, which happens to be not only an outer envelope but also an important interface with the world.

References

Aly A, Pace D, Cram A (2006) Brachioplasty in the patient with massive weight loss. Aesthet Surg J 26:76–84. doi:10.1016/j.asj.2005.11.004

Banja J (2004) Obesity, responsibility, and empathy. Case Manag 15:43–46. doi:doi.org/10.1016/j.casemgr.2004.09.005

Boriani F, Taveggia A, Cravero L (2005) Obesity and body contouring: contemporary lessons from a historical example. Obes Surg 15:1218. doi:10.1381/0960892055002392

Cintra W, Modolin MLA, Gemperli R, Gobbi CIC, Faintuch J, Ferreira MC (2008) Quality of life after abdominoplasty in women after bariatric surgery. Obes Surg 18:728–732. doi:10.1007/s11695-007-9280-2

Gusenoff JA, Messing S, O'Malley W, Langstein HN (2008) Patterns of plastic surgical use after gastric bypass: who can afford it and who will return for more. Plast Reconstr Surg 122:951–958. doi:10.1097/PRS.0b013e3181811c55

Heneghan HM, Heinberg L, Windover A, Rogula T, Schauer PR (2012) Weighing the evidence for an association between obesity and suicide risk. Surg Obes Relat Dis 8:98–107. doi:10.1016/j.soard.2011.10.007

Highton L, Ekwobi C, Rose V (2012) Post-bariatric surgery body contouring in the NHS: a survey of UK bariatric surgeons. J Plast Reconstr Aesthet Surg 65:426–432. doi:10.1016/j.bjps.2011.09.047

Iglesias M, Butron P, Abarca L, Perez-Monzo MF, de Rienzo-Madero B (2010) An anthropometric classification of body contour deformities after massive weight loss. Ann Plast Surg 65:129–134. doi:10.1097/SAP.0b013e3181c9c336

Kenkel JM (2006) Body contouring surgery after massive weight loss. Marking and operative techniques. Plast Reconstr Surg 117(Suppl 1):47–73

Kitzinger HB, Abayev S, Pittermann A, Karle B, Bohdjalian A, Langer FB, Prager G, Frey M (2012) After massive weight loss: patients' expectations of body contouring surgery. Obes Surg 22:544–548. doi:10.1007/s11695-011-0551-6

Kramer RS, Ward R (2010) Internal facial features are signals of personality and health. J Exp Psychol (Hove) 63:2273–2287. doi:10.1080/17470211003770912

Labardi L, Gentile P, Gigliotti S, Marianetti M, Colicchia GM, Pascali M, Brinci L, Cervelli V (2012) Medial thighplasty: horizontal and vertical procedures after massive weight loss. J Cutan Aesthet Surg 5:20–25. doi:10.4103/0974-2077.94330

Modolin ML, Cintra W, Faintuch J, Silva MM, Moraes LB, Gemperli R, Ferreira MC (2011) Improved double-fuse technique for post-bariatric brachioplasty. Rev Col Bras Cir 38:217–222. doi:doi.org/10.1590/S0100-69912011000400003

Modolin M, Cintra W, Gobbi CIC, Ferreira MC (2003) Circumferential abdominoplasty for sequential treatment after morbid obesity. Obes Surg 13:95–100. doi:10.1381/096089203321136665

Olbers T (2011) Bariatric surgery or the fairy tale about the ugly duckling. Lakartidningen 108:2570–2573

Orpheu SC, Coltro PS, Scopel GP, Gomez DS, Rodrigues CJ, Modolin ML, Faintuch J, Gemperli R, Ferreira MC (2010) Collagen and elastic content of abdominal skin after surgical weight loss. Obes Surg 20:480–486. doi:10.1007/s11695-009-0019-0

Rubin JP, Gusenoff JA, Coon D (2009) Dermal suspension and parenchymal reshaping mastopexy after massive weight loss: statistical analysis with concomitant procedures from a prospective registry. Plast Reconstr Surg 123:782–789. doi:10.1097/PRS.0b013e31819ba1a8

Sclafani AP (2005) Restoration of the jawline and the neck after bariatric surgery. Facial Plast Surg 21:28–32. doi:10.1055/s-2005-871760

Sell H, Eckel J (2010) Adipose tissue inflammation: novel insight into the role of macrophages and lymphocytes. Curr Opin Clin Nutr Metab Care 13:366–370. doi:10.1097/MCO.0b013e32833aab7f

Shermak MA, Chang D, Magnuson TH, Schweitzer MA (2006) An outcomes analysis of patients undergoing body contouring surgery after massive weight loss. Plast Reconstr Surg 118:1026–1031. doi:10.1097/01.prs.0000232417.05081.db

Song AY, Rubin JP, Thomas V, Dudas JR, Marra KG, Fernstrom MH (2006) Body image and quality of life in post massive weight loss body contouring patients. Obesity (Silver Spring) 14:1626–1636. doi:10.1038/oby.2006.187

Stevens WG, Freeman ME, Stoker DA, Quardt SM, Cohen R, Hirsch EM (2007) One-staged mastopexy with breast augmentation: a review of 321 patients. Plast Reconstr Surg 120:1674–1679. doi:10.1097/01.prs.0000282726.29350.ba

Whitaker IS, Karoo RO, Spyrou G, Fenton OM (2007) The birth of plastic surgery: the story of nasal reconstruction from the Edwin Smith Papyrus to the twenty-first century. Plast Reconstr Surg 120:327–336. doi:10.1097/01.prs.0000264445.76315.6d

Chapter 15
New Technologies for Improving the Psychological Treatment

Giuseppe Riva, Pietro Cipresso, Fabrizia Mantovani, Antonios Dakanalis and Andrea Gaggioli

Abstract Most clinicians and patients consider obesity just as a problem of energy balance: more energy input than expenditure. However, the epidemiological data and the clinical practice clearly show that obesity is more complex than described by this simple equation. In particular, the chapter underlined the possible role of negative body image in the etiology of this disturbance. Specifically, we suggested that unhealthful weight-control behaviors, that induce significant weight gain over time, may be driven by an allocentric negative body image that is no more updated by contrasting egocentric representations driven by perception (*allocentric lock*). In other words, subjects are locked to an allocentric negative representation of their body that their sensory inputs are no more able to update even after the dramatic body changes following a successful diet or bariatric surgery. In the chapter, we also discussed the possible role of virtual reality (VR) in addressing this problem within an integrated treatment approach. Specifically, we suggested that adding a 10-session experiential protocol based on the free NeuroVR (http://www.neurovr.org) virtual reality software to the classical treatment of

G. Riva (✉) · P. Cipresso · A. Gaggioli
Applied Technology for Neuro-Psychology Lab, IRCSS Istituto Auxologico Italiano, Via Pelizza da Volpedo 41, 20149 Milan, Italy
e-mail: giuseppe.riva@unicatt.it

P. Cipresso
e-mail: p.cipresso@auxologico.it

A. Gaggioli
e-mail: andrea.gaggioli@auxologico.it

G. Riva · P. Cipresso · A. Gaggioli
Department of Psychology, Università Cattolica del Sacro Cuore, Largo Gemelli 1, 2013 Milan, Italy

F. Mantovani
CESCOM, Università Milano-Bicocca, Piazza della Scienza, 3, 20126 Milan, Italy
e-mail: fabrizia.mantovani@unimib.it

A. Dakanalis
Department of Psychology, University of Pavia, P.za Botta 6, 27100 Pavia, Italy
e-mail: antonios.dakanalis@unipv.it

P. Capodaglio et al. (eds.), *Disabling Obesity*, DOI: 10.1007/978-3-642-35972-9_15,

obesity may improve its long-term outcome. The characteristics of the protocol and the outcome of two controlled clinical trials used to test it (VEPSY UPDATED - ISRCTN59019572, and AVATOB-NCT01394393) are detailed and discussed.

15.1 Introduction

Weight-related disorders are becoming a critical problem in both Western and developing countries. During the past 20 years, there has been a dramatic increase in both obesity and eating disorders in most countries and rates remain high. Unfortunately, the treatment of weight-related disorders is not easy: the typical treatment includes a combination of diet, exercise, cognitive and/or behavior modification, and the final choice depends on the overall health condition, level of disorders, and willingness to address the problem.

The raising prevalence of weight-related disorders is pushing eating disorder and obesity researchers to start a collaboration between the fields to address them. In particular, their effort is focused on the identification of risk factors that are shared between these weight-related disorders (Haines and Neumark-Sztainer 2006): apparently, unhealthful weight-control behaviors—such as fasting (going without eating for 24 h for weight control), vomiting, or laxative abuse—are the common antecedents of both obesity and eating disorders (Haines and Neumark-Sztainer 2006; Neumark-Sztainer et al. 2006; Neumark-Sztainer 2009; Johnston 2004; Stice et al. 2005, 2008). For example, Neumark-Sztainer and colleagues (Neumark-Sztainer et al. 2006) discussed the results of the Project EAT II (Eating Among Teens), a longitudinal study involving 2,516 ethnically and socioeconomically diverse adolescents. They report that, 5 years later, the use of unhealthful weight-control behaviors increased six times the risk for binge eating with loss of control, three times the risk for being overweight, and two to five times the risk for extreme weight-control behaviors such as the use of diet pills and self-induced vomiting. A similar result was found by Stice and colleagues (2008): in a different longitudinal study fasting was the best predictor for the future onset, 5 years later, of binge eating and bulimia nervosa.

It is well-known from epidemiological studies that childhood obesity has different ethnic, socioeconomic (compared with affluent white children, the poor Hispanic, white, and black children have 2.7, 1.9, and 3.2 times higher odds of obesity), and behavioral risk factors (Singh et al. 2008). Between the behavioral variables higher television viewing, and higher physical inactivity levels were all independently associated with higher obesity prevalence.

However, in a 4-year longitudinal study on 496 adolescent girls, Stice and colleagues (2005) studied the psychological and behavioral risk factors able to predict the onset of obesity in adolescent girls. Their data show that participants who were on a weight-loss diet, or who used maladaptive compensatory behaviors for weight control at T1 of the study showed, 4 years later, an increased risk for obesity onset. A more recent 10-year longitudinal study confirmed this datum (Neumark-Sztainer et al. 2012). As stated by the authors:

> Findings clearly indicate that dieting and unhealthy weight control behaviors, as reported by adolescents, predict significant weight gain over time (p. 80).

These data have an important clinical implication: the evidence that youths practicing unhealthful weight-control behaviors are at higher risk for obesity implies that prevention and treatment interventions should also focus on the causes of these behaviors. In other words, why do adolescents decide to start such radical weight-control behaviors? In a recent letter to the *Yahoo Answer* site an adolescent girl wrote:

> I hate my body so much. My top half (arms, stomach) is fine. I have a very flat stomach, my arms are great (my back is even bony looking) I wear a XS, S in shirts. but my lower half (butt, thighs) are huge, seriously I mean how can I loose weight there? (online: http://answers.yahoo.com/question/index?qid=20120504232627AAPkxrl).

The words of the girl clearly explain her behavior: she wants going on diet because she does not like her body (Riva et al. 2000c). A study by Kostanski and Gullone (1999) with a sample of 431 Australian pre-adolescent children (7–10 years) confirms this interpretation: pre-adolescents as young as 7 years of age are unsatisfied with their body appearance and deliberately engage in restrictive eating behaviors. More, a recent study (Friend et al. 2012) showed that in adolescents frequent self-weighing is associated with lower body satisfaction and higher rates of unhealthy and extreme weight-control behaviors.

This is even true for morbid obesity, a chronic condition that is hard to treat with diet, exercise, and psychological treatment alone. For this reason, in contrast to the nonsurgical treatment of obesity, bariatric surgery is becoming the treatment of choice for morbid obesity (van Hout and van Heck 2009).

Nevertheless, all roses have thorns, and this is true for bariatric surgery, too. In fact, this approach does not lead to equal results in every patient (van Hout et al. 2003): the long-term efficacy is strongly influenced by compliance to adequate dietary rules in which psychosocial factors and behavioral changes can play a major role.

In particular, a critical problem is the expected outcome of the treatment. As underlined by Kaly and colleagues (2008) there is a significant difference between the weight loss clinicians consider successful following bariatric surgery and the weight loss potential patients expect to achieve. As a general guideline, bariatric surgery is considered successful when 50 % of excess weight is lost and the weight loss is sustained up to 5 years. However, most obese patients have different expectations: in the previous study, patients declared to be "happy" after a 77 ± 9 % excess body weight loss and considered "acceptable" a 67 ± 10 % excess body weight loss. A 49 ± 14 % excess body weight loss, the gold standard for clinicians, was considered "disappointing".

The situation does not always change after the treatment: a problem affecting some patients is body image dissatisfaction (Pecori et al. 2007; Adami et al. 1999; Morrow et al. 2008): patients with 200 pounds overweight still have body disparagement after losing 100 pounds. For example, in a recent post in a blog, some months after bariatric surgery Stella writes:

> I feel like I have somewhat of a "dysmorphic" view of my body I don't really see myself as a size 4 or a size 2 (only if they run big since I gained a few lbs) I still see myself as fat and the damn skin on my arms does not help the matters. I have such a hard time finding shirts cause of it and I can almost never find a dress cause I don't want my arm cheese to show. I want to be normal, is this normal??? (online: http://www.shrinkingstella.com/).

The words of the woman clearly explain her situation: the weight loss was unable to modify her negative experience of the body. This expectation was linked to the common belief that the best way to improve one's body image is to lose weight. However, recent studies have questioned this belief: dietary intervention, even if accompanied by significant weight loss, may be ineffective in reducing total body dissatisfaction (Rosen et al. 1995; Rosen 1996a). And unfortunately body disparagement has a negative effect on long-term follow-up of obesity, even the one treated using bariatric surgery: uncontrolled eating and grazing after surgery are associated with poorer and elevated psychological distress (Colles et al. 2008).

Given the importance of body image satisfaction for the quality of life of obese persons, these findings argue for the potential benefits of treatment strategies for improving appearance satisfaction for obese individuals, to improve the success of their weight-management efforts (Thompson et al. 1999). Unfortunately, obesity researchers have not yet systematically added body image interventions in their programs. Almost 20 years ago in a review on the behavioral obesity treatment literature, Rosen (1996b) did not find any study including psychological techniques specifically designed to modify body image. But today, the situation is not radically different. In their new cognitive behavioral approach to the treatment of obesity (Cooper et al. 2010) Cooper and Fairburn address the experience of the body only as potential obstacle to the acceptance of weight maintenance (Cooper et al. 2003) and not as a one of the possible causes of the disorder. Unfortunately, the results of the proposed approach are similar to the old one (Cooper et al. 2010):

> Both of the main treatments resulted in an average weight loss of about ten percent of initial weight… The participants were subsequently followed-up for three years post-treatment. The great majority regained almost all the weight that they had lost with the new treatment being no better than the behavioural treatment in preventing weight regain (p. 706).

We do not agree with this vision (Riva et al. 2006). Following the emerging "embodied cognition" approach (Bermúdez et al. 1995; Clark 1997; Clancey 1997; Gallagher 2003) we consider body image as an integral part of the subjects' identity. As noted by Gallagher (1995) the body experience is not neutral but it places constraints on intentional consciousness:

> changes or distortion introduced at the level of body schema result in changes or distortions in intentional consciousness (p. 239).

In particular in this chapter we will introduce and discuss the "allocentric lock theory" (Riva 2007, 2012; Riva et al. 2012b; Riva and Gaudio 2012).

Specifically, we will suggest that an allocentric negative body image that is no more updated by contrasting egocentric representations driven by perception is the common antecedent of both obesity and eating disorders (Riva 2011). In other words, these patients are locked to an allocentric (*observer view*) negative representation of their body that their sensory inputs are no more able to update even after dramatic body changes, including the one following either a diet or bariatric surgery. Moreover, we will discuss the possible role of virtual reality (VR) in addressing it. Specifically, we suggest that adding a 10-session experiential protocol based on virtual reality to the treatment of obesity may improve its long-term outcome.

15.2 Unlocking the Allocentric Lock

15.2.1 The Allocentric Lock Theory

Psychology and neuroscience indicate that our spatial experience, including the experience of the body, involves the integration of different sensory inputs within two different reference frames: *egocentric* and *allocentric* (Klatzky 1998; Mou et al. 2004):

- *Egocentric frame*: it is referred to the body of the observer and allows him/her to locate objects relative to the body center. When we adopt an egocentric stance we represent the object relative to ourselves.
- *Allocentric frame*: it is referred to space external to the perceiver. When we adopt an allocentric stance the object is represented independently of our own current relation with it.

As suggested by Byrne and Becker (2007), the transformation from egocentric to allocentric representations of space is done by neurons in different medial temporal lobe structures. If, for some reasons, this transformation is impaired, the subjects cannot use anymore the sensory inputs to update the contents of the allocentric representation of their body. This is what may be behind the body dissatisfaction experienced by many obese patients even after a significant weight loss: an altered somato representation—that is not updated by contrasting egocentric parietal representations driven by perception—priming the processing of any further body-related experience (Riva 2011). In simpler words, the egocentric perception-driven experience of the real body does not modify the allocentric memory-driven experience of a negative body: these patients are locked to an allocentric negative representation of their body (Riva 2010). However, the impossibility of using sensory inputs for updating the allocentric representation of the body—patients hate their body even after the surgery or significant weight loss—locks the patients into an unsatisfying body that may explain their depression and low quality of life (Omalu et al. 2007; Masheb et al. 2007).

15.2.2 Unlocking the Virtual Body Using Virtual Reality

The evolution of technology is providing new tools and methods for health care (Riva et al. 2004b). Between them, an emerging trend is the use of virtual reality (VR) (Satava and Jones 2002; Riva and Gamberini 2000; Riva and Gaggioli 2008).

VR consists of a three-dimensional (3D) graphical environment where a user can interface with the environment through a variety of computer peripheral devices. Using visual, aural, or haptic devices, the user can experience the environment as if it were a part of the real world (Riva and Davide 2001).

In surgery, for instance, VR is used in simulation training (Fried et al. 2010). In fact, VR allows the surgeon to interact efficiently with 3D computerized databases of medical images in real-time using his natural senses and skills (Tanoue et al. 2010). In sum, for physicians, and surgeons, the ultimate goal of VR is the presentation of virtual objects to all of the human senses in a way identical to their natural counterpart (Székely and Satava 1999).

In clinical psychology, the ultimate goal is different. VR is used to offer a new human-computer interaction paradigm in which patients are no longer simply external observers of images on a computer screen but are active participants within a computer-generated 3D virtual world (Riva et al. 2002). Moreover, VR can be considered an "embodied technology" for its effects on body perceptions (Spagnolli and Gamberini 2005): it is possible the use of VR for inducing controlled changes to the experience of the body (Riva et al. 2000a; Riva 1997; Slater et al. 2010; Lenggenhager et al. 2007).

On the one side, different authors showed that is possible to use VR both to induce illusory perceptions—e.g. a fake limb (Slater et al. 2009)—by altering the normal association between touch and its visual correlate. It is even possible to generate a body transfer illusion (Slater et al. 2010): Slater and colleagues substituted the experience of male subjects' own bodies with a life-sized virtual human female body.

On the other side, it is also possible to use VR to improve body image (Riva 1998a, b), even in patients with eating disorders (Riva et al. 1999, 2000b, 2003) or obesity (Riva et al. 2000a, 2006).

As noted by Gallagher (1995), "[different] studies indicate that changes in various aspects of body schemata have an effect on the way subjects perceive their own body" (p. 237). Following this vision it is possible the use of VR to induce a controlled sensory rearrangement that facilitates an update of the locked allocentric representation of the body.

A possible strategy toward this goal is the adaptation to virtual reality of the imagery rescripting method developed for the treatment of post-traumatic stress disorders (Smucker et al. 1995; Riva 2011). Specifically, Riva developed a specific body image rescripting protocol based on VR that is included as part of the experiential cognitive therapy (ECT) described below.

15.2.2.1 The Proposed Approach: The Experiential Cognitive Therapy

Developed by Giuseppe Riva and his group (Riva et al. 1998, 2000a), ECT is a relatively short term (15-session in 6 weeks), patient-oriented approach that focuses on individual discovery (Riva et al. 2004a, 2006). As in the case of cognitive behavioral therapy (CBT), ECT uses a combination of nutritional, cognitive, and behavioral procedures to help the patient identify and change the maintaining mechanisms in obesity and eating disorders. However, ECT differs from the typical CBT approach in the use of VR, in its focus on empowerment and in its focus on the negative emotions related to the body:

- *Its focus on the body experience*. A major reason patients want to lose weight (Rosen 1996b). The experience of the body is not addressed only as potential obstacle to the acceptance of weight maintenance (Cooper et al. 2003). As we underlined before, in line with the emerging "embodied cognition" approach (Bermúdez et al. 1995; Clark 1997; Clancey 1997; Gallagher 2003), we consider body image as an integral part of the subjects' identity.
- *Its focus on the empowerment process*. We consider a critical goal for the long-term efficacy of bariatric surgery, the ability of the patient in defining a realistic target weight range, in monitoring eating behaviors and in managing the frustration related to weight fluctuations. However, we do not consider these abilities as behavioral skills only. Here we agree with the DiClemente position (1986) that describes them as part of a broader individual dimension, defined "control self-efficacy": "an individual's ability to control the addictive behavior in a variety of provocative situations" (p. 303). Bandura (1989, 1997) proposed the concept of self-efficacy as an explanation of behavior and behavior change. People tend to avoid activities they believe exceed their coping abilities and undertake those they consider themselves capable of handling. Following this approach, the main strategy to raise self-efficacy is "empowerment", the process of helping people feel a sense of control over their lives. Within this process there are three critical dimensions to address: (Menon 1999)
  - *Perceived control*. Includes beliefs about authority, decision-making skills, availability of resources, autonomy in the scheduling and performance of work, etc.;
  - *Perceived competence*. Reflects role-mastery, which besides requiring the skillful accomplishment of one or more assigned tasks, also requires successful coping with non-routine role-related situations;
  - *Goal internalization*. This dimension captures the energizing property of a worthy cause or exciting vision.

- *The use of virtual reality (VR)*. VR helps the therapist in providing the two above features of our approach: body experience treatment and empowerment. The use of a VR treatment makes it possible to induce a controlled sensory rearrangement (see Table 15.1) that facilitates an update of the locked allocentric representation of the body (Riva 1998a, c). Further, VR has the right features to support empowerment, since it is a special, sheltered setting where patients can start to explore and act without feeling threatened (Botella et al. 1998).

ECT has been tested in different case studies and two controlled trials with 211 obese patients (Riva et al. 2006) and 36 binge eating patients (Riva et al. 2003). The two controlled trials show that ECT provides better results in the follow-up than competing approaches, including both nutritional and cognitive behavioral therapy.

15.2.2.2 The Protocol

The protocol (Riva 2011) includes five weekly group sessions aimed at improving motivation to change and assertiveness, and 10 biweekly virtual reality sessions.

The first VR session is used to assess any stimuli that could elicit abnormal eating behavior. Specifically, the attention is focused on the patient's concerns about food, eating, shape, and weight. This assessment is normally part of the Temptation Exposure with Response Prevention protocol (Schlundt and Johnson 1990). At the end of the first VR session the therapist uses the *miracle question*, a typical approach used by the solution-focused brief therapy (deShazer 1985; McFarland 1995). According to this approach, the therapist asks the patient to imagine what life would be like without her/his complaint. Answering this question in writing the patient constructs her/his own solution, which then guides the therapeutic process (deShazer 1988). According to de Shazer (1988) this approach is useful for helping patients establish goals that can be used to verify the results of the therapy. Using VR to experience the effects of the miracle, the patient is more likely not only to gain an awareness of her need to do something to create change but also to experience a greater sense of personal efficacy.

The next nine VR sessions are used to assess and modify:

- *The expectations and emotions related to food and weight.* This is done both by integrating different cognitive-behavioral methods: Countering, Alternative Interpretation, Label Shifting, Deactivating the Illness Belief (see Table 15.1).
- *The strategies used to cope with difficult interpersonal and potential maintenance situations.* This is done both by using the Temptation Exposure with Response Prevention (Schlundt and Johnson 1990; Riva 1998c)—and by working on these three empowering dimensions (Menon 1999): *perceived control, perceived competence* and *goal internalization.*
- *The body experience of the subject.* To do this the virtual environment integrates the therapeutic methods used by Butter and Cash (1987) and Wooley and Wooley (1985), and the body image rescripting protocol based on the Allocentric Lock hypothesis (see Table 15.1) (Smucker et al. 1995).

15.2.2.3 The Virtual Reality Experience

The VR sessions are based on the free NeuroVR software (http://www.neurovr.org). NeuroVR is an enhanced version of the original Virtual Reality for Body Image

Modification (VEBIM) immersive virtual environment, previously used in different preliminary studies on non-clinical subjects (Riva 1997, 1998a).

NeuroVR is composed of 14 virtual environments, used by the therapist within a 60-minute session with the patient. The environments present critical situations related to the maintaining/relapse mechanisms (e.g., Home, Supermarket, Pub, Restaurant, Swimming Pool, Beach, and Gymnasium) and two body image comparison areas.

Using the NeuroVR Editor (see Fig. 15.1), the psychological stimuli/stressors appropriate for any given scenario can be chosen from a rich database of 2D and 3D objects, and easily placed into the pre-designed virtual scenario by using an icon-based interface (no programming skills are required).

In addition to static objects, the NeuroVR Editor allows both to add audio object and to overlay on the 3D scene video composited with a transparent alpha channel.

The editing of the scene is performed in real time, and effects of changes can be checked from different views (frontal, lateral and top).

The edited scene is then visualized and experienced using the NeuroVR Player.

Through the VR experience, the patients practice both eating/emotional/relational management and general decision-making and problem-solving skills. By directly practicing these skills within the VR environment, the patient is helped in developing specific strategies for avoiding and/or coping with these.

Specifically, in the VR sessions the therapist uses the "20/20/20 rule". During the first 20 min, the therapist focuses on getting a clear understanding of the patient's current concerns, level of general functioning, and the experiences related to food. This part of the session tends to be characterized by patients doing most of the talking, although therapist guides with questions and reflection to get a sense of the patient's current status. The second 20 min is devoted to the virtual reality experience. During this part of the session the patient enters the virtual environment

Fig. 15.1 The NeuroVR Editor

and faces a specific critical situation. Here, the patient is helped in developing specific strategies for avoiding and/or coping with it. In the final 20 min, the therapist explores the patient's understanding of what happened in VR and the specific reactions—emotional and behavioral—to the different situations experienced. If needed, some new strategies for coping with the VR situations are presented and discussed. The different techniques used in the VR sessions are detailed in Table 15.1.

Table 15.1 The VR body image rescripting protocol (Adapted from Riva 2011)

Phase 1: Interview	During a clinical interview the patient is asked to relive the contents of the allocentric negative body image and the situation/s in which it was created and/or reinforced (e.g. being teased by my boyfriend at home) in as much detail as possible. The meaning of the experience for the patient was also elicited
Phase 2: Development of the VR scene	The clinician reproduces the setting of the identified situation (e.g. the corridor of the classroom where my boyfriend teased me) using one of the different scenes available in the free NeuroVR software (http://www.neurovr.org)
Phase 3: Egocentric Experience of the VR scene	The patient is asked to reexperience the event in VR from a first person perspective (the patient does not see his/her body in the scene) expressing and discussing his/her feelings. The patient is then asked what was needed to happen change the feelings in a positive direction The main cognitive techniques used in this phase, if needed, are: Countering: Once a list of distorted perceptions and cognitions is developed, the process of countering these thoughts and beliefs begins Label Shifting: The patient first tries to identify the kinds of negative words she uses to interpret situations in her life, such as bad, terrible, obese, inferior, and hateful. The situations in which these labels are used are then listed. The patient and therapist replace each emotional label with two or more descriptive words
Phase 4: Allocentric Experience of the VR scene	The patient is asked to reexperience the event in VR from a third person perspective (the patient sees his/her body in the scene) intervening both to calm and reassuring his/her virtual avatar and to counter any negative evaluation. The therapist follows the Socratic approach, for example "What would need to happen for you to feel better? How does it look through the eyes of a third person? Is there anything you as a third person like to do? How do the other people respond?" The main cognitive techniques used in this phase, if needed, are: Alternative Interpretation: The patient learns to stop and consider other interpretations of a situation before proceeding to the decision-making stage Deactivating the Illness Belief: The therapist first helps the client list her beliefs concerning weight and eating

15.3 Conclusions

Most clinicians and patients consider obesity just as a problem of energy input and expenditure: more energy input than expenditure. However, the clinical practice and epidemiological data clearly show that obesity is more complex than expected by this simple equation. In particular, this chapter underlined the possible role of negative body image in the etiology of this disturbance.

Specifically, we suggested that body disparagement may be produced by an allocentric negative body image that is no more updated by contrasting egocentric representations driven by perception. In other words, these subjects are locked to an allocentric (*observer view*) negative representation of their body that their sensory inputs are no more able to update even after dramatic body changes. The impossibility of using sensory inputs for updating the allocentric representation of the body—patients hate their body even after a significant weight loss—locks the patients into an unsatisfying body that may explain their depression, low quality of life and difficulty in maintaining an effective eating behavior (Omalu et al. 2007; Masheb et al. 2007).

How can we unlock this virtual body? This chapter suggested as possible answer an exciting new technology: virtual reality (VR). VR is well-known by surgeons: it is used in surgical training and allows the surgeon to interact efficiently with 3D computerized databases of medical images in real-time using his natural senses and skills. However, VR can be considered an "embodied technology" for its effects on body perceptions: VR can be used for inducing controlled changes to the experience of the body. In sum, it is possible the use of VR to induce a controlled sensory rearrangement that facilitates an update of the locked allocentric representation of the body.

Specifically, a specific body image rescripting protocol based on VR is part of the ECT described in this chapter. ECT is a relatively short term (five weekly group sessions aimed at improving motivation to change and assertiveness, and 10 biweekly virtual reality sessions in 6 weeks), patient-oriented approach that focuses on individual discovery. As cognitive behavioral therapy (CBT), ECT uses a combination of nutritional, cognitive, and behavioral procedures to help the patient identify and change the maintaining mechanisms in obesity and eating disorders. However, ECT differs from the typical CBT approach in the use of VR, in its focus on empowerment and in its focus on the negative emotions related to the body (Villani et al. 2012).

In the VR sessions the therapist uses the "20/20/20 rule". During the first 20 min, the therapist focuses on getting a clear understanding of the patient's current concerns, level of general functioning, and the experiences related to food. The second 20 min are devoted to the virtual reality experience. During this part of the session, the patient enters the virtual environment and faces a specific critical situation. In the final 20 min, the therapist explores the patient's understanding of what happened in VR and the specific reactions—emotional and behavioral—to the different situations experienced.

ECT has been tested in different case studies and two controlled trials with 211 obese patients and 36 binge eating patients (VEPSY UPDATED-ISRCTN59019572). The two controlled trials show that ECT provides better results in the follow-up than competing approaches. This positive experience is being replicated in Mexico. The "Laboratorio de Enseñanza Virtual y Ciberpsicología" at the School of Psychology of the Universidad Nacional Autonoma de Mexico, in cooperation with the Obesity Unit of the Medica Sur Hospital in México City have recently started a controlled clinical trial, recently approved by the US ClinicalTrial.gov database (Virtual Environments For Supporting Obesity Treatment—AVATOB-NCT01394393). The trial, that will include 60 morbid obese patients both treated with bariatric surgery and without it, started its work in June 2011 and is expected to complete in 2013.

Even if the final results are not available yet, recently we reported (Riva et al. 2012a, b) the clinical case of Patricia, a 44-year-old woman, who entered in the bariatric protocol with a weight of 114 kg., 1.55 m. tall and a B.M.I. of 47. Even if at the start of the protocol her weight was 80.2 kg (a 30 % body weight loss and a 62 % excess body weight loss), the weight loss achieved after the surgery was unable to modify her negative experience of the body: she expressed the need to improve self-esteem, be more attractive, and also be more attractive to others. Moreover, the clinical data underlined a moderate level of depression matched by a high level of body dissatisfaction. The clinical data after the treatment showed a significant improvement in all the psychological variables matched both by an improvement in the subjective physical and emotional well-being, and by relevant behavioral changes in the personal and social daily life.

In conclusion, the available clinical data suggest the added value of ECT as part of an integrated obesity treatment based on the experiential approach allowed by virtual reality. Longer follow-up data and multi-centric trials are required to investigate the possible effects of the behavioral and body image changes on the long-term maintenance of the weight loss.

References

Adami GF, Meneghelli A, Bressani A, Scopinaro N (1999) Body image in obese patients before and after stable weight reduction following bariatric surgery. J Psychosom Res 46(3):275–281. doi:S0022399998000944

Bandura A (1989) Social cognitive theory. In: Vasta R (ed) Annals of child development, vol 6., Six theories of child developmentJAI Press, Greenwich, pp 1–60

Bandura A (1997) Self-efficacy: The exercise of control. W H Freeman and Co., New York

Bermúdez J, Marcel AJ, Eilan N (1995) The body and the self. MIT Press, Cambridge

Botella C, Perpiña C, Baños RM, Garcia-Palacios A (1998) Virtual reality: a new clinical setting lab. Studies in Health Technology and Informatics 58:73–81

Butters JW, Cash TF (1987) Cognitive-behavioral treatment of women's body image satisfaction: a controlled outcome-study. J Consult Clin Psychol 55:889–897

Byrne P, Becker S, Burgess N (2007) Remembering the past and imagining the future: a neural model of spatial memory and imagery. Psychol Rev 114(2):340–375

Clancey WJ (1997) Situated cognition: on human knowledge and computer representation. Cambridge University Press, Cambridge

Clark A (1997) Being there: putting brain body and world together again. MIT Press, Cambridge

Colles SL, Dixon JB, O'Brien PE (2008) Grazing and loss of control related to eating: two high-risk factors following bariatric surgery. Obesity (Silver Spring) 16(3):615–622. doi:oby20071 0110.1038/oby.2007.101

Cooper Z, Doll HA, Hawker DM, Byrne S, Bonner G, Eeley E, O'Connor ME, Fairburn CG (2010) Testing a new cognitive behavioural treatment for obesity: a randomized controlled trial with three-year follow-up. Behav Res Ther 48(8):706–713. doi:10.1016/j.brat.2010.03.008

Cooper Z, Fairburn CG, Hawker DM (2003) Cognitive-behavioral treatment of obesity: a clinician's guide. Guilford Publications, Inc., New York

deShazer S (1985) Keys to solutions in brief therapy. W.W. Norton, New York

deShazer S (1988) Clues: investigating solutions in brief therapy. Norton, New York

DiClemente CC (1986) Self-efficacy and the addictive behaviors. J Soc Clin Psychol 4:302–315

Fried MP, Sadoughi B, Gibber MJ, Jacobs JB, Lebowitz RA, Ross DA, Bent JP 3rd, Parikh SR, Sasaki CT, Schaefer SD (2010) From virtual reality to the operating room: the endoscopic sinus surgery simulator experiment. Otolaryngol Head Neck Surg 142(2):202–207. doi:S0194-5998(09)01772-010.1016/j.otohns.2009.11.023

Friend S, Bauer KW, Madden TC, Neumark-Sztainer D (2012) Self-weighing among adolescents: associations with body mass index, body satisfaction, weight control behaviors, and binge eating. J Acad Nutr Diet 112(1):99–103. doi:10.1016/j.jada.2011.08.036

Gallagher S (1995) Body schema and intentionality. In: Bermúdez JL, Marcel A, Eilan N (eds) The body and the self, vol 225–244. MIT Press, Cambridge, MA

Gallagher S (2003) Bodily self-awareness and object perception. Theoria et Historia Scientiarum: International Journal for Interdisciplinary Studies 7 (1):online: http://www2.canisius.edu/~gallaghr/theoria03.html

Haines J, Neumark-Sztainer D (2006) Prevention of obesity and eating disorders: a consideration of shared risk factors. Health Educ Res 21(6):770–782

Johnston JM (2004) Eating disorders and childhood obesity: who are the real gluttons? CMAJ 171(12):1459–1460. doi:171/12/145910.1503/cmaj.1041111

Kaly P, Orellana S, Torrella T, Takagishi C, Saff-Koche L, Murr MM (2008) Unrealistic weight loss expectations in candidates for bariatric surgery. Surg Obes Relat Dis 4(1):6–10. doi:S1550-7289(07)00715-010.1016/j.soard.2007.10.012

Klatzky RL (1998) Allocentric and egocentric spatial representations: definitions, distinctions and interconnections. In: Freska C, Habel C, Wender KF (eds) Spatial Cognition: an interdisciplinary approach to representation and processing of spatial knowledge. Springer-Verlag, Berlin, pp 1–17

Kostanski M, Gullone E (1999) Dieting and body image in the child's world: conceptualization and behavior. J Genet Psychol 160(4):488–499

Lenggenhager B, Tadi T, Metzinger T, Blanke O (2007) Video ergo sum: manipulating bodily self-consciousness. Science 317(5841):1096–1099. doi:317/5841/109610.1126/science.1143439

Masheb RM, White MA, Toth CM, Burke-Martindale CH, Rothschild B, Grilo CM (2007) The prognostic significance of depressive symptoms for predicting quality of life 12 months after gastric bypass. Compr Psychiatry 48(3):231–236. doi:S0010-440X(07)00014-410.1016/j.comppsych.2007.01.005

McFarland B (1995) Brief therapy and eating disorders. Jossey-Bass Publishers, San Francisco

Menon ST (1999) Psychological empowerment: definition, measurement, and validation. Canadian Journal of Behavioural Science 31(3):161–164

Morrow J, Gluck M, Lorence M, Flancbaum L, Geliebter A (2008) Night eating status and influence on body weight, body image, hunger, and cortisol pre- and post- Roux-en-Y Gastric Bypass (RYGB) surgery. Eat Weight Disord 13(4):e96–e99 5219 [pii]

Mou W, McNamara TP, Valiquette CM, Rump B (2004) Allocentric and egocentric updating of spatial memories. J Exp Psychol Learn Mem Cogn 30(1):142–157

Neumark-Sztainer D (2009) Preventing obesity and eating disorders in adolescents: what can health care providers do? J Adolesc Health 44(3):206–213. doi:S1054-139X(08)00663-010.1016/j.jadohealth.2008.11.005
Neumark-Sztainer D, Wall M, Guo J, Story M, Haines J, Eisenberg M (2006) Obesity, disordered eating, and eating disorders in a longitudinal study of adolescents: how do dieters fare 5 years later? J Am Diet Assoc 106(4):559–568. doi:S0002-8223(06)00004-610.1016/j.jada.2006.01.003
Neumark-Sztainer D, Wall M, Story M, Standish AR (2012) Dieting and unhealthy weight control behaviors during adolescence: associations with 10-year changes in body mass index. J adolesc health: off publ Soc adolesc Med 50(1):80–86. doi:10.1016/j.jadohealth.2011.05.010
Omalu BI, Ives DG, Buhari AM, Lindner JL, Schauer PR, Wecht CH, Kuller LH (2007) Death rates and causes of death after bariatric surgery for Pennsylvania residents, 1995–2004. Arch Surg 142 (10):923–928; discussion 929. doi:142/10/923 [pii] 10.1001/archsurg.142.10.923
Pecori L, Serra Cervetti GG, Marinari GM, Migliori F, Adami GF (2007) Attitudes of morbidly obese patients to weight loss and body image following bariatric surgery and body contouring. Obes Surg 17(1):68–73
Riva G (1997) The virtual environment for body-image modification (VEBIM): Development and preliminary evaluation. Presence, Teleoperators, and Virtual Environments 6(1):106–117
Riva G (1998a) Modifications of body image induced by virtual reality. Percept Mot Skills 86:163–170
Riva G (1998b) Virtual environment for body-image modification: virtual reality system for the treatment of body image disturbances. Comput Hum Behav 14(3):477–490
Riva G (1998c) Virtual Reality vs. Virtual Body: The use of virtual environments in the treatment of body experience disturbances. Cyber Psychol Behavior 1(2):129–137
Riva G (2007) Virtual Body, real pain: The Allocentric Lock Hypothesis. In: Zampini M, Pavani F (eds) Body Representation Workshop, Rovereto, Italy, 8–10 October 2007. Università degli Studi di Trento, Rovereto, Italy, p online: http://www.cimec.unitn.it/events/brw/Poster/107_abs_GIUSEPPE_RIVA.pdf
Riva G (2010) Neuroscience and Eating Disorders: The role of the medial temporal lobe. Nature Precedings, online: http://precedings.nature.com/documents/4235/. Accessed July, 2012
Riva G (2011) The key to unlocking the virtual body: virtual reality in the treatment of obesity and eating disorders. J Diabetes Sci Technol 5(2):283–292
Riva G (2012) Neuroscience and eating disorders: The allocentric lock hypothesis. Medical Hypotheses 78:254–257. doi:S0306-9877(11)00566-410.1016/j.mehy.2011.10.039
Riva G, Gamberini L (2000) Virtual reality in telemedicine. Telemedicine Journal 6(3):325–338
Riva G, Davide F (eds) (2001) Communications through virtual technologies: identity, community and technology in the communication age. emerging communication: Studies on new technologies and practices in communication. Ios Press. Online: http://www.emergingcommunication.com/volume1.html, Amsterdam
Riva G, Gaggioli A (2008) Virtual clinical therapy. Lecture Notes in Computer Sciences 4650:90–107
Riva G, Gaudio S (2012) Allocentric lock in anorexia nervosa: new evidences from neuroimaging studies. Med Hypotheses 79(1):113–117
Riva G, Bacchetta M, Baruffi M, Rinaldi S, Molinari E (1998) Experiential cognitive therapy in anorexia nervosa. Eating and Weight Disorders 3:141–150
Riva G, Bacchetta M, Baruffi M, Rinaldi S, Molinari E (1999) Virtual reality based experiential cognitive treatment of anorexia nervosa. J Behav Therap Exp Psychiatr 30(3):221–230
Riva G, Bacchetta M, Baruffi M, Rinaldi S, Vincelli F, Molinari E (2000a) Virtual reality based experiential cognitive treatment of obesity and binge-eating disorders. Clin Psychol Psychotherap 7(3):209–219
Riva G, Bacchetta M, Baruffi M, Rinaldi S, Vincelli F, Molinari E (2000b) Virtual reality-based experiential cognitive treatment of obesity and binge-eating disorders. Clin Psychol Psychotherap 7(3):209–219
Riva G, Marchi S, Molinari E (2000c) Body image and eating restraint: a structural modeling analysis. Eat Weight Disord 5(1):38–42

Riva G, Bacchetta M, Cesa G, Conti S, Molinari E (2003) Six-month follow-up of in-patient experiential-cognitive therapy for binge eating disorders. CyberPsychol Behav 6(3):251–258

Riva G, Botella C, Légeron P, Optale G (eds) (2004a) Cybertherapy: Internet and Virtual Reality as Assessment and Rehabilitation Tools for Clinical Psychology and Neuroscience. IOS Press; Online: http://www.cybertherapy.info/pages/book3.htm, Amsterdam

Riva G, Vatalaro F, Davide F, Alcañiz M (eds) (2004b) Ambient intelligence: the evolution of technology, communication and cognition towards the future of human-computer interaction. IOS Press. On-line: http://www.emergingcommunication.com/volume6.html, Amsterdam

Riva G, Bacchetta M, Cesa G, Conti S, Castelnuovo G, Mantovani F, Molinari E (2006) Is severe obesity a form of addiction? Rationale, clinical approach, and controlled clinical trial. CyberPsychol Behav 9(4):457–479

Riva G, Cárdenas-López G, Duran X, Torres-Villalobos GM, Gaggioli A (2012a) Virtual reality in the treatment of body image disturbances after bariatric Surgery: a clinical case. Stud Health Technol Inform 181:54–59

Riva G, Castelnuovo G, Cesa G, Gaggioli A, Mantovani F, Molinari E (2012b) Virtual reality for enhancing the cognitive behavioral treatment of obesity: a controlled Study with one-year follow-up. In: Medicine 2.0′12 Boston

Riva G, Molinari E, Vincelli F (2002) Interaction and presence in the clinical relationship: virtual reality (VR) as communicative medium between patient and therapist. IEEE Trans Inf Technol Biomed 6(3):198–205

Rosen JC (1996a) Body image assessment and treatment in controlled studies of eating disorders. Int J Eat Disord 19:341–343

Rosen JC (1996b) Improving body image in obesity. In: Thompson JK (ed) Body image, eating disorders and obesity. American Psychological Association, Washington, pp 425–440

Rosen JC, Orosan P, Reiter J (1995) Cognitive behavior therapy for negative body image in obese women. Behav Ther 26:25–42

Satava RM, Jones SB (2002) Medical applications of virtual reality. In: Stanney KM (ed) Handbook of virtual environments: design, implementation, and applications. Lawrence Erlbaum Associates, Inc., Mahwah, pp 368–391

Schlundt DG, Johnson WG (1990) Eating disorders: assessment and treatment. Allyn and Bacon, Needham Heights

Singh GK, Kogan MD, Van Dyck PC, Siahpush M (2008) Racial/Ethnic, Socioeconomic, and Behavioral Determinants of Childhood and Adolescent Obesity in the United States: Analyzing Independent and Joint Associations. Ann Epidemiol 18(9):682–695

Slater M, Perez-Marcos D, Ehrsson HH, Sanchez-Vives MV (2009) Inducing illusory ownership of a virtual body. Front Neurosci 3(2):214–220. doi:10.3389/neuro.01.029.2009

Slater M, Spanlang B, Sanchez-Vives MV, Blanke O (2010) First person experience of body transfer in virtual reality. PLoS ONE 5(5):e10564. doi:10.1371/journal.pone.0010564

Smucker MR, Dancu C, Foa EB, Niederee JL (1995) Imagery rescripting: a new treatment for survivors of childhood sexual abuse suffering from posttraumatic stress. J Cogn Psychotherap 9:3–17

Spagnolli A, Gamberini L (2005) A place for presence. understanding the human involvement in mediated interactive environments. Psychology Journal 3 (1):6-15. On-line: http://www.psychnology.org/article801.htm

Stice E, Presnell K, Shaw H, Rohde P (2005) Psychological and behavioral risk factors for obesity onset in adolescent girls: a prospective study. J Consult Clin Psychol 73(2):195–202

Stice E, Davis K, Miller NP, Marti CN (2008) Fasting increases risk for onset of binge eating and bulimic pathology: a 5-year prospective study. J Abnorm Psychol 117(4):941–946. doi:2008-16252-01910.1037/a0013644

Székely G, Satava RM (1999) Virtual reality in medicine. Bmj 319(7220):1305

Tanoue K, Uemura M, Kenmotsu H, Ieiri S, Konishi K, Ohuchida K, Onimaru M, Nagao Y, Kumashiro R, Tomikawa M, Hashizume M (2010) Skills assessment using a virtual reality simulator, LapSim, after training to develop fundamental skills for endoscopic surgery. Minim Invasive Ther Allied Technol 19(1):24–29. doi:10.3109/13645700903492993

Thompson JK, Heinberg LJ, Altabe M, Tantleff-Dunn S (1999) Exacting beauty: Theory, assessment and treatment of body image disturbance. American Psychological Association, Washington DC

van Hout GC, van Heck GL (2009) Bariatric psychology, psychological aspects of weight loss surgery. Obes Facts 2(1):10–15. doi:00019356410.1159/000193564

van Hout GC, Leibbrandt AJ, Jakimowicz JJ, Smulders JF, Schoon EJ, Van Spreeuwel JP, van Heck GL (2003) Bariatric surgery and bariatric psychology: general overview and the Dutch approach. Obes Surg 13(6):926–931. doi:10.1381/096089203322618795

Villani D, Gatti E, Confalonieri E, Riva G (2012). Am I my avatar? A tool to investigate virtual body image representation in adolescence. CyberPsych Behav Soc Netw 15(8):435-440

Wooley SC, Wooley OW (1985) Intensive out-patient and residential tratment for bulimia. In: Garner DM, Garfinkel PE (eds) Handbook of psychotherapy for anorexia and bulimia. Guilford Press, New York, pp 120–132

Chapter 16
Chronic Care Models for Obesity Management

Maria Letizia Petroni

Abstract Because of the high prevalence of obesity, a major challenge for national health services worldwide is to elaborate effective, efficacious, and sustainable models for assistance and rehabilitation of the complicated obese patient. Two main chronic care models for obesity management have so far been described: the British "Triple tier model for weight management", and more recently the Italian "Hub and Spoke model for obesity management". The triple tier model was primarily designed for increasing efficacy for weight management in primary care; emphasis is put on weight control rather than on disease management. The hub and spoke model describes a service network with different but integrated levels of care, with the patient at the center. It identifies highly specialized hospitals (hub) for diagnosis and treatment of morbid obesity and of its complications, patients being referred to and from peripheral centers (spokes), for initiation/continuation of their treatment and rehabilitation program. Regardless of the model chosen, the development of clinical pathways for obesity—from primary care to rehabilitation—for morbid and complicated obesity requires first of all an appropriate organizational, structural, and educational setting. They should clearly define the multidisciplinary integration between different levels of care and between professionals, identifying roles, and responsibilities. Obese patients should be guaranteed to be cared for in hospitals meeting defined standards for accommodation, patients' handling, equipment by trained medical, and nursing personnel. Telemedicine has the potential to further improve chronic care management of the obese patients, both for weight management and management of the complications.

M. L. Petroni (✉)
Clinical Nutrition Laboratory, IRCCS Istituto Auxologico Italiano,
Piancavallo, Verbania, Italy
e-mail: marialetizia.petroni@gmail.com

M. L. Petroni
Obesity Rehabilitation and Clinical Nutrition Services, Ospedale Privato Accreditato,
Villa Igea, Forlì, Italy

P. Capodaglio et al. (eds.), *Disabling Obesity*, DOI: 10.1007/978-3-642-35972-9_16,

16.1 Introduction

Chronic care models were started to be developed in the latter part of the twentieth century with the aim to improve the assessment and treatment of chronically ill patients and to limit unnecessary costs (Wagner 1998). The model described by Wagner and colleagues is based on six elements i.e., the community, the health system, self-management support, delivery system design, decision support, and clinical information systems. In this model, patients take an active part in their care and provide feedback to professionals and to health administrators. The Chronic Care Model can be applied to a variety of chronic illnesses, health care settings, and target populations (Wagner 1998).

Despite the huge burden represented by disabling obesity on healthcare costs there is a surprisingly shortage of published research or debate on chronic care models for obesity management in the scientific literature.

Most papers have focused on cost-effectiveness treatments for weight loss (Fuller et al. 2012; Rucker et al. 2007; Colquitt et al. 2009; Picot et al. 2009), comparing dietary and behavioral interventions with pharmacological or surgical treatment of obesity. In some of such trials, the presence of "disabling diseases" represents an exclusion criteria; in the majority of the remaining trials improvements on obesity-related disability are seldom assessed, and if so assessment is carried out indirectly through questionnaires on health-related quality of life rather than on direct functional assessment.

It is not surprising therefore that even within the in-depth analysis carried out within the framework of Health Technology Assessment Programme of the British National Health Service (Avenell et al. 2004) on the long-term effects and economic consequences of obesity treatments there is no mention of disability. The authors have developed economic models of treatments for obesity in terms of reduction of cardiovascular risk factors and on quality-adjusted life years (QALYs). Unfortunately, the instruments on which the calculation of QALYs have been based present with a number of limitations, such as not being restricted to functional status, but including a conflation of symptom, function, and health-perception scales not specific to obesity, or being excessively psychosocial oriented or being not empirically validated (Donini et al. 2011).

Only recently, Yaskin and colleagues have dealt with the concept of chronic care models in obesity management (Yaskin et al. 2009). They pointed out in their review that as with other chronic diseases, improved outcomes are associated with effective self-management of obesity across the life span, and therefore obesity fits squarely within the disease management and chronic care models (Yaskin et al. 2009). They conclude that—with the limitations due to the uneven quality of available studies—the combination approaches i.e. surgical or pharmacologic, combined with a behavioral intervention, were most likely to be effective. Management of the co-morbidities within this model was not specifically addressed.

Only a small part of the obese population worldwide has access to adequate services to the severity of the condition (including diagnostic, treatment, and rehabilitation of complications).

The big question is whether it is possible to identify one ore more models for assistance and rehabilitation of the complicated obese patient, which would prove in the long term to be effective, efficacious, and sustainable. Despite the scarce evidence from the literature, nevertheless, the present chapter will try and describe possible chronic care models for obesity, especially disabling obesity. Details will be given about the Italian forerunner experience in this sector.

16.2 Theoretical Models

16.2.1 The Triple-Tier Model for Weight Management

In 2001, the United Kingdom National Audit Office document *Tackling Obesity in England* (National Audit Office 2001) described the program offered by a specialist obesity clinic under the National Health Service, consisting in a combination of: group behavior modification, low calorie liquid diet program, appointments with the dietician and the doctor, clinical psychology assessment, pharmacological intervention for prescription of anti-obesity drugs, referral to other units for surgical intervention, and recruitment into clinical research programs where appropriate. This survey did not mention any experience on rehabilitative care for complicated obesity.

Mainly based on the findings and recommendations of the National Audit Office Document, in 2005 Maryon-Davis described a triple-tier pathway for weight management (Maryon-Davis 2005) from the brief intervention in primary care (Tier 1) through community-based lifestyle program (Tier 2) to hospital-based treatment (Tier 3) (Fig. 16.1).

This model was primarily designed for increasing efficacy for weight management in primary care. Tiers 1 and 2 of the pathway are described in detail, including organizational aspects and lines of evidence. The upper level of care (Tier 3) is just mentioned as "specialist hospital-based service, involving outpatient or in-patient dietary management and behavioural therapy, with or without exercise. It would also include possible referral for surgery" and no further detail is given.

16.2.2 The "Hub and Spoke" Model for Obesity Management

The "Hub and spoke" model put forward for the management of chronic conditions of special healthcare and economic interest, involves the concentration of tailored care, in centers of excellence (hub) and the referral of patients to peripheral centers (spokes), for the continuation/integration of their treatment/rehabilitation program.

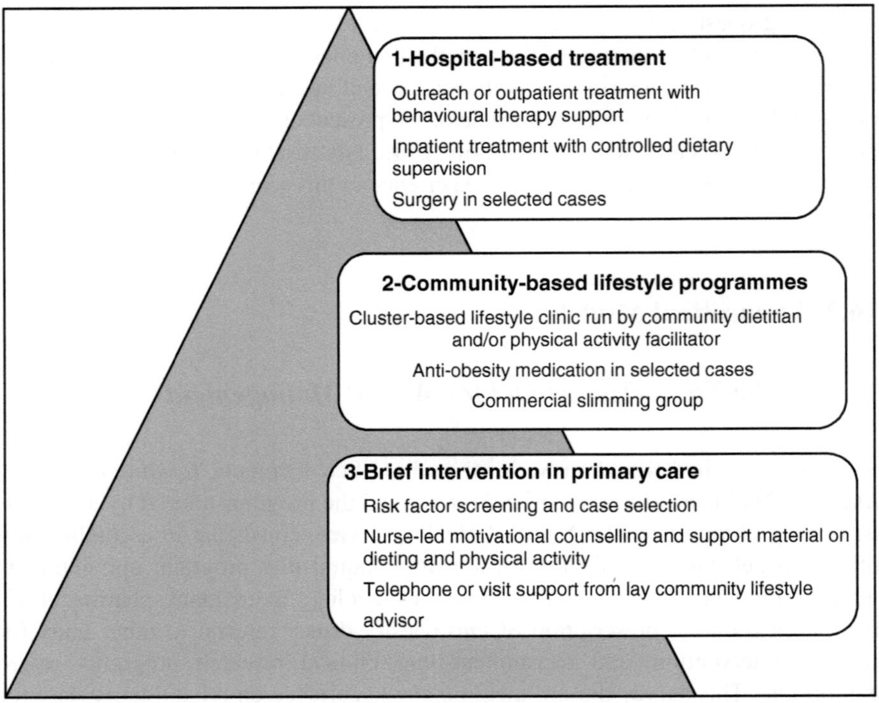

Fig. 16.1 The triple-tier for weight management (modified from Maryon-Davis 2005)

Networks created using this approach aim to guarantee a coordinated action of intervention, thus assuring patients optimum healthcare in the best-suited facility in terms of clinical and organizational appropriateness (Nobilio and Ugolini 2003). Until recently, in Italy, the concept of "dedicated care" for very obese patients in specialized centers organized into networks has never been introduced into National or regional legislation (Italian Ministry of Health 2011a). The consequence was (and unfortunately still is) an unacceptable heterogeneity between hospitals and between different Regions on the level of services offered to patients and on access to care, some of the few hospitals offering rehabilitation programs specific to severe obesity being set in a "legal vacuum" then at risk of inappropriateness according to regional guidelines in force that favor the post-acute rehabilitation rather than organ-specific and multidimensional rehabilitation, despite the recognition of a full dignity to obesity (metabolic) rehabilitation at the national level (Italian Ministry of Health 2011b).

In 2011, the Italian Ministry of Health published a document on the "Clinical, instrumental, technological and operative appropriateness criteria for the prevention, diagnosis and treatment of diabetes and obesity" (Italian Ministry of Health 2011a). This document proposes for the first time an "hub and spoke" model for obesity management, emphasizing the importance of community as a setting for the analysis of the needs and the integration between the different levels of care, with the patient at the center. Since this is a chronic condition, the service network should take the patient into charge and follow it over time, inserting her/him in pathways of care with various levels of diagnostic and therapeutic intensity, always part of a program of shared follow-up.

The document describes the *Hub Centre*—Hospital Center of High Specialization—as units with medical and nursing professionals with specific skills and experience covering the different clinical aspects of obesity and comorbidities related to it, with a role of excellence in the field of internal medicine (in particularly in the field of metabolism, endocrinology, and clinical nutrition) and bariatric surgery. According to this document, to ensure an adequate level of service and justify the use of dedicated resources with 24 h a day operation, the Hub must provide presumably at least 600 new cases/year, 150 bariatric surgery operations, 3,000 outpatients' visits/year of morbid and complicated obese patients.

The *Spoke Centre* is described as a center responsible for: (a) the management of patients with varying degrees of obesity that presents complex clinical conditions; (b) the implementation of therapeutic measures (rehabilitation/physical reconditioning/nutrition) which are most appropriate for the management of patients in the subacute phase; (c) the transfer to higher level of care for the patients in critical clinical conditions.

Depending on the need, and the number of cases dealt with, the Spoke center could optionally activate a bariatric surgery unit.

In this model, access to any of the network nodes should be able to occur through the general practitioners (GPs) or the specialist centers, either public or private accredited with the National Health Service (outpatients'clinics, day hospital, day service, and inpatient admissions) who are to assist obese patients.

The requirements common to the two levels of support (Hub and Spoke) are:

- Presence of a multidisciplinary team for obese patient management formed by experienced doctors and nurses;
- Systematic collection of clinical information through medical records in order to share the patient care process and to evaluate quality of care;
- Electronic network to allow teleconsultations between the network centers;
- Activation of early rehabilitation as needed;
- Weekly multidisciplinary meetings;
- Close collaboration between nursing and other forms of multidisciplinary care.

The Hub also provides educational programs and training of highly specialized personnel in Spoke hospitals, with the aim to improve quality standards of knowledge and expertise in the fields of internal medicine, endocrinology, clinical nutrition, general and bariatric surgery, and health management of obesity.

The Hub is responsible for collecting and processing clinical data of outcome from Spoke hospitals, promoting, and continuously updating the Regional and National epidemiological registry of morbid obesity.

The contact with the Hub center should be easy and immediate, with the appointment of medical/paramedical figures which play the role of "referents of

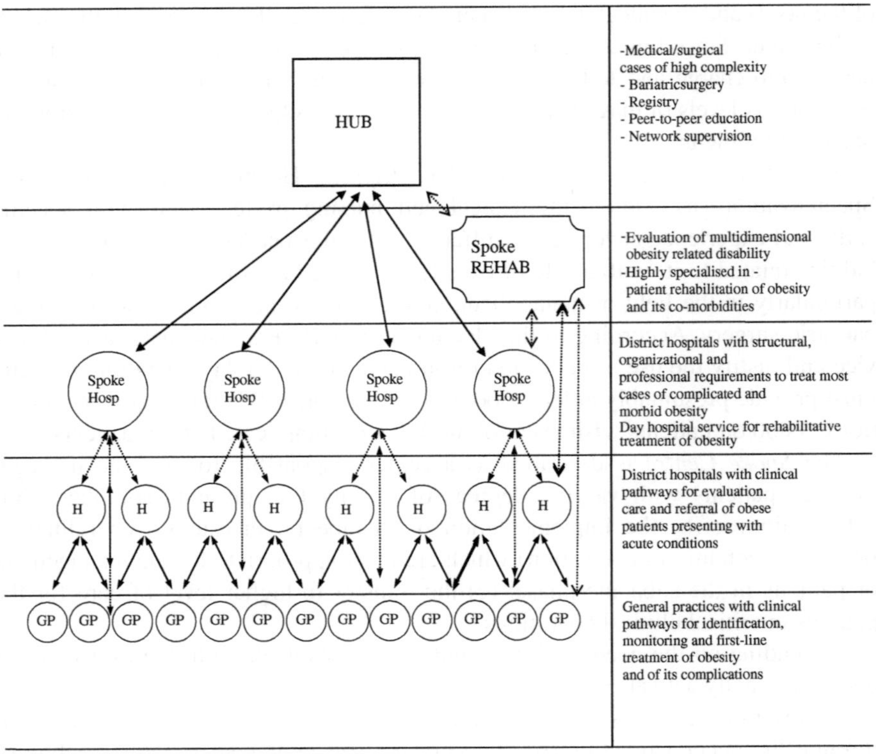

Fig. 16.2 Regional networks for obesity management based on the Hub and Spoke model

access" to the Center. Obese patients with major disabilities, frailty and difficulties, and with little or no family support must be able to rely on an integrated home care treatment and rehabilitation plan, which is managed and coordinated by the GP. The local services must guarantee intermediate care in-between the follow-up visits at the specialist Center in view of a better continuity with the treatment and rehabilitation program of the obese patient (medical monitoring, nutritional, psychological, psychiatric, physiotherapy treatment, etc.). The service network must also be articulated both longitudinally and transversely: (a) longitudinally because the same patient may need, over the years, different types of care in relation to the course of the disease and the exacerbation of complications; (b) transversely because different obese patients, in relation to the degree of comorbidities-frailty-disability can take advantage of different nodes of the service network (Fig. 16.2).

As far as planning of rehabilitative services are concerned, a reasonable estimate for implementing the model could be to provide one in-patient rehabilitation bed in a highly specialized center per 1,000 severely or complicated obese patients, and 2 to 4 day-care rehabilitation seats in peripheral spoke centers per every 1,000 severely or complicated obese patients. In order to ensure adequate standards of care and experience, a specialized obesity rehabilitation unit should run at least 20 beds for inpatient treatment. Also considering the minimum requirements for Hub centers, regional health services with less than 20,000 severely obese or complicated patients in their territory should identify interregional Hub centers and rehabilitation centers.

The first Italian Region to setup a regional Hub and Spoke obesity network has been the Veneto Region. The Hub center is located in the Padua University Hospital, to whose medical and surgical wards obese patients with life threatening conditions are referred, and where a bariatric surgery center operates and outpatients' endocrinology and nutrition clinics dedicated to morbid obesity are run. The regional in-patient Obesity Rehabilitation spoke center is located at Garda (Verona); appropriateness criteria for access under the regional health service have been disciplined (Veneto Region 2012). At a national level, the scientific institute Istituto Auxologico Italiano Ospedale San Giuseppe at Piancavallo (VB) which was born as a Foundation in 1963, has represented and still represents the main hub center of referral for medical and rehabilitative management of morbid obesity from all over the country (Zaninelli 2008).

16.3 Integrated Pathways for Clinical Management of Obesity

16.3.1 From Primary Care to Rehabilitation

The Italian Ministry of Health 2011 document stated that "the particular characteristics of morbid obesity as a chronic disease, its comorbidities and consequential disabilities that negatively impact both quality of life and health expenditure calls for an approach that also involves rehabilitation and not just treatment alone. It is therefore important to devise and develop pathways of care based on a multidisciplinary approach that not only deal with the weight issue over a longer period, but, above all, prevent and treat its complications. Regional obesity prevention, diagnosis and treatment networks are structured into various levels including both community and hospital facilities" (Italian Ministry of Health 2011a).

Integrated care pathways are structured multidisciplinary care plans which detail essential steps in the care of patients with a specific clinical problem. Clinical pathways appeared as a result of the adaptation of the documents used in industrial quality management, the so-called standard operating procedures. They have been proposed as an aid to promote the translation of national guidelines into local protocols and their subsequent application to clinical practice; to improve multidisciplinary communication and care planning, including with primary care; to improve quality standards, clinician–patient communication, and patient satisfaction (Campbell et al. 1998). Integrated care pathways describe, for a specific clinical condition, the tasks to be carried out together with the timing and sequence of these tasks and the discipline involved in completing the task. In its simplest form, they consist of a single multidisciplinary record, which is part of the patient's clinical record together with a patient summary sheet.

Experience has so far been greater in surgical services, where care is more easily predictable and set routines of practice were already in place. Not surprisingly, so far clinical pathways in obesity have only been reported for bariatric surgery. They have been reported to improve several indicators of perioperative quality of care (Ronellenfitsch et al. 2012), and to reduce both variability in professional care patterns and hospital costs (Campillo-Soto et al. 2008). For its complex pathogenesis, the high variability of response to treatment and the several comorbidities, obesity as a whole represents a condition which cannot be framed in a few standard operating procedures. The development of pathways of care—from primary care to rehabilitation—for morbid and complicated obesity requires first of all an appropriate organizational, structural, and educational setting.

From the organization point of view, the first step is to identify referents (medical and nursing area) for the quality of assistance to obese patient within each setting (territorial primary care; district hospital; highly specialized hospital; rehabilitation hospitals). As next steps, a mapping of all structures and teams more frequently involved with the care of obese people, the identification of key persons ("facilitators") within each team to be involved in drawing up and diffusing the protocols, a survey to identify adequacy and pitfalls of existing care, and the involvement of patients' organizations.

About structural requirements, highly specialized centers must have large premises accessible to people with disabilities, divided into: a waiting area proportionate to influx; suitable and equipped clinics that have chairs, beds, toilets, and lifts able to support weight at least up to 250 kg, surgical ward should have operation tables, equipment, and stretches able to support superobese patients; the same for intensive care units (ICUs) and hospital wards dedicated to treat medical and surgical complications of obese patients. Radiology Departments of regional centers should have dedicated equipment for morbidly obese and superobese (e.g., open MR) (Italian Ministry of Health 2011a). Each obesity clinic must have: computer, height rod, scale weighting up to 300 kg, sphygmomanometer with large armband, ophthalmoscope, reflectometer for glucose, and educational material. The area dedicated to the clinics must have a bioelectrical impedance equipment and possibly dual energy X-ray absorptiometry (DEXA) for body composition and indirect calorimeter for measurement of resting energy expenditure.

With regard to training, it is essential for GPs as well as hospital medical (nonobesity specialists) and nursing staff to be correctly and timely informed of all the diagnostic and therapeutic aspects related to obesity. It is of paramount importance that the primary care physician and the staff of district services are able to recognize the factors predisposing to complications, to recognize and monitor alarm signals, to direct patients to the appropriate level of care, thus avoiding unnecessary delay in diagnosis and treatment. Moreover, the training of surgeons who regularly treat obese patients must not merely include continuing updates on surgical techniques and tools, but also dedicate greater attention to the concept of chronicity and long-term management (Italian Ministry of Health 2011a).

Table 16.1 Tasks for primary care setting in clinical pathways for obesity management

Periodic evaluations	Monitoring and supervision
Weight, waist circumference	Basic assessment of obesity-related disabilities
Blood pressure, heart rate, and peripheral oxygen saturation	Basic assessment of obesity-related medical comorbidities (including psychiatric)
Fasting glucose, lipid profile, uric acid, glycated haemoglobin, and microalbuminuria (as applicable)	Referral to superior levels of treatment and rehabilitation according to shared clinical pathway
Smoking habits	Two-way communication with hospital specialists
Physical activity levels	
Alcohol consumption	Monitoring of assumption of pharmacological treatment for obesity complications and of nutritional supplements following bariatric surgery
Tailored screening for obesity-related neoplasm (endometrial, breast, and colorectal cancer)	
Early detection of symptoms suggestive of cardio-respiratory complication including obstructive sleep apnoea syndrome	Monitoring of side-effects of pharmacological treatments for obesity (if prescribed by the specialist) and for its complications
Assessment of pain and risk of wounds	Referral for and supervision of home-care assistance
Availability of social and familiar support for proper nutrition and adhesion to therapies	
Therapies	**Further treatment option (if locally available)**
Pharmacological treatment of cardiovascular and metabolic complications of obesity	Referral to a community exercise specialist for adapted physical activity
Wound care	Referral to a community dietician for dietary care plan, nutrition education, and monitoring
Brief motivational intervention on reducing sedentary behavior	
Brief motivational intervention on smoking cessation	Referral to a community psychologist for individual/group support and cognitive-behavioral therapy for obesity
Brief nutritional counseling	
Prescription of physical therapy for osteoarticular complication	

A number of indicators of process and of outcome in the disease management of obesity, from GPs to hospital management and bariatric surgery have been described (Italian Ministry of Health 2011a). Periodical evaluation of disability (e.g. by using the TSD.OC score as described in Chap. 6) could help to define the appropriate level of care in individual patients.

Last, but not least, medical and paramedical personnel should be trained to have an emphatic and nondiscriminating approach toward obese patients, thus avoiding to perpetuate anti-obesity stigma which is detrimental to patients' adhesion to care (Sikorski et al. 2011).

Clinical pathways for obesity should clearly define the multidisciplinary integration between different levels of care and between professionals, identifying roles, and responsibilities.

Primary care practitioners i.e. GPs, practice nurses, health visitors, community nurses, community pharmacists, and, if available, exercise specialists, community dieticians, and psychologists can provide first-line management of obesity in high-risk patients. Table 16.1 lists the main tasks that could be carried out within a primary care setting.

In the minimum set of clinical pathways for obesity in *district hospitals* there should be included the emergency treatment of the morbid obese patient; well-defined anesthesiological procedures to treat surgical conditions in obese patients; the prompt identification of cases to be referred to the Hub center or to other Spoke hospitals in the network with the appropriate setting to treat critical medical and surgical complications in morbid obesity; the appropriate drug dosage of available hospital medications for obesity; nursing of the obese patient. District hospitals which represent "Spoke" hospitals in the network should develop—in close co-operation with the Hub hospital—additional clinical pathways, and make them available to nonspoke hospitals (Table 16.2).

Rehabilitative centres should set multidisciplinary pathways covering evaluation, rehabilitative programs, and outcome measures in the areas of physical, nutritional, psychological, social, nursing rehabilitation, and plus those related to

Table 16.2 Suggested minimum set of clinical pathways for obesity management in district hospitals (to be revised for adaptation to obese patients or to be added)

District hospital (non-spoke)	District hospital (spoke)
	Same as for non-spoke hospital plus those listed above
Medical and surgical areas	
Peripheral and central venous cannulation	Diagnostic and therapeutic protocols and procedures for the medical complications of obesity in the following areas:
Main medical and surgical conditions included in current clinical pathways	Metabolic
Drug dosage in obesity	Endocrine
Anaesthetic procedures	Cardiovascular
Management of the obese patient within the Hub and Spoke network	Respiratory
	Gastrointestinal and hepatological
	Renal
	Osteo-articular
	Psychiatric
	Geriatric medicine
	Bariatric surgery
	Assessment (multidisciplinary)
	Surgical procedures
	post-surgical follow up (multidisciplinary)
	Obesity rehabilitation (or referral protocols if located elsewhere)
Nursing area	
Accidents and emergency triage	Evaluation of obesity-related disability and frailty (multidisciplinary)
Mobilization and transfers of obese patients	
Patient review instrument (PRI)	
Pression ulcer prevention	
Wound care	
Others	
Radiological procedures	Psychological assessment and cognitive-behavioral treatment
	Nutritional assessment and dietary treatment

medical complications of obesity. Integrated pathways should be drawn for pre- and post-surgical rehabilitation of patients undergoing bariatric surgery.

These pathways are described in detail in other chapters of the present book.

The *Hub centre* should ensure that all clinical pathways developed within the network comply with stringent standards of care, are evidence-based whenever possible, and are being regularly updated. Also, as already mentioned, they are responsible for keeping the regional Registry of morbid obesity and for the continuous medical education programs on obesity under the Regional Health Service.

16.4 Surgery or Rehabilitation? Are They Really Conflicting Approaches?

Some degree of theoretical conflict does exist in the whole literature between surgical and nonsurgical approaches to obesity, and probably explains why publications are scarce regarding chronic care models. One gap is the relative lack of dialog between surgeons and internists, as perhaps none of the pathways are truly complete and fully balanced. Surgical models and protocols address surgical needs, which are essentially acute and short term. Clinical pathways are mainly directed toward lifelong care, based on the different assumption that obesity is a permanent condition.

In surgical circles, though obesity continues to be defined as a chronic incurable disease, pragmatically it is viewed as practically the opposite. If somebody has lost more than 50 % of excess body weight after bariatric intervention he or she is considered compensated, if not cured. A not much different concept applies to diabetes and metabolic surgery, where outcomes like long-term remission and even cure are beginning to be utilized. Surgeons do not dismiss the need for lifelong care, however, they believe it should be mostly periodical assessment, not full-fledged assistance. In case of recurrence they prefer to re-operate and start the process all over again. One must admit much optimism and enthusiasm underlying the bariatric approach, and indeed surgeons have great trust in their procedures, believing that they can remove disease with their hands. Indeed, it should be considered that a significant proportion of surgically managed patients do get rid of a few comorbidities and some others improve, at least for up to 10–15 years, as longer follow-ups are not available yet (Sjostrom et al. 2012).

Nevertheless, only a fraction of those with severe or complicated obesity are potentially candidates to surgery, and not all candidates are suitable to proceed to surgery (Neff and le Roux 2012).

A chronic care model that includes rehabilitation not only can help to improve co morbidities and quality of life in these patients, but also can make candidates for surgery patients previously not eligible. Also, although obesity surgery can significantly reduce disability in a large proportion of patients, in others residual disability (especially psychiatric) and co morbidities can persist or worsen despite weight loss (Tindle et al. 2010; Mirabelli et al. 2011; Neff and le Roux 2012) while in others it can reappear following weight regain (Neff and le Roux 2012). These all represent strong arguments in favor of including bariatric surgery in chronic care models. Those patients with a long-lasting efficacy of bariatric surgery on comorbidities and quality of life could eventually be managed in primary care setting under specialist supervision, while the remaining ones will require settings of higher complexity, which can include rehabilitation coupled to repeated surgery where deemed appropriate.

16.5 Future Approaches: Telemedicine

A new promising method for granting continuity of care to wide populations of patients at low costs is telemedicine: the term "telemedicine" has a wide definition—medicine practised at a distance—uses information and communication technologies (ICT) in order to exchange information useful for the diagnosis, treatment, rehabilitation, and prevention of diseases (Castelnuovo et al. 2003). Main roles for telemedicine to support integrated care in chronic disease management have been in providing education (to improve self-management), in enabling information transfer (e.g. telemonitoring), in facilitating contact with health professionals (e.g. telephone support and follow-up), and in improving electronic records. That is, telemedicine has been used in both the process of care and the outcome of care (Wootton 2012).

Care-at-a-distance (telecare) can be carried out with tools such as websites, e-mail, chat lines, videoconference, telephone, and mobile phones. Compared to no intervention or minimal interventions (pamphlets, usual care), *interactive computer-based* interventions (mainly behavioral treatments) are an effective intervention for weight loss and weight maintenance (Wieland et al. 2012). However, compared to in-person interventions, interactive computer-based interventions result in smaller weight losses and lower levels of weight maintenance (Wieland et al. 2012). *Mobile technologies* as potential interventions to promote weight loss showed strong evidence to occurrence of weight loss in the short term, but only moderate evidence for the medium term, with little or no data on long-term follow-up, cost-effectiveness, and patient acceptability (Bacigalupo et al. 2012).

Also, novel medical technologies can be interfaced to communication technologies in order to allow at-distance medical monitoring of obesity complications, like type 2 diabetes, respiratory insufficiency, obstructive sleep apnoea, arterial hypertension, and coronary heart disease. A US-based company offers the TelehealthLink™ obesity package currently including a weight scale, a blood glucose monitor, and a small communication gateway box (http://www.telehealthlink.com). The weight scale and blood glucose monitor connects to the communication gateway which sends the data to company's servers real-time via an Internet connection. Additional healthcare devices for hypertension and asthma can be added into the package according to patient monitoring needs. Centralized real-time data collection allows to process vital signals automatically and simultaneously informs caregivers or/and family members if the patient needs immediate medical attentions. Web interface enables physicians, caregivers, and family members to monitor patients' health status. No research data specifically addressed to obesity using this system has yet been published.

One possible reason why telemedicine programs for obesity so far tested have not given fully satisfactory results could be the lack of involvement of primary care providers. In a controlled study, where intervention provided patients with weight-loss support remotely—through the telephone, a study-specific website, and e-mail but also where primary care providers reinforced participation at routinely scheduled visits—was effective in promoting weight loss at 24 months as the intervention providing in-person support (Appel et al. 2011).

Web-based strategies for enforcing and monitoring prescribed physical activity could play an important role for enforcing long-term weight loss but also long-term care of obesity. A promising tool is the Mywellnesskey® accelerometer (Technogym, Italy), devised to measure and reinforce the adherence to physical activity recommendations, which has been validated and tested in patients discharged from cardiac rehabilitation program (Guiraud et al. 2012a). This tool has a visual display to give immediate feedback to the patient on amount of daily physical activity prescribed and that already carried out, and can be interfaced to remote control by the prescribing physician through the patient's personal computer. In a controlled study, telephone support based on accelerometer recordings proved to be an effective strategy to improve adherence to physical activity in noncompliant cardiac patients following rehabilitation (Guiraud et al. 2012b).

Another tool which couples accelerometer, pedometer, galvanic skin response, skin temperature, and heat flux (Bodymedia® Sensewear Armband) is being tested for measuring physical activity and energy expenditure in obese patients with type 2 diabetes within a multidisciplinary telemedicine approach (Castelnuovo et al. 2010, 2011).

A word of caution about easy enthusiasm for telemedicine in the management of chronic diseases comes from a recent analysis by Wootton of 141 randomized controlled trials on telemedicine interventions for asthma, COPD, diabetes, heart failure, and hypertension (Wootton 2012). The author underlines that firstly, most studies have reported positive effects, while almost none have reported negative effects, suggesting publication bias; secondly, that there were no major differences in the value of the telemedicine intervention between the disease types; thirdly, that most studies have been relatively short term, although it seems unlikely that in a chronic disease, any intervention can have much effect unless applied for a long period. And finally, that there have been very few studies of cost-effectiveness and patients' acceptance. Thus, the author concludes that evidence base for the value of telemedicine in managing chronic diseases is on the whole weak and contradictory, and further higher quality research is warranted (Wootton 2012).

Acknowledgments The author wish to thank Professor Antonio Liuzzi for his pivotal work about clinical pathways in obesity and Professor Joel Faintuch for his comments which largely contributed to Sect. 16.4.

References

Appel LJ, Clark JM, Yeh HC, Wang NY, Coughlin JW, Daumit G, Miller ER 3rd, Dalcin A, Jerome GJ, Geller S, Noronha G, Pozefsky T, Charleston J, Reynolds JB, Durkin N, Rubin RR, Louis TA, Brancati FL (2011) Comparative effectiveness of weight-loss interventions in clinical practice. N Engl J Med 365:1959–1968

Avenell A, Broom J, Brown TJ, Poobalan A, Aucott L, Stearns SC, Smith WC, Jung RT, Campbell MK, Grant AM (2004) Systematic review of the long-term effects and economic consequences of treatments for obesity and implications for health improvement. Health Technol Assess 8:1–182

Bacigalupo R, Cudd P, Littlewood C, Bissell P, Hawley MS, Buckley Woods H (2012) Interventions employing mobile technology for overweight and obesity: an early systematic review of randomized controlled trials. Obes Rev. doi:10.1111/obr.12006 [EPub ahead of print]

Campbell H, Hotchkiss R, Bradshaw N, Porteous M (1998) Integrated care pathways. BMJ 316:133–137

Campillo-Soto A, Martín-Lorenzo JG, Lirón-Ruíz R, Torralba-Martínez JA, Bento-Gerard M, Flores-Pastor B, Aguayo-Albasini JL (2008) Evaluation of the clinical pathway for laparoscopic bariatric surgery. Obes Surg 18:395–400

Castelnuovo G, Gaggioli A, Mantovani F, Riva G (2003) From psychotherapy to e-therapy: the integration of traditional techniques and new communication tools in clinical settings. Cyberpsychol Behav 6:375

Castelnuovo G, Manzoni GM, Cuzziol P, Cesa GL, Corti S, Tuzzi C, Villa V, Liuzzi A, Petroni ML, Molinari E (2011) TECNOB study: ad interim results of a randomized controlled trial of a multidisciplinary telecare intervention for obese patients with type-2 diabetes. Clin Pract Epidemiol Ment Health 7:44–50

Castelnuovo G, Manzoni GM, Cuzziol P, Cesa GL, Tuzzi C, Villa V, Liuzzi A, Petroni ML, Molinari E (2010) TECNOB: study design of a randomized controlled trial of a multidisciplinary telecare intervention for obese patients with type-2 diabetes. BMC Public Health 10:204

Colquitt JL, Picot J, Loveman E, Clegg AJ (2009) Surgery for obesity. Cochrane Database Syst Rev 2:CD003641

Donini LM, Brunani A, Sirtori A, Savina C, Tempera S, Cuzzolaro M, Spera G, Cimolin V, Precilios H, Raggi A, Capodaglio P (2011) SIO-SISDCA task force. Assessing disability in morbidly obese individuals: the Italian Society of Obesity test for obesity-related disabilities. Disabil Rehabil 33:2509–2518

Fuller NR, Colagiuri S, Schofield D, Olson AD, Shrestha R, Holzapfel C, Wolfenstetter SB, Holle R, Ahern AL, Hauner H, Jebb SA, Caterson ID (2012) A within-trial cost-effectiveness analysis of primary care referral to a commercial provider for weight loss treatment, relative to standard care-an international randomised controlled trial. Int J Obes. doi: 10.1038/ijo.2012.139 [EPub ahead of print]

Guiraud T, Granger R, Gremeaux V, Bousquet M, Richard L, Soukarie L, Babin T, Labrunee M, Bosquet L, Pathak A (2012a) Accelerometer as a tool to assess sedentarity and adherence to physical activity recommendations after cardiac rehabilitation program. Ann Phys Rehabil Med 55:312–321

Guiraud T, Granger R, Gremeaux V, Bousquet M, Richard L, Soukarié L, Babin T, Labrunée M, Sanguignol F, Bosquet L, Golay A, Pathak A (2012b) Telephone support oriented by accelerometric measurements enhances adherence to physical activity recommendations in noncompliant patients after a cardiac rehabilitation program. Arch Phys Med Rehabil. doi:S0003-9993(12)00534-5

Information from the company's website http://www.telehealthlink.com at 15 November 2012

Italian Ministry of Health (2011) Appropriateness criteria for the prevention, diagnosis and treatment of diabetes and obesity (Appropriatezza clinica, strutturale, tecnologica e operativa per la prevenzione, diagnosi e terapia dell'obesità e del diabete mellito). Quaderni della Salute, 10 (http://www.quadernidellasalute.it/archivio-quaderni/10-luglio-agosto-2011.php)

Italian Ministry of Health (2011) The centrality of the individual in rehabilitation: new organizational and management models (La centralità della Persona in riabilitazione:nuovi modelli organizzativi e gestionali). Quaderni della Salute, 8 (http://www.quadernidellasalute.it/archivio-quaderni/8-marzo-aprile-2011.php)

Maryon-Davis A (2005) Weight management in primary care: how can it be made more effective? Proc Nutr Soc 64:97–103

Mirabelli D, Petroni ML, Ferrante D, Merletti F (2011) Risk of suicide and bariatric surgery. Am J Med 124:e17

National Audit Office (2001) Tackling Obesity in England. Report by the Controller and Auditor General. The Stationery Office (http://www.nao.org.uk/publications/0001/tackling_obesity_in_england.aspx)

Neff KJH, le Roux CW (2012) Bariatric surgery: a best practice article. J Clin Pathol [EPub ahead of print]

Nobilio L, Ugolini C (2003) Selective referrals in a 'hub and spoke' institutional setting: the case of coronary angioplasty procedures. Health Policy 63:95–107

Picot J, Jones J, Colquitt JL, Gospodarevskaya E, Loveman E, Baxter L, Clegg AJ (2009) The clinical effectiveness and cost-effectiveness of bariatric (weight loss) surgery for obesity: a systematic review and economic evaluation. Health Technol Assess 13(41):1–190, 215–357

Ronellenfitsch U, Schwarzbach M, Kring A, Kienle P, Post S, Hasenberg T (2012) The effect of clinical pathways for bariatric surgery on perioperative quality of care. Obes Surg 22:732–739

Rucker D, Padwal R, Li SK, Curioni C, Lau DC (2007) Long term pharmacotherapy for obesity and overweight: updated meta-analysis. BMJ 335:1194–1199

Sikorski C, Luppa M, Kaiser M, Glaesmer H, Schomerus G, König HH, Riedel-Heller SG (2011) The stigma of obesity in the general public and its implications for public health—a systematic review. BMC Public Health 11:661

Sjostrom L, Peltonen M, Jacobson P, Sjostrom D, Karason K, Wedel H et al (2012) Bariatric surgery and long-term cardiovascular events. JAMA 307:56–65

Tindle HA, Omalu B, Courcoulas A, Marcus M, Hammers J, Kuller LH (2010) Risk of suicide after long-term follow-up from bariatric surgery. Am J Med 123:1036–1042

Veneto Region (2012) Regional ordinance 99, 9 July 2012

Wagner EH (1998) Chronic disease management: what will it take to improve care for chronic illness? Effective Clin Pract 1:2–4

Wieland LS, Falzon L, Sciamanna CN, Trudeau KJ, Brodney S, Schwartz JE, Davidson KW (2012) Interactive computer-based interventions for weight loss or weight maintenance in overweight or obese people. Cochrane Database Syst Rev 8:CD007675

Wootton R (2012) Twenty years of telemedicine in chronic disease management—an evidence synthesis. J Telemed Telecare 18:211–220

Yaskin J, Toner RW, Goldfarb N (2009) Obesity management interventions: a review of the evidence. Popul Health Manag 12:305–316

Zaninelli S (2008) Istituto auxologico italiano 1958–2008. Cinquant'anni di ricerca e cura. Franco Angeli, Milano

Index

P. Capodaglio et al. (eds.), *Disabling Obesity*, DOI: 10.1007/978-3-642-35972-9,

MIX
Papier aus verantwortungsvollen Quellen
Paper from responsible sources
FSC® C105338

If you have any concerns about our products,
you can contact us on
ProductSafety@springernature.com

In case Publisher is established outside the EU,
the EU authorized representative is:
Springer Nature Customer Service Center GmbH
Europaplatz 3, 69115 Heidelberg, Germany

Printed by Libri Plureos GmbH
in Hamburg, Germany